Methods in Molecular Biology™

For further volumes:
http://www.springer.com/series/7651

Imaging Gene Expression

Methods and Protocols

Edited by

Yaron Shav-Tal

The Mina and Everard Goodman Faculty of Life Sciences and Institute of Nanotechnology, Bar-Ilan University, Ramat Gan, Israel

Editor
Yaron Shav-Tal
The Mina and Everard Goodman Faculty
of Life Sciences and Institute of Nanotechnology
Bar-Ilan University
Ramat Gan, Israel

Additional material to this book can be downloaded from http://extras.springer.com

ISSN 1064-3745 ISSN 1940-6029 (electronic)
ISBN 978-1-62703-525-5 ISBN 978-1-62703-526-2 (eBook)
DOI 10.1007/978-1-62703-526-2
Springer New York Heidelberg Dordrecht London

Library of Congress Control Number: 2013942490

Printed on acid-free paper

Humana Press is a brand of Springer
Springer is part of Springer Science+Business Media (www.springer.com)

Dedication

To my mentors, Dov Zipori and Robert Singer.

Preface

Gene expression can be a term of many meanings, depending on the type of research field one belongs to. The activity of genes has been measured since the days when radioactive labeling of nucleic acids became possible, and ever since the field has been ploughed through by scores of biochemical and molecular analyses, and recently, bioinformatics and genomics approaches. Imaging of gene expression has also made an impact on our understanding of the gene expression pathway, particularly since the field uniquely deals with gene activity within single cells rather than taking a population view like the above-mentioned techniques. As imaging technologies and approaches have evolved, the scope of certain imaging techniques has moved far beyond the production of purely illustrative images or appealing time-lapse movies to providing the scientist with a rich range of ways to measure and quantify the biological process and outcome of gene expression. Close interactions between biologists and physicists trying to better understand the mechanistics of gene activity have allowed the import of techniques and terminology once solely used in physics, thereby bringing to light the emerging field of biophysics.

This book aims to offer up-to-date approaches and protocols that scientists in the field have developed, which would benefit the broader scientific community. The methods not only describe the technical design of the experiments but also focus on the expected outcome and provide tips and personal insights for the benefit of the user, which we all know can sometimes tip the outcome between the failure and success of a technique. This volume is divided into three parts. The first section deals with the output of a gene, namely the RNA molecules that are transcribed from the gene and the way by which these molecules can be tracked or quantified in fixed or living cells. The second section provides protocols that focus on the gene, DNA, or chromatin. The third portrays a variety of ways by which nuclear processes intertwined with gene expression can be followed and quantified in living cells as well as approaches for studying several subnuclear structures found in eukaryotic cells. Finally, since cells make up tissues, it is imperative to be able to follow these processes in whole tissues, or even better, in the actual living organism. Therefore imaging in the context of a complete organism is given special notice in the book. The chapters have been contributed by both well-established and young scientists, and I am ever grateful to all the authors who have participated and invested time, thought, and energy in the bringing this collection to light.

Ramat Gan, Israel *Yaron Shav-Tal*

Contents

Contributors

DAVID P. BAZETT-JONES • *Program in Genetics and Genome Biology, The Hospital for Sick Children, Toronto, ON, Canada*
ANDREW S. BELMONT • *Department of Cell and Developmental Biology, University of Illinois, Urbana, IL, USA*
QIAN BIAN • *Department of Molecular and Cell Biology, University of California, Berkeley, CA, USA*
IRENA BRONSHTEIN BERGER • *Physics Department & Institute for Nanotechnology, Bar-Ilan University, Ramat-Gan, Israel*
SEBASTIAN BULTMANN • *Department of Biology, Center for Integrated Protein Science Munich (CIPSM), Ludwig Maximilians University Munich, Munich, Germany*
DANIELLE CANNON • *Department of Cell and Developmental Biology and MRC Laboratory for Molecular Cell Biology, University College London, London, UK*
M. CRISTINA CARDOSO • *Department of Biology, Technische Universität Darmstadt, Darmstadt, Germany*
CORELLA S. CASAS-DELUCCHI • *Department of Biology, Technische Universität Darmstadt, Darmstadt, Germany*
JONATHAN R. CHUBB • *MRC Laboratory for Molecular Cell Biology, Department of Cell and Developmental Biology, University College London, London, UK*
ADAM M. CORRIGAN • *MRC Laboratory for Molecular Cell Biology, Department of Cell and Developmental Biology, University College London, London, UK*
THOMAS CREMER • *Department Biology II, Anthropology and Human Genetics, Biocenter, Ludwig-Maximilians-University (LMU), Martinsried, Germany*
MIROSLAV DUNDR • *Department of Cell Biology, Chicago Medical School, Rosalind Franklin University of Medicine & Science, North Chicago, IL, USA*
LIRON EVEN-FAITELSON • *Program in Genetics and Genome Biology, The Hospital for Sick Children, Toronto, ON, Canada*
MATTHEW L. FERGUSON • *Center for Cancer Research, National Cancer Institute, National Institutes of Health, Bethesda, MD, USA*
EDEN FUSSNER • *Program in Genetics and Genome Biology, The Hospital for Sick Children, Toronto, ON, Canada*
SOURAV GANGULY • *Laboratory of Receptor Biology and Gene Expression, National Cancer Institute, National Institutes of Health, Bethesda, MD, USA*
YUVAL GARINI • *Physics Department & Institute for Nanotechnology, Bar-Ilan University, Ramat-Gan, Israel*
ROGER A. GREENBERG • *Department of Cancer Biology, Abramson Family Cancer Research Institute, Basser Research Center for BRCA1/2, Perelman School of Medicine, University of Pennsylvania, Philadelphia, PA, USA; Department of Pathology, Abramson Family Cancer Research Institute, Basser Research Center for BRCA1/2, Perelman School of Medicine, University of Pennsylvania, Philadelphia, PA, USA*

GORDON L. HAGER • *Laboratory of Receptor Biology and Gene Expression, National Cancer Institute, National Institutes of Health, Bethesda, MD, USA*
ARIGELA HARIKUMAR • *Department of Genetics, The Alexander Silberman Institute of Life Sciences, The Hebrew University of Jerusalem, Jerusalem, Israel*
EDITH HEARD • *Mammalian Developmental Epigenetics Group, Institut Curie, CNRS UMR3215, INSERM U934, Paris, France*
DANIÈLE HERNANDEZ-VERDUN • *Nuclei and Cell Cycle and Macromolecular Complexes in Live Cells, Institut Jacques Monod – UMR 7592 CNRS - Université Paris Diderot-Paris 7, Paris, France*
BARBARA HÜBNER • *Department Biology II, Anthropology and Human Genetics, Biocenter, Ludwig-Maximilians-University (LMU), Martinsried, Germany*
PINHAS KAFRI • *The Mina & Everard Goodman Faculty of Life Sciences & Institute of Nanotechnology, Bar-Ilan University, Ramat Gan, Israel*
ALON KALO • *The Mina & Everard Goodman Faculty of Life Sciences & Institute of Nanotechnology, Bar-Ilan University, Ramat Gan, Israel*
TIM P. KAMINSKI • *Institute of Physical and Theoretical Chemistry, Rheinische Friedrich-Wilhelms-University Bonn, Bonn, Germany*
DAVID A. KELLY • *The Wellcome Trust Centre for Cell Biology, University of Edinburgh, Edinburgh, UK*
ELDAD KEPTEN • *Physics Department & Institute for Nanotechnology, Bar-Ilan University, Ramat-Gan, Israel*
NIMISH KHANNA • *Department of Cell and Developmental Biology, University of Illinois, Urbana, IL, USA*
ULRICH KUBITSCHECK • *Institute of Physical and Theoretical Chemistry, Rheinische Friedrich-Wilhelms-University Bonn, Bonn, Germany*
CHRISTIAN LANCTÔT • *First Faculty of Medicine, Institute of Cellular Biology and Pathology, Charles University in Prague, Prague, Czech Republic*
DANIEL R. LARSON • *Center for Cancer Research, National Cancer Institute, National Institutes of Health, Bethesda, MD, USA*
HEINRICH LEONHARDT • *Department of Biology, Center for Integrated Protein Science Munich (CIPSM), Ludwig Maximilians University Munich, Munich, Germany*
REN LI • *Program in Genetics and Genome Biology, The Hospital for Sick Children, Toronto, ON, Canada*
EMILIE LOUVET • *Laboratory of Plasma Membrane and Nuclear Signaling, Graduate School of Biostudies, Kyoto University, Kyoto, Japan*
DAVIDE MAZZA • *Center for Experimental Imaging, Istituto Scientifico Ospedale San Raffaele e Universita' Vita-Salute San Raffaele, Bethesda, MD, USA*
JAMES G. MCNALLY • *Laboratory of Receptor Biology and Gene Expression, National Cancer Institute, National Institutes of Health, Bethesda, MD, USA*
PETER MEISTER • *Institute of Cell Biology, University of Bern, Bern, Switzerland*
ERAN MESHORER • *Department of Genetics, The Alexander Silberman Institute of Life Sciences, The Hebrew University of Jerusalem, Jerusalem, Israel*
TINA B. MIRANDA • *Laboratory of Receptor Biology and Gene Expression, National Cancer Institute, National Institutes of Health, Bethesda, MD, USA*
TETSUYA MURAMOTO • *Laboratory for Cell Signaling Dynamics, Quantitative Biology Center, RIKEN, Osaka, Japan*
ELEONORA MURO • *Randall Division of Cell and Molecular Biophysics, King's College London, London, UK*

JÜRGEN NEUMANN • *Department Biology II, Human Biology and BioImaging, Biocenter, Ludwig-Maximilians-University (LMU), Martinsried, Germany*
CLAUDIO NIETZEL • *Institute of Physical and Theoretical Chemistry, Rheinische Friedrich-Wilhelms-University Bonn, Bonn, Germany*
THORU PEDERSON • *Department of Biochemistry and Molecular Pharmacology and Program in Cell and Developmental Dynamics, University of Massachusetts Medical School, Worcester, MA, USA*
TRISTAN PIOLOT • *Institut Curie, CNRS UMR3215, INSERM U934, Paris, France*
MATT PLUTZ • *Department of Cell and Developmental Biology, University of Illinois, Urbana, IL, USA*
JOAN C. RITLAND POLITZ • *Basic Sciences Division, Fred Hutchinson Cancer Research Center, Seattle, WA, USA*
TIM POLLEX • *Mammalian Developmental Epigenetics Group, Institut Curie, CNRS UMR3215, INSERM U934, Paris, France*
SAMIR RAHMAN • *Département de Biochimie, Faculté de Médecine, Université de Montréal, Montréal, QC, Canada*
MARIUS REINHART • *Department of Biology, Technische Universität Darmstadt, Darmstadt, Germany*
ANDREA SCHARF • *IUF – Leibniz Research Institute for Environmental Medicine at Heinrich-Heine University Düsseldorf, Düsseldorf, Germany*
ERIC C. SCHIRMER • *The Wellcome Trust Centre for Cell Biology, University of Edinburgh, Edinburgh, UK*
NIRAJ M. SHANBHAG • *Department of Cancer Biology, Abramson Family Cancer Research Institute, Basser Research Center for BRCA1/2, Perelman School of Medicine, University of Pennsylvania, Philadelphia, PA, USA*
YARON SHAV-TAL • *The Mina and Everard Goodman Faculty of Life Sciences and Institute of Nanotechnology, Bar-Ilan University, Ramat Gan, Israel*
JAN PETER SIEBRASSE • *Institute of Physical and Theoretical Chemistry, Rheinische Friedrich-Wilhelms-University Bonn, Bonn, Germany*
KARI-PEKKA SKARP • *Institute of Biotechnology, University of Helsinki, Helsinki, Finland*
JAN-HENDRIK SPILLE • *Institute of Physical and Theoretical Chemistry, Rheinische Friedrich-Wilhelms-University Bonn, Bonn, Germany*
MICHELLE STEVENSE • *Institut für Physiologische Chemie, Medizinische Fakultät, TU Dresden, Dresden, Germany*
MIKE STRAUSS • *Department of Biological Chemistry and Molecular Pharmacology, Harvard Medical School, Boston, MA, USA*
MARIA K. VARTIAINEN • *Institute of Biotechnology, University of Helsinki, Helsinki, Finland*
ANNA VON MIKECZ • *IUF – Leibniz Research Institute for Environmental Medicine at Heinrich-Heine University Düsseldorf, Düsseldorf, Germany*
TY C. VOSS • *Laboratory of Receptor Biology and Gene Expression, National Cancer Institute, National Institutes of Health, Bethesda, MD, USA*
DANIEL ZENKLUSEN • *Département de Biochimie, Faculté de Médecine, Université de Montréal, Montréal, QC, Canada*
NIKOLAJ ZULEGER • *The Wellcome Trust Centre for Cell Biology, University of Edinburgh, Edinburgh, UK*

Part I

Imaging Gene Expression and RNA Dynamics in Cells and Organisms

Chapter 1

High-Throughput Fluorescence-Based Screen to Identify Factors Involved in Nuclear Receptor Recruitment to Response Elements

Tina B. Miranda, Ty C. Voss, and Gordon L. Hager

Abstract

The glucocorticoid receptor is an inducible transcription factor which plays important roles in many physiological processes. Upon activation, GR interacts with regulatory elements and modulates the expression of genes. Although GR is widely expressed in multiple tissues, its binding sites within chromatin and the genes it regulates are tissue specific. Many accessory proteins and cofactors are thought to play a role in dictating GR's function; however, mechanisms involved in targeting GR to specific sites in the genome are not well understood. Here we describe a high-throughput fluorescence-based method to identify factors involved in GR loading at response elements. This screen utilizes a genetically engineered cell line that contains 200 repeats of a glucocorticoid response promoter and expresses GFP-tagged GR. Upon treatment with corticosteroids, GFP–GR forms a steady-state distribution at the promoter array, and its concentration at this focal point can be quantitatively determined. This system provides a novel approach to identify activities important for GR loading at its response element using siRNA libraries to target factors that enhance or inhibit receptor localization.

Key words Nuclear receptor, Glucocorticoid receptor, siRNA screen, Chromatin

1 Introduction

The glucocorticoid receptor (GR) belongs to a class of ligand-inducible transcription factors, which control a wide spectrum of important biological processes including metabolism, development, and inflammatory responses. Other members of the nuclear receptor superfamily including the estrogen receptor (ER), progesterone receptor (PR), androgen receptor (AR), and thyroid receptor (TR) also play prominent roles in physiological processes. In the absence of hormone, GR resides in the cytoplasmic compartment of the cell. Once bound to its ligand, the receptor translocates into the nucleus where it binds either to GR response elements (GREs) present on the DNA as a homodimer or to other regulatory sequences through tethering with additional

Yaron Shav-Tal (ed.), *Imaging Gene Expression: Methods and Protocols*, Methods in Molecular Biology, vol. 1042, DOI 10.1007/978-1-62703-526-2_1, © Springer Science+Business Media, LLC 2013

proteins [1]. GR can induce a positive outcome on transcription by recruiting components of the basic transcriptional machinery or can negatively regulate gene expression. Although GR is expressed in most tissues, the genes it regulates are tissue specific [2]. The chromosomal architecture at these response elements has been shown to be an important mediator for nuclear receptor binding at these sites and contributes to the cell-type-specific activity of these regulatory elements. In addition, many accessory proteins and cofactors are thought to play a role in regulating the function of GR.

As a general approach to the identification of factors that affect the ability of GR to bind to GREs, we have conducted an image-based high-throughput screen to identify chromatin modifiers important for GR loading at GREs (Miranda et al., manuscript in preparation). Although our initial screen focused on chromatin modifiers, other siRNA libraries can be screened using the same system. Direct GR binding to response elements in living cells can be visualized using a genetically engineered cell line containing 200 repeats of the glucocorticoid responsive promoter, MMTV, and expressing GFP-tagged GR [3, 4]. Upon treatment with dexamethasone, these cells form a steady-state distribution of GFP–GR at the promoter arrays, and this localized binding appears as a focal structure in the nucleus of hormone-treated cells (Fig. 1). Using previously described computer algorithms [5, 6], these structures can be automatically detected, and the levels of

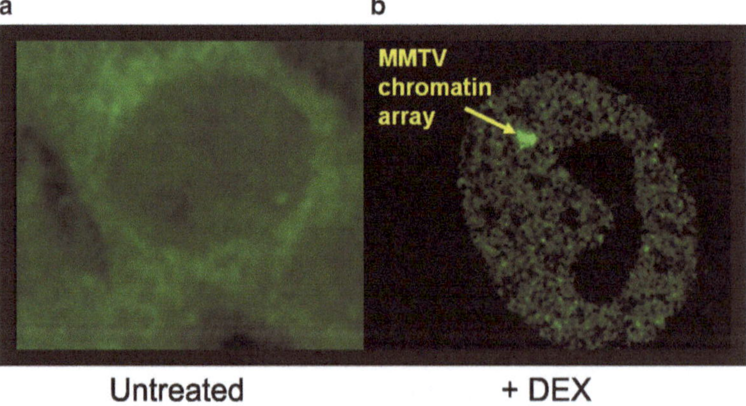

Fig. 1 Cells stably expressing GFP–GR and containing 200 copies of the MMTV promoter were treated either with the vehicle (untreated) or with dexamethasone for 30 min. Cells were then imaged using the Opera Imaging System (PerkinElmer). In the untreated cells (**a**), GFP–GR is localized to the cytoplasm. Upon treatment with dexamethasone (**b**), GFP–GR localizes to the nucleus and forms on the array of MMTV repeats

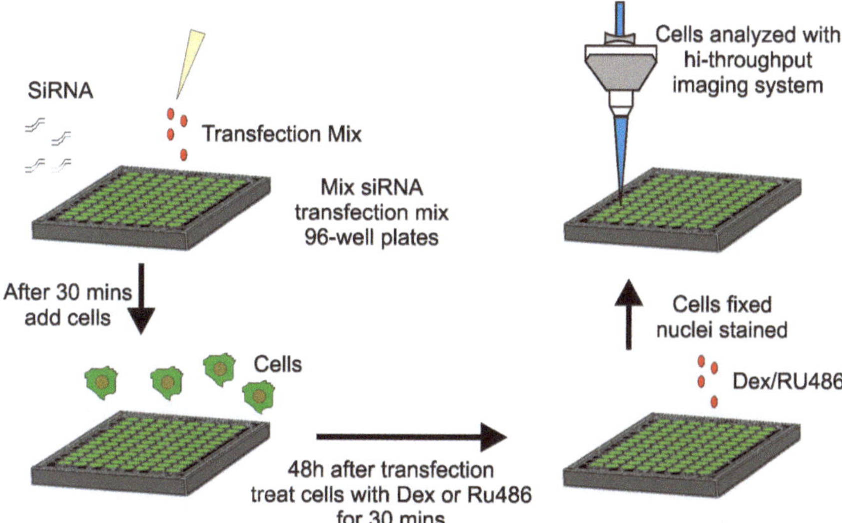

Fig. 2 Scheme of the siRNA screening procedure. Briefly, 3,617 cells are transfected with siRNA for 48 h. Cells are then treated with corticoid steroids for 30 min and then fixed with paraformaldehyde. Nuclei are stained and then cells are imaged using a high-throughput imaging system

GFP–GR concentrated at the amplified response elements can be quantitatively determined. The imaging data obtained has been previously shown to be comparative to ChIP data for GR loading at this array [6].

In this chapter we provide a detailed protocol that can be used to screen for factors that affect GR loading at its response elements (Fig. 2). This screen can be adapted to other nuclear receptors and/or transcription factors. Using multiple robotic systems (High-Throughput Imaging Facility, NCI), cells are reverse transfected with a siRNA library targeting 300 different chromatin modifiers. Forty-eight hours after transfection cells are treated with either dexamethasone (DEX) or the antagonist RU486. RU486 is a type II antagonist that induces GR to translocate into the nucleus but causes a decrease in GR loading at the array [6–8]. Cells transfected with a scrambled siRNA are treated with RU486 as a control for decreases in array loading (Fig. 3a). Cells transfected with siRNA to GAPDH are used to determine transfection efficiency (Fig. 3b, c). After treatment with the corticoid steroids for 30 min, cells are fixed and nuclei are stained with a fluorescent dye. Cells are then imaged using a high-throughput imaging system and images are analyzed using previously defined algorithms. Although we screened for factors that affect GR loading at a response element, other parameters, such as screening for factors that are involved in translocation of GR or GR turnover, can be examined with this system by incorporating modifications to the image analysis algorithms.

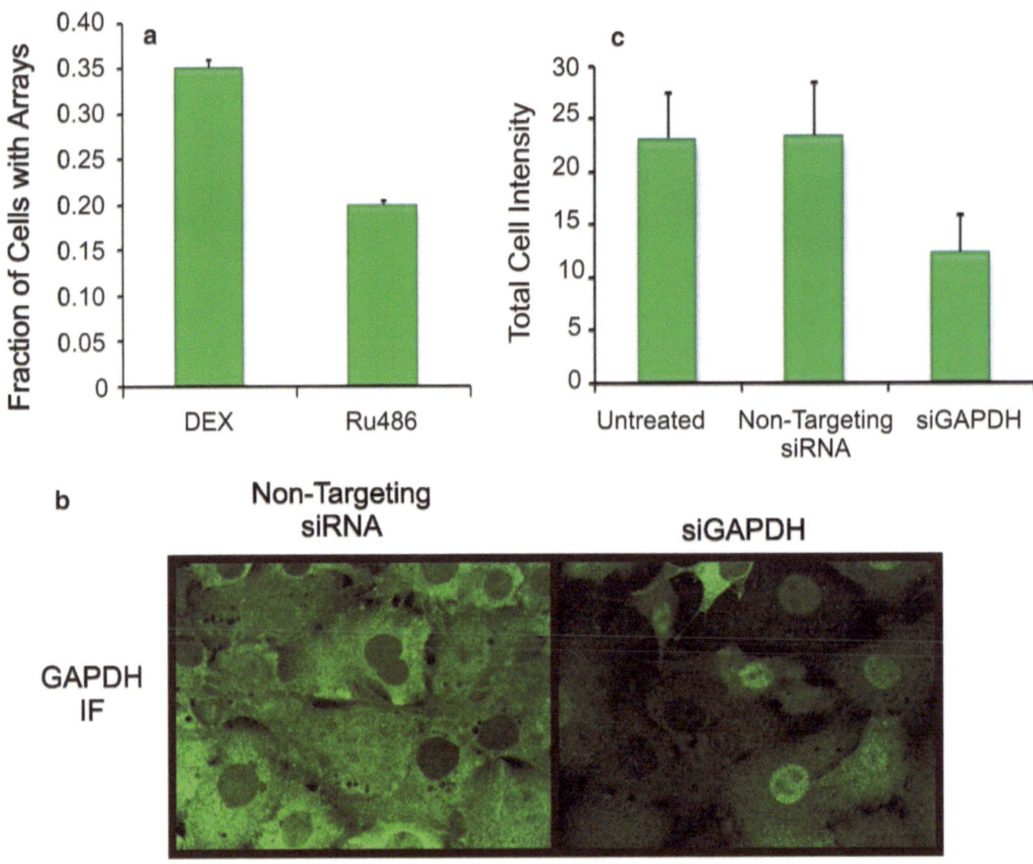

Fig. 3 Controls used in screen. (**a**) Upon treatment of cells with RU486 or dexamethasone (DEX), GR translocates into the nucleus. However, the fraction of arrays formed by GFP–GR binding to the MMTV repeats is lower in cells treated with RU486 than cells treated with DEX. (**b**) Immunofluorescence for GAPDH in cells transfected with nontargeting siRNA or cells transfected with siGAPDH. (**c**) Quantification of total cell intensity for GAPDH immunofluorescence. Numbers are an average from six 96-well plates with four wells per plate containing either nontargeting siRNA or siRNA to GAPDH

2 Materials

2.1 Cell Culture

1. A genetically engineered mouse mammary cell line containing 200 repeats of the glucocorticoid responsive promoter, MMTV, and expressing GFP-tagged GR (3,617 cells) [3, 4] (*see* **Note 1**).
2. Complete medium: Dulbecco's Modified Eagle's Medium (DMEM) supplemented with 10 % fetal calf serum, sodium pyruvate, nonessential amino acids, 2 mM glutamine, penicillin/streptomycin, and 5 μg/mL tetracycline to repress expression of the tet-regulated fusion proteins is used for maintenance of cells.

3. 0.25 % trypsin.
4. Phosphate-buffered saline (PBS).
5. Hemacytometer.

2.2 siRNA Screening

1. DharmaFECT 2 transfection reagent (Dharmacon) (*see* **Note 2**).
2. JANUS Automated Workstation (PerkinElmer 8-tip MPD).
3. Multidrop Combi (Thermo Scientific).
4. MatriPlate 96-well glass bottom (Matrical Bioscience; cat. No. MGB096-1-1-LG-BC).
5. DMEM.
6. A siRNA library (Dharmacon siGENOME siRNA pools).
7. Control siRNAs including a siRNA targeting GAPDH (Dharmacon ON-TARGETplus pool; cat. No. D-001830-20-05) and a nontargeting siRNA control (Dharmacon ON-TARGETplus control nontargeting pool; cat. No. D-001810-10-05).
8. Inducible medium: DMEM growth medium supplemented with 10 % charcoal–dextran-treated serum, sodium pyruvate, nonessential amino acids, 2 mM glutamine, and penicillin/streptomycin (*see* **Note 3**).

2.3 Treatment of Cells with Dexamethasone/RU486 and Fixation

1. 100 μM dexamethasone (DEX) in ethanol (store at −20 °C).
2. 100 μM RU486 in ethanol (store at −20 °C).
3. Inducible medium.
4. PBS.
5. 12 % paraformaldehyde (w/v) in PBS.
6. Bio Tek liquid handler (BioTek Instruments EL406).

2.4 Staining of the Nuclear Compartment

1. PBS.
2. DRAQ5 (BioStatus Limited; 5 mM; cat. No. DR50200) or DAPI diluted 1:5,000 in PBS.

2.5 Immunofluorescence Labeling of GAPDH

1. GAPDH antibody.
2. Texas Red dye conjugated AffiniPure donkey anti-mouse IgG (H + L) antibody.
3. PBS.
4. 0.5 % Triton X-100 in PBS.
5. 5 % bovine serum albumin (w/v) in PBS. Filter through a 0.45 μM filter.
6. 0.5 % Tween 20 in PBS.

2.6 Imaging and Imaging Analysis

1. Opera High Content Screening System (PerkinElmer).
2. Acapella imaging software (PerkinElmer).
3. Image analysis algorithms described previously [5, 6].

3 Methods

3.1 Cell Culture

1. For maintenance, grow 3,617 cells in DMEM supplemented with 10 % fetal calf serum, sodium pyruvate, nonessential amino acids, 2 mM glutamine, penicillin/streptomycin, and 5 μg/mL tetracycline at 37 °C and 5 % CO_2.
2. Passage the cells using standard tissue culture practices. Cells are usually split 1:6 every 2 days.

3.2 siRNA Screening

1. Dissolve all the siRNAs to a final concentration of 500 nM with siRNA buffer provided by the manufacturer. Manufacturer provides siRNAs in a 96-well plate.
2. Aliquot 100 μL/well of each siRNA into individual 96-well plates (*see* **Note 4**). At least four wells should contain nontargeting siRNA and two wells should contain siRNA against GAPDH as controls (*see* **Note 5**). We use the JANUS Automated Workstation for all siRNA pipetting to avoid any errors. Each well contains a pool of four siRNAs per gene. Small library screens are done independently at least three times with two technical replicates each time.
3. Prepare a transfection mix: Dilute DharmaFECT 2 in DMEM to a concentration of 0.1 μL/10 μL (*see* **Note 6**). DMEM should be at room temperature.
4. Using the JANUS automated system, take 10 μL from the siRNA daughter plate and add to the 96-well glass bottom MatriPlate.
5. Then take 10 μL of the transfection mix and add to the siRNA on the MatriPlate 96-well glass bottom. Mixing of transfection mix and siRNA is done by pipetting up and down five times. Incubate at room temperature for at least 30 min to allow transfection complexes to form.
6. Meanwhile, wash flasks containing the cells three times with PBS. Add the appropriate volume of 0.25 % trypsin and incubate the cells at room temperature until they detach from the bottom of the flask (approximately 1–2 min). Suspend the cells in inducible medium. Determine the concentration by counting the cells using a hemacytometer. Dilute cells in the inducible medium to a concentration of 10,000 cells/80 μL.

7. After transfection complexes have formed, add 80 μL of cells per each well of the 96-well glass bottom MatriPlate using the Multidrop Combi.
8. Incubate the plates for 48 h at 37 °C and 5 % CO_2 before treating the cells with DEX or RU486.

3.3 Treatment of Cells with DEX/RU486 and Fixation

1. Dilute 100 μM of DEX stock solution and 100 μM of RU486 stock solution in inducible medium so that the final concentration is 200 nM.
2. Using the Multidrop Combi, add 100 μL of the diluted RU486 (final concentration is 100 nM) to the control wells and 100 μL of the diluted DEX (final concentration is 100 nM) to every other well of the 96-well glass bottom MatriPlate. Incubate the cells at 37 °C for 30 min.
3. Using the BioTek liquid handler, add 100 μL of 12 % paraformaldehyde to each well (*see* **Note 7**). Incubate plate at room temperature for 15 min.
4. Wells are then washed six times with 200 μL of PBS using the BioTek liquid handler. Plates can be stored at this time at 4 °C in PBS.

3.4 Staining of the Nuclear Compartment

1. Using the BioTek liquid handler, the liquid is aspirated from the wells.
2. Then 100 μL of DRAQ5 (diluted 1:5,000 in PBS) is added to each well. Plates are incubated at room temperature for 15 min.
3. Wells are then washed six times with 200 μL of PBS using the BioTek liquid handler (plates can be stored at this time at 4 °C in PBS wrapped in aluminum foil).

3.5 Immunofluorescence Labeling of GAPDH

In order to confirm that the transfections worked, control cells transfected with siRNA against GAPDH should be included on every plate. Changes in the levels of GAPDH protein upon knockdown can be observed by conducting immunofluorescence using an antibody against GAPDH (Fig. 3b, c). The levels of GAPDH in cells transfected with nontargeting siRNA are then compared with wells transfected with the siRNA targeting GAPDH. Levels of cellular GAPDH can be quantified using the Acapella software. Under our conditions we observe a 50 % knockdown of GAPDH (*see* **Note 8**):

1. For control wells that contained siRNA to GAPDH or nontargeting siRNA, remove the PBS and add 100 μL of 0.5 % Triton X-100 to each well to permeabilize the cells. Incubate the plates on ice for 10 min.
2. Then wash the wells twice with 100 μL of 5 % BSA.

3. To block the cells, add 100 μL of 5 % BSA and incubate for 20 min at room temperature. Then remove the blocking solution.
4. Dilute the GAPDH antibody 1:1,000 in 5 % BSA. Then add 100 μL of antibody to each well and incubate for 1–2 h at room temperature.
5. Wash the wells three times with 0.5 % Tween 20.
6. Dilute the Texas Red donkey anti-mouse antibody 1:400 in 5 % BSA.
7. Remove the 0.5 % Tween 20 from the wells and add 100 μL of the diluted antibody.
8. Incubate at room temperature for 1 h.
9. Then wash the wells six times with 200 μL of PBS. Plates can be stored at 4 °C for about 1 week in PBS wrapped in aluminum foil.

3.6 Imaging and Imaging Analysis

1. The Opera High Content Screening System is used to image the plates. Twelve fields per well should be imaged with five z-stacks 1 μM apart.
2. Three exposures are imaged using the following wavelengths: 488 nM (images GFP–GR), 561 nM (images GAPDH immunofluorescence), and 640 nM (images DRAQ5).
3. Nuclear segmentation is then done using the Acapella imaging software as per manufacturer's instructions. Levels of GAPDH in the cells can also be analyzed using Acapella.
4. The tandem array of genes with GFP–GR is detected using previously described algorithms [5, 6].

3.7 Secondary Screen

To confirm the results from the primary screen, a secondary screen is conducted using siRNA against the top 20 hits from the primary screen. This time ON-TARGETplus siRNA SMART pools (Dharmacon) is used. These are siRNA that have been chemically modified to reduce off-target effects. The screen is conducted exactly as described for the primary screen except that the secondary screen is performed independently at least four times with three technical replicates each time, in order to produce solid statistical data.

3.8 Deconvolution of siRNA

After completion of the secondary screen, the siRNA pools are separated into individual siRNAs and re-screened in order to further eliminate off-target effects. For targets that were confirmed by the secondary screen, the screen is repeated using each of the individual siRNAs from the SMART siRNA pools. If more than two siRNAs per target from the SMART pools give the same effect, then it is less likely that the effect is an off-target effect. In this screen the individual siRNAs are randomly placed on a 96-well plate (nine wells per siRNA) to account for variation due to well position. This screen is conducted independently for two times, with three technical replicates each time.

3.9 Other Potential Validation Experiments

Many other types of experiments can be conducted in order to validate the siRNA screen. Immunofluorescence can be performed using antibodies raised against the top hits to confirm that the target localizes to the array of response elements. Chromatin immunoprecipitation (ChIP) studies to examine GR binding can be done in cells in which a top hit has been knocked down, in order to confirm the loss of binding at the array using a different method. RNA can be isolated from cells transfected with siRNA from a top hit, and qPCR can be conducted on the cDNA to confirm knockdown of the siRNA target.

4 Notes

1. In order to conduct the screen using other receptors or transcription factors, new cell lines would have to be engineered to contain an array of response elements specific to the factor being analyzed. That factor would also have to be tagged with a fluorescent tag.
2. Transfection reagents should be stored at 4 °C and kept on ice while working with them in order to prevent a decrease in it efficiency.
3. Antibiotics can be used with 3,617 cells, as it does not affect the health of these cells during transfection. This keeps the cells from becoming contaminated during the transfection process. However, it should be noted that the health of other cell types may be affected by the presence of antibiotics in the medium during the transfection procedure.
4. Diluted siRNA stocks should be aliquoted into several individual plates and stored at −80 °C to avoid more than two freeze–thaw cycles. This will help maintain the integrity of the siRNA for an extended period of time.
5. Control siRNAs should be included on all plates. This includes nontargeting siRNA (for both DEX or RU486 treatments) and siRNA to GAPDH.
6. If transfections are done in other cell types, different transfection reagents should be tested and concentrations optimized.
7. Fresh paraformaldehyde is in PBS usually prepared before each experiment. However, we find that using 16 % paraformaldehyde (Electron Microscopy Sciences cat. No. 15710) diluted to 12 % on the day of the experiments works well.

8. It should be noted that optimization of the system should be conducted carefully. Optimal transfection conditions can cause a decrease in array formation. Therefore, a fine balance needs to be found between transfection conditions and cell health.

Acknowledgements

This research was supported, in part, by the Intramural Research Program of the National Institutes of Health, National Cancer Institute, Center for Cancer Research. T. B. M. was supported, in part, by a National Institute of General Medical Sciences Pharmacological Research and Training Fellowship.

References

1. Biddie SC, Hager GL (2009) Glucocorticoid receptor dynamics and gene regulation. Stress 12:193–205
2. Wiench M, Miranda TB, Hager GL (2011) Control of nuclear receptor function by local chromatin structure. FEBS J 278: 2211–2230
3. Kramer P, Fragoso G, Pennie WD et al (1999) Transcriptional state of the mouse mammary tumor virus promoter can effect topological domain size in vivo. J Biol Chem 274:28590–28597
4. Walker D, Htun H, Hager GL (1999) Using inducible vectors to study intracellular trafficking of GFP-tagged steroid/nuclear receptors in living cells. Methods 19:386–393
5. Voss TC, Schiltz RL, Sung MH et al (2011) Dynamic exchange at regulatory elements during chromatin remodeling underlies assisted loading mechanism. Cell 146:544–554
6. Voss TC, John S, Hager GL (2006) Single cell analysis of glucocorticoid receptor action reveals that stochastic post-chromatin association mechanisms regulate ligand-specific transcription. Mol Endocrinol 20:2641–2655
7. Schulz M, Eggert M, Baniahmad A et al (2002) RU486-induced glucocorticoid receptor agonism is controlled by the receptor N terminus and by corepressor binding. J Biol Chem 277: 26238–26243
8. Szapary D, Huang Y, Simons SS Jr (1999) Opposing effects of corepressor and coactivators in determining the dose–response curve of agonists, and residual agonist activity of antagonists, for glucocorticoid receptor-regulated gene expression. Mol Endocrinol 13:2108–2121

Chapter 2

Live-Cell Imaging Combined with Immunofluorescence, RNA, or DNA FISH to Study the Nuclear Dynamics and Expression of the X-Inactivation Center

Tim Pollex, Tristan Piolot, and Edith Heard

Abstract

Differentiation of embryonic stem cells is accompanied by changes of gene expression and chromatin and chromosome dynamics. One of the most impressive examples for these changes is inactivation of one of the two X chromosomes occurring upon differentiation of mouse female embryonic stem cells. With a few exceptions, these events have been mainly studied in fixed cells. In order to better understand the dynamics, kinetics, and order of events during differentiation, one needs to employ live-cell imaging techniques. Here, we describe a combination of live-cell imaging with techniques that can be used in fixed cells (e.g., RNA FISH) to correlate locus dynamics or subnuclear localization with, e.g., gene expression. To study locus dynamics in female ES cells, we generated cell lines containing TetO arrays in the X-inactivation center, the locus on the X chromosome regulating X-inactivation, which can be visualized upon expression of TetR fused to fluorescent proteins. We will use this system to elaborate on how to generate ES cell lines for live-cell imaging of locus dynamics, how to culture ES cells prior to live-cell imaging, and to describe typical live-cell imaging conditions for ES cells using different microscopes. Furthermore, we will explain how RNA, DNA FISH, or immunofluorescence can be applied following live-cell imaging to correlate gene expression with locus dynamics.

Key words Live-cell imaging, X chromosome inactivation, Fluorescent in situ hybridization (FISH), Embryonic stem cells, Tet operator, Tet repressor

1 Introduction

Differentiation of embryonic stem cells (ESCs) is accompanied by changes in chromatin structure and content. Live-cell imaging of chromosomes and chromatin is a very powerful means of studying the dynamics, kinetics, and order of events that occur during ESC differentiation, at the single-cell level. One of the hallmarks of differentiation in female ESCs is random X chromosome inactivation (XCI), the process ensuring dosage compensation in female mammals, whereby either the maternal or paternal X chromosome becomes transcriptionally silenced during the first few days of ESC

Yaron Shav-Tal (ed.), *Imaging Gene Expression: Methods and Protocols*, Methods in Molecular Biology, vol. 1042, DOI 10.1007/978-1-62703-526-2_2, © Springer Science+Business Media, LLC 2013

differentiation. Random XCI is regulated by an X-linked region known as the X-inactivation center (Xic) [1] which contains the genes encoding for the long noncoding RNA Xist [2–4] and its antisense transcript Tsix [5] as well as numerous long-range regulatory elements [6]. Monoallelic upregulation of *Xist* is the first step during random XCI. It is regulated by multiple dynamic events occurring as soon as ES cell differentiation begins. The upregulation of *Xist* from only one of the two X chromosomes is partly ensured by the monoallelic downregulation of its antisense repressor, *Tsix*. The transient downregulation of *Tsix* on one allele has been linked to transient homologous associations between the two Xics at this locus during early ES cell differentiation. Xic trans-interactions or "pairing" events have been proposed to be involved in both sensing the number of X chromosomes that are present [7] and establishing asymmetric *Tsix* expression, such that monoallelic *Tsix* downregulation leads to monoallelic *Xist* upregulation [7–10].

Dissecting the order and function of these dynamic and transient events represents a perfect illustration of the need for live-cell imaging techniques in ESCs and their differentiated derivatives. Several questions including the duration of the pairing process, as well as the extent to which close proximity is actually followed by complete overlap between the two Xic loci, can be addressed using live-cell imaging of Xic dynamics during ESC differentiation. Live-cell imaging has also enabled Xic pairing events to be linked with a functional outcome. When combined with subsequent RNA FISH, it was found that pairing of the two Xics is followed by transient, asymmetric downregulation of *Tsix* [10]. Based on this combination of live and fixed cell analyses, it has been concluded that Xic pairing is likely to be an important upstream event for symmetry-breaking between the two Xics.

In order to visualize Xic dynamics in living ESCs, we have exploited the bacterial Tet operator (TetO)/Tet repressor (TetR) system. The TetO array can be visualized in living cells by transient or stable expression of TetR fused to fluorescent proteins (e.g., mCherry) which can bind with high affinity to the TetO array. In our study, a multicopy array of 224 single TetO sites was inserted by homologous recombination into a region adjacent to *Tsix* within the Xic in female mESCs (Fig. 1). An ES cell line in which both X chromosomes were tagged in this way was then derived [10] and a TetR-mCherry transgene was used to follow the dynamics of the Xic in living cells, using a DeltaVision microscope. This was the first report of live-cell imaging of an endogenous locus in ESCs and their early differentiated derivatives [10].

In this chapter we present a detailed description of the current methodology that we use to investigate nuclear and chromosome dynamics in ES cells, using the previously described female ESC line which carries two TetO-/TetR-tagged Xic loci. Several aspects

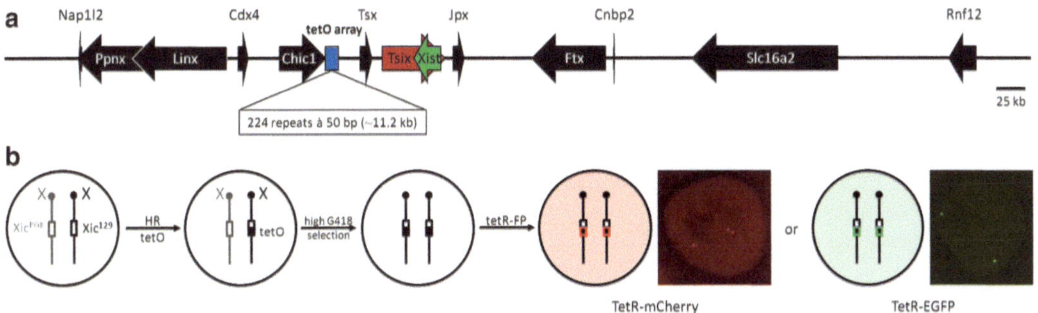

Fig. 1 (**a**) Schematic representation of the X-inactivation center (Xic), adapted from ref. 10. A TetO array containing 224 repeats was integrated by homologous recombination close to the Chic1 gene in the X-inactivation center approximately 40 kb upstream of the 5′ end of Tsix. (**b**) Integration of the TetO array resulted in a heterozygous cell line which has been treated with high concentrations of G418 to generate ES cell lines homozygous for the integrated TetO array [10]. TetR-FP binds the TetO arrays in the Xic and can be detected by live-cell imaging

will be covered, including (a) rapid acquisition live-cell imaging at high resolution in time, to study the dynamics and kinetics of fast processes; (b) imaging over more extended periods with less frequent acquisitions, to assess events over entire cell cycles and during differentiation; and (c) combination of live-cell imaging and fixed cell analysis (e.g., RNA FISH, immunofluorescence), for example, to follow Xic pairing dynamics and correlate this with *Tsix/Xist* expression patterns using RNA FISH [10]. We also mention recent advances that we have made concerning longer-term live-cell imaging of ESC using the OMX microscope which provides increased sensitivity and decreased phototoxicity for live-cell imaging.

2 Materials

2.1 Tissue Culture

1. ESC medium: Dulbecco's Modified Eagle's Medium (DMEM) supplemented with 15 % fetal calf serum (batch tested FCS for ES cells), 0.2 % (v/v) 2-mercaptoethanol, and 1,000 U/ml leukemia inhibitory factor (LIF).
2. Differentiation medium: DMEM supplemented with 10 % FCS, 0.2 % (v/v) 2-mercaptoethanol, and 100 nM all-*trans*-retinoic acid (RA).
3. Freezing medium: 10 % dimethyl sulfoxide (DMSO) in FCS.
4. 6, 24, 96-well tissue culture plates.
5. Tissue culture flasks.
6. 10-cm tissue culture dishes.
7. 10× phosphate-buffered saline (PBS).

8. Double-processed tissue culture water.
9. 0.1 % gelatin in 1× PBS.
10. Alternative pre-coat for live-cell imaging: poly-d-lysine or human plasma fibronectin-purified protein.
11. 0.05 % trypsin-EDTA (1×) with phenol red.
12. 50 mg/ml hygromycin B.
13. Lipofectamine 2000 transfection reagent.

2.2 Live-Cell Imaging

1. 12, 18, and 30 mm #1.5 glass round coverslips.
2. MatriPlate 96-well glass-bottom microwell plate (MatriCal Bioscience).
3. Ludin chamber type 3 (Life Imaging Services).
4. Non-coat 35-mm petri dish with #1S 12 mm coverslip with grid (Matsunami Glass Ind.).
5. Non-coat 35-mm petri dish with #1.5 glass coverslip (MatTek Corp).
6. Fine forceps.
7. Paraffin.
8. DeltaVision Core (Applied Precision) equipped with a CoolSNAP HQ2 camera and an incubator chamber with CO_2 perfusion system (Okolab).
9. DeltaVision OMX V3 (Applied Precision) equipped with three Evolve back illuminated EMCCD cameras (Photometrics), three laser lines (405, 491, and 561 nm, powered with 200 mV, respectively). Speckles are removed by shaking of the multimode fiber at 4 kHz; OMX is equipped with an objective heater system (Applied Precision) and a heating stage (Applied Precision). Glass-bottom dishes with a 0.17 mm thickness (MatTek) are placed in a cell observation room (adapted from Applied Precision Instrument) and placed on a heating sample holder (custom made). The observation room includes a heating lid (adapted from Bioptechs) and a temperature probe (adapted from Bioptechs) which can be placed directly in the medium to monitor the temperature in the culture dish. CO_2 is perfused through the heating lid using a LIS CO_2 controller system.

2.3 Immunofluorescence/RNA/DNA FISH

1. Fixation solution: 3 % paraformaldehyde in PBS (freshly prepared, pH adjusted).
2. Permeabilization solution: 0.5 % Triton X-100 in PBS supplemented with 2-mM vanadyl ribonucleoside complex (VRC).

3. Antibody dilution solution: 1 % bovine serum albumin (BSA) in 1× PBS supplemented with 500 U/ml RNase inhibitor (e.g., RiboLock 40 U/μl).
4. 70 % Ethanol (v/v).
5. Nick translation kit (Abbott) containing nick translation enzyme, dNTP solutions, and 10× nick translation buffer.
6. Spectrum green dUTP (Abbott).
7. Spectrum red dUTP (Abbott).
8. Cy-5 dUTP.
9. Formamide (FA). Upon opening, aliquot immediately and keep at −20 °C.
10. Mouse Cot-1 DNA (Invitrogen).
11. DNA, molecular biology grade from fish sperm (Roche).
12. 3 M sodium acetate pH 5.2.
13. 10 mg/ml RNase A.
14. 5,000 U/ml RNase H.
15. 20× SSC buffer concentrate.
16. 2× Hybridization buffer: 40 % (w/v) sodium dextran sulfate, 20× BSA, 400-mM VRC in 4× SSC.
17. Wash buffer: 50 % formamide/2× SSC pH 7.2–7.4.
18. DNA counterstaining solution: 4′,6-diamidino-2-phenylindole dihydrochloride (DAPI) in 2× SSC.
19. Mounting solution: 90 % (v/v) glycerol, 0.1 % (w/v) *p-phenylenediamine*, pH 9.0 in PBS.
20. Eppendorf Centrifuge 5417R.
21. Eppendorf Concentrator plus.
22. Eppendorf Thermomixer comfort.
23. Liquiport Liquid pump.
24. Shake'N'Bake Hybridization Oven.

2.4 Plasmids

1. pBROAD3-TetR-ICP22-mCherry.
2. pBROAD3-TetR-ICP22-EGFP.
3. pBROAD3-mCherry-PCNA.

2.5 ES Cell Lines

1. PGK 31′-60 (PGK 2TetO TetR-mCherry).
2. PGK 31′-T12 TetR-ICP22-EGFP 12C (PGK 2TetO TetR-EGFP).

3 Methods

3.1 General Tissue Culture

All of the protocols described here apply to ESCs that grow in feeder-free conditions. Prior to live-cell imaging, ESCs or their differentiated derivatives should be growing optimally (good colony morphology, minimal cell death). Suboptimal cells should never be used for live-cell imaging, as they will rapidly die or show low/aberrant fluorescence. Usually it can take several days, and sometimes more than one passage for ESCs, to reach optimal conditions. Furthermore, imaging should be performed a minimum of 24-h post-plating of the cells onto the coverslip/in the chamber.

3.1.1 Thawing and Culturing ESCs

1. Thaw cells from liquid nitrogen in a water bath at 37 °C and resuspend them in an appropriate amount of ESC medium.
2. Pellet cells for 5 min at 180 × *g* at room temperature in a 15-ml Falcon tube.
3. Resuspend the cell pellet in an appropriate amount of medium and distribute to a well or flask pre-coated with 0.1 % gelatin. Cells are grown as an adherent culture in wells or flasks and are split every 2–3 days in ESC medium.

3.1.2 Passaging ESCs and Plating for Imaging

1. Wash 70–80 % confluent cells once with pre-warmed 1× PBS (37 °C).
2. Incubate the cells for 8–10 min at 37 °C in 0.05 % trypsin-EDTA.
3. Stop trypsinization by addition of ESC medium and resuspend cells by vigorous pipetting with a 5-ml pipette (foaming indicates optimal dissociation). Split cells 1:5 to 1:10 (depending on the ESC line) into gelatin-coated wells or flasks. For live-cell imaging, count and distribute appropriate cell numbers onto pre-prepared, gelatin-coated petri dishes or coverslips.

3.1.3 Freezing Cells

1. Centrifuge cells after trypsinization and resuspension at room temperature at 180 × *g* for 5 min.
2. Aspirate medium and resuspend cells in freezing medium and distribute to freezing tubes.
3. Freeze cells gradually at −80 °C, then transfer to liquid nitrogen the next day for long-term storage.

3.2 Generation of Stably Transgenic ESC Lines for Live-Cell Imaging Purposes

Imaging of chromosome dynamics can be performed by transient expression of TetR-FP in ES cell lines containing TetO arrays. However, it should be noted that optimal imaging results can only be obtained using ESC lines stably expressing TetR-FP. The following section describes in brief the general procedure for the derivation of ESC lines stably expressing a TetR-FP fusion protein. The example given involves co-transfection of plasmids expressing the TetR-FP fusion proteins, as well as a plasmid expressing a selectable marker.

3.2.1 Transfection

1. Grow ESCs to 70–80 % confluence in a well of a 6-well plate.
2. Change ES medium 2 h prior to transfection.
3. Perform transfection with Lipofectamine 2000, following the manufacturer's instructions, using 3 μg of the TetR-FP expression vector (e.g., pBROAD3-TetR-ICP22-mCherry [10]) mixed with 0.2 μg selection marker expression plasmid (e.g., hygromycin resistance) per transfected well.
4. Incubate overnight.
5. Change medium and trypsinize the transfected cells the following morning and plate out 10 % and 90 % of the cells onto two pre-gelatinized 10-cm tissue culture dishes and culture the cells in ES medium for another day.
6. On the following day, change ES medium and start selection by adding the appropriate drug (e.g., hygromycin B; *see* **Note 1**).
7. After selection begins, medium should be changed every day until clone picking. Nonresistant cells should die during the next 1–5 days depending on the drug used.

3.2.2 ESC Clone Picking

1. Discrete ESC colonies should become visible by eye after approximately 7–10 days of culture. Colonies should be picked when they are still smaller than 1 mm and should not be allowed to grow too large or become necrotic in the center as this will increase the chances of differentiation and loss.
2. Wash the dish once with 10 ml 1× PBS. Fill the dish with 5 ml PBS.
3. Pick single colonies in 30 μl PBS using a 200-μl pipette tip and pipette them into individual wells of a 96-well plate.
4. Add 30 μl of 0.05 % trypsin-EDTA per well to the picked colonies and incubate at 37 °C for 8 min.
5. Add 150-μl ES medium to stop trypsinization and dissociate the cells of the colony into a single-cell suspension by pipetting using a multi-pipette and 200-μl tips.
6. Distribute the dissociated cells to a pre-gelatinized 96-well plate. Change medium daily until cells grow to a confluence of 60–70 %.

3.2.3 Screening of Drug-Resistant Clones for the Presence and Expression of the TetR-FP Transgene

1. Split cells 1:3 into a pre-gelatinized 96-well plate (for future propagation) and a pre-gelatinized 96-well plate with glass-bottom dish (for screening).

Table 1
Typical numbers of ESCs seeded at day −1 for live-cell imaging during ESC differentiation with LIF withdrawal and addition retinoic acid in PGK derived cell lines

Day of differentiation	Number of seeded cells
Day 0	$\geq 5 \times 10^4$ cm^{-2}
Day 0.5	~4×10^4 cm^{-2}
Day 1	~3×10^4 cm^{-2}
Day 1.5	~2×10^4 cm^{-2}
Day 2	~1.75×10^4 cm^{-2}
Day 2.5	~1.5×10^4 cm^{-2}
Day 3	~1.25×10^4 cm^{-2}
Day 3.5	~1×10^4 cm^{-2}
Day 4	$\leq 1 \times 10^4$ cm^{-2}

2. The clones can be screened in the 96-well plate with glass bottom using the DeltaVision microscope to detect expression of TetR-FP and visibility of TetO loci (*see* **Note 2**).

3.2.4 Propagation of Selected Clones

1. When clones have reached 70–80 %, confluence in the remaining 96-well plate transfers them into single, pre-gelatinized wells of a 48-well or 24-well plate by trypsinization. Each clone is likely to have its own growth kinetics and should only be passaged when it appears optimal in terms of confluence and morphology.
2. Repeat the procedure to transfer clones from the 24-well plate to 6-well plates. Cells can be frozen down and stored at this stage (6-well)-or further propagated in 25-cm^2 flasks after culturing them in 6-well plates.

3.3 Live-Cell Imaging of ESCs

3.3.1 Preparation of ESCs for Imaging Using a Ludin Chamber

1. Sterilize 12-mm coverslips (*see* **Note 3**) by flaming with ethanol and distribute sterile coverslips in the wells of a 6-well tissue culture plate (three coverslips per well). Pre-coat coverslips with 0.1 % gelatin in PBS (*see* **Note 4**). Aspirate the pre-coating solution before distribution of the cells.
2. Seed cells at day −1 according to a scheme that provides subconfluent cell density (*see* **Note 5**) on the day of the imaging experiment. Typical cell densities at day −1 are depicted in Table 1 (as mentioned above, ESCs should be in optimal condition for live-cell imaging).

3. If differentiating cells are required, start differentiation (day 0) by washing out ESC medium and adding differentiation medium. Between the medium changes, wash the cells three times with pre-warmed 1× PBS to fully remove any traces of ESC medium.
4. Change medium 1–2 h before imaging (*see* **Note 6**).
5. Mount the coverslip into the Ludin live-cell chamber (*see* **Notes** 7 and **8**): Take the coverslips from the 6-well plate with a pair of fine forceps and place it into the bottom of the Ludin chamber with cells facing up. Place a 12 mm rubber ring on top of the coverslip. Carefully screw the top parts of the Ludin chamber on the bottom part. Fill the Ludin chamber with 0.5–1-ml pre-warmed ESC medium or differentiation medium. Place a 30-mm glass coverslip on top of the chamber to prevent evaporation.
6. Place the Ludin chamber on the objective table of the DeltaVision microscope after putting immersion oil on the objective. Place the CO_2 perfusion system on the objective table (*see* **Note 9**). Focus the cells in the bright-field mode using the eyepiece.
7. Start imaging (*see* Subheading 3.3.3).

3.3.2 Preparation for Imaging Using Plain or Gridded 35-mm Petri Dishes (See Note 10)

1. Pre-coat the surface of the petri dish with 0.1 % gelatin in PBS (*see* **Note 4**).
2. Follow **steps 2–4** of Subheading 3.3.1.
3. Place the petri dish on the objective table of the DeltaVision microscope after putting immersion oil to the objective.
4. Note the orientation of the petri dish on the objective table (*see* **Note 11**).
5. Place the CO_2 perfusion system on the objective table (*see* **Note 8**). Focus the cells in the bright-field mode using the eyepiece.
6. Note the exact position of the imaged cells and colonies on the gridded coverslip in the petri dish using the bright-field mode and the numbered grid on the coverslip (also *see* Subheading 3.4) and memorize these positions using the microscope software.
7. Start imaging (*see* Subheading 3.3.3).

3.3.3 Live-Cell Imaging of Chromosome Dynamics During Differentiation Using the DeltaVision Microscope

1. Live-cell imaging conditions with any given microscope need to be adjusted, depending on the nature of the experiment and on the ESC line being used. Conditions need to be optimized to avoid phototoxicity issues and to ensure that the signal-to-noise ratios of the fluorescence signals are adequate. This can vary a lot depending on the ESC line and fluorescent protein being used. For example, in the case of imaging of TetO/TetR loci, taking only snapshots of cells to determine the position of

the TetO array can be done with rather high exposure times. Long-term live-cell imaging requires different imaging conditions (see below). The ESCs to be imaged can be located and focused on using the eyepiece and the bright-field mode of the microscope. Once cells are in focus, fluorescent light can be used to determine optimal imaging conditions (*see* **Note 12**). Avoid unnecessary exposure of cells with fluorescent light, which can induce phototoxicity and bleaching of the signal.

2. Determining optimal imaging conditions is crucial for the quality of the imaging experiment and the viability of the cells. The exposure time for live-cell imaging needs to be adjusted to minimize bleaching and phototoxicity. Depending on the excitation light and light source as well as the FPs used and their expression levels, phototoxic effects are positively correlated with the total exposure to light. The number of Z stacks will depend on the size of the object being imaged but should be minimized to reduce the total amount of light that the cells are exposed to during the acquisition. For our purposes (a single TetO array locus on the X chromosome), a step size of 0.3 μm was sufficient to faithfully detect the TetO array and measure 3D distances between the two loci on different X chromosomes. Depending on the size of the array and the signal-to-noise ratio in the cell line used, the step size might need to be adjusted. However, decreasing the step size means more images per Z stack, which will result in a higher amount of light exposure.
3. Use very low exposure times and test for phototoxicity (*see* Subheading 3.3.5) when performing long-term time-lapse live-cell imaging. Mouse ESCs are very sensitive to light and their differentiated derivatives even more so (*see* **Note 10**). The intervals between the acquisitions should be adapted to the total time of imaging. Typical examples of live-cell imaging with the DeltaVision are described below.
4. Figure 2a depicts a typical time-lapse acquisition performed over 3 h with 10-min intervals between acquisitions at the DeltaVision microscope. Acquired were Z stacks with a total number of 34 planes spaced by 300 nm (total stack size 10.2 μm). Acquisitioln time was 60 ms and the transmission filter was set to 32 %.
5. Figure 2b depicts a typical time-lapse acquisition performed over 30 min with 1-min intervals.

3.3.4 Long-Term Live-Cell Imaging Using the OMX Microscope

In order to increase speed of image acquisition, decrease the exposure time, and decrease phototoxicity during live-cell imaging of ESCs, OMX microscopy presents several advantages over the previous DeltaVision system we used. The DeltaVision is equipped with a standard metalloid lamp producing (depending on the filters used) a polychromatic excitation light. The OMX uses laser

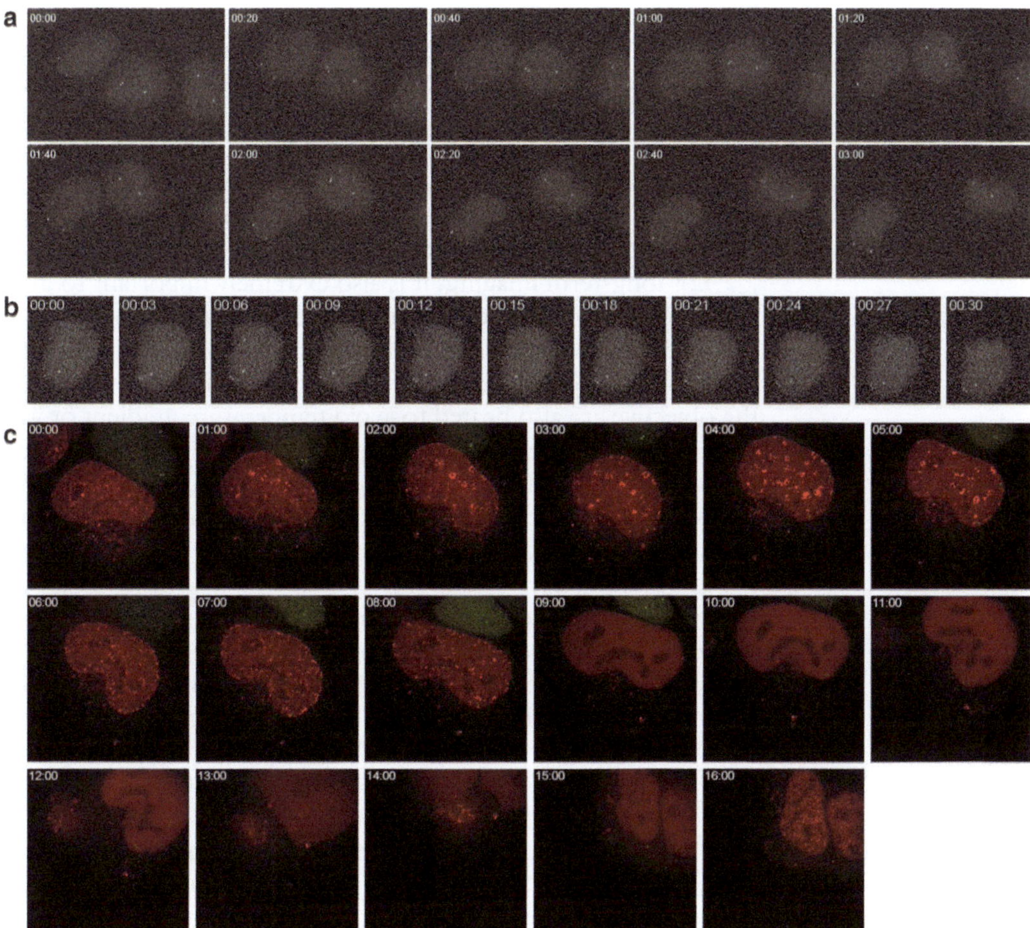

Fig. 2 (**a**) PGK 2TetO TetR-EGFP (undifferentiated) ESCs imaged for 3 h using the DeltaVision microscope with 10-min intervals (19 time points in total, shown is only every second time point). The displayed images are Z projections of stacks containing 34 planes with 300-nm spacing (total stack size 10.2 μm). The exposure time was set to 60 ms and the transmission filter set to 32 %. Nuclear shape can be determined from the TetR-EGFP background signal. The TetR-EGFP bound to one TetO array is represented by one spot. Depending on the cell cycle stage, two to four individual spots can be detected representing either unreplicated arrays, replicated arrays, or replicated arrays with two separated sister chromatids. The loci and the nuclei are highly dynamic. (**b**) PGK 2TetO TetR-EGFP (undifferentiated) imaged for 30 min using the DeltaVision microscope with 1-min intervals (31 time points in total, shown is only every third time point). The displayed images are Z projections of stacks containing 34 planes with 300 nm spacing (total stack size 10.2 μm). The exposure time was set to 60 ms and the transmission filter set to 32 %. Even during short-term acquisitions, one can observe highly dynamic sister chromatids and loci. Position and shape of ES cell nuclei also change during short-term acquisitions. (**c**) PGK 2TetO TetR-EGFP (undifferentiated) transfected with pBROAD3-mCherry-PCNA imaged for 16 h using the OMX microscope with 10-min intervals (96 time points, shown is only every sixth time point). The displayed images are Z projections of stacks containing 47 planes with 300-nm spacing (total stack size 14.1 μm). Acquisitions were made in the conventional mode using simultaneous acquisition. Exposure time was set to 20 ms for the green channel (488-nm laser, 0.1 % laser power) and 10 ms for the red channel (568-nm laser, 0.1 % laser power). The cell in the center of the images was in mid-S phase during the start of the acquisition, enters late S phase (time points: 2–8 h) before entering G_2 (time point 9 h). Cytokinesis occurs at time points 13–14 h. S phase is reinitiated in the daughter cells between time points 15 and 16 h

illumination, resulting in monochromatic excitation. In addition, the OMX contains multiple cameras (see specifications in Subheading 2) allowing for parallel multiwavelength acquisition in the different color channels. The OMX system is extremely stable, being encased in a vibration and dust-free environment, thus rendering far better conditions for long-term imaging. Parallel acquisition with multiple cameras and laser excitation allow reduction in exposure times and faster speed of acquisition using this system.

For long-term imaging, it is also extremely important to guarantee optimal and stable temperature and CO_2 conditions. To this end, the glass-bottom petri dish is covered with a heating lid in the observation room of the OMX. The sample holder, the stage, and the objective are also heated to avoid temperature gradient or variation. Temperature should be monitored using a temperature probe (*see* Subheading 2). Temperature fluctuations larger than 0.1 K from the optimal temperature should be avoided. Optimal CO_2 perfusion needs to be ensured using a gas perfusion system (*see* Subheading 2).

Figure 2c depicts a typical time-lapse two-color acquisition using the OMX, performed for 16 h with 10-min intervals. Z stacks consist of 48 planes, spaced by 300 nm (total stack size 14 μm). Exposure time was 20 ms for the green channel (488 nm) and 10 ms for the red channel (561 nm). Transmission was set to 0.1 % and the EMCCD camera gain to 300.

3.3.5 Assessing the Cell State After Long-Term Live-Cell Imaging

The color of the culture medium and cell morphology should always be recorded at the beginning and end of any live-cell imaging experiment. Useful indicators of possible phototoxicity include cell morphology, cell cycle, and cell division perturbations. In the ESC line we use as a model system, the nucleoli have a different background staining than the rest of the nucleus. Significant changes in nucleolar morphology can also be used an indicator of phototoxic effects. Rounding up of the cells can indicate initiation of cell death. Creation of micronuclei is another indicator of phototoxicity. Prolongation of the cell cycle can also be indicative of phototoxic effects (induction of checkpoint). To assess for the degree of cell viability after imaging, the cells can either be fixed or assessed for DNA damage or cell cycle markers by immunofluorescence, or else they should be left on the microscope following imaging for several hours and checked for continued viability.

3.4 RNA FISH Following Live-Cell Imaging

Depending on the purpose of the experiment, cells can be fixed either directly after live-cell imaging or else following a given period of time. The protocols outlined below are described in more detail in ref. 11. An example of live-cell imaging of ESCs and their differentiated counterparts followed by fixation and RNA FISH is given in Fig. 3. Position of the cells was memorized using a petri dish with gridded coverslip (Fig. 3a; *see* also Subheading 3.3.2).

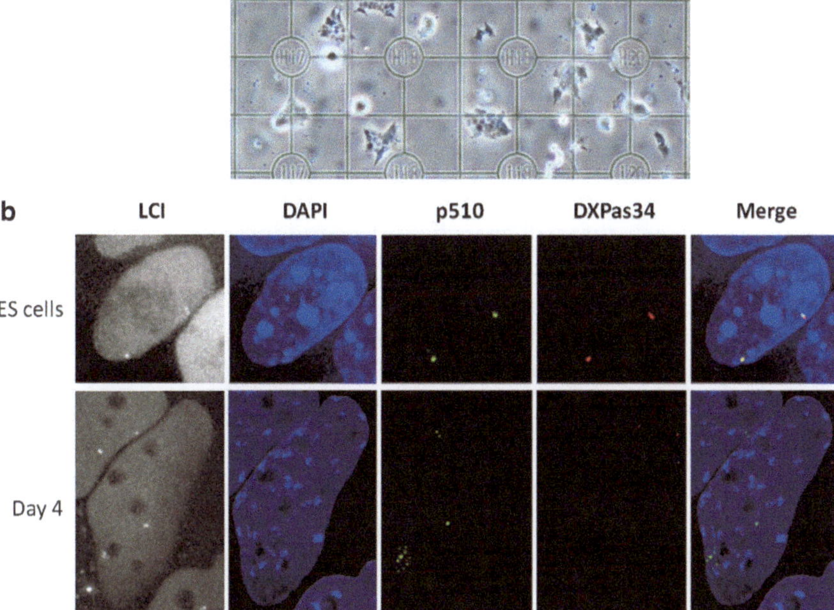

Fig. 3 (**a**) Small colonies (5–10 cells each) of the feeder-independent PGK12.1 ESC line in a 35-mm petri dish with 12-mm coverslip with grid. Positions and orientation of the colonies and cells in live-cell imaging can be memorized using the grid with the indicated letters and numbers. (**b**) Live-cell imaging (LCI) of PGK 2TetO TetR-mCherry cells with subsequent fixation and RNA FISH. ESCs or their differentiated counterparts (day 4 after LIF withdrawal) were imaged in a gridded petri dish directly before fixation. The same cells were imaged again after performing RNA FISH using a probe hybridizing to *Xist/Tsix* (p510) and *Tsix* only (DXPas34). In undifferentiated ES cells, Tsix is expressed biallelically from both X chromosomes (*upper panel*). At day 4 after LIF withdrawal, *Xist* expression has been monoallelically upregulated from one of the X chromosomes (*lower panel*)

3.4.1 Fixation and Permeabilization

1. Wash cells (on coverslips or in petri dishes) in pre-warmed 1× PBS.
2. Fix cells for 10 min in fixation solution at room temperature.
3. Wash cells three times with 1× PBS.
4. Permeabilize cells for 4–5 min in ice-cold permeabilization solution on ice.
5. Wash cells three times with 70 % ethanol (only for RNA and DNA FISH).
6. Store coverslips in 6-well plates or petri dishes in 70 % ethanol at −20 °C (*see* **Note 13**).

3.4.2 FISH DNA Probe Labeling

1. For a 50-μl reaction mix, 1–2 μg of plasmid or BAC/plasmid DNA is mixed with water to 17.5 μl, 2.5 μl of 0.2 mM SR-, SG-, Cy5-dUTP, 10 μl 10 mM each dNTP mix (dGTP, dATP,

dCTP), 5 μl 10 mM dTTP, 5 μl 10× nick translation buffer, and 8–10 μl nick translation enzyme.

2. Incubate for 16 h at 15 °C in the dark.
3. Inactivate the reaction by freezing at −20 °C. Probes can be stored for up to a few weeks at −20 °C.

3.4.3 Probe Preparation

1. 0.1 μg probe per coverslip and Cot-1 DNA (competition is required for most probes as they can contain repeat sequences that will cross hybridize and increase background unless competed away prior to hybridization) are precipitated by addition of 10 μg salmon sperm DNA, 1/10 volume 3 M sodium acetate pH 5.2, and three volumes ethanol at 16,100 × *g* and 4 °C for 30 min.
2. Wash precipitate with 75 % ethanol and spin down at 16,100 × *g* and 4 °C for 5 min.
3. Dry for 2 min in a concentrator/speed vac.
4. Resuspend in an appropriate volume of formamide (half the volume required for hybridization).
5. Incubate for at least 30 min at 37 °C and 1,400 rpm in a Thermomixer.
6. Denature for 10 min at 75 °C in a Thermomixer.
7. Quench on ice, or if competition is to be performed, put directly at 37 °C for at least 30 min in a Thermomixer.
8. Mix the probe solution with an equal volume of 2× hybridization solution.

3.4.4 Hybridization and Washes

1. Dehydrate coverslips (in wells or petri dishes) by sequential washing in 1× 80 %, 1× 95 %, and 2× 100 % ethanol for 5 min.
2. Dry coverslips or petri dishes completely.
3. Hybridization using coverslips: Spot the 12-μl probe hybridization mix (*see* **step 8**, Subheading 3.4.3), without making bubbles, onto a slide and lower the coverslip, with cells facing into the hybridization mix. Continue with **step 5**.
4. Hybridization in live-cell imaging petri dishes: Spot an appropriate amount of hybridization solution directly onto the coverslip in the bottom of the petri dish. Place a 12-mm glass coverslip on top of the hybridization solution to prevent evaporation. Continue with **step 5**.
5. Place slides or petri dishes in a humid chamber (tissue paper soaked in 50 % FA/2× SSC) and incubate at 37 °C overnight.
6. Post-hybridization washes (slides): Add 1 ml of 50 % FA/2× SSC onto the coverslip on the slide to loosen it, remove the coverslip carefully from the slide (taking care not to scrape the cells), and place it with cells facing upwards, into a 6-well plate containing 50 % FA/2× SSC. Continue with **step 8**.

7. Post-hybridization washes (petri dishes): Add 1 ml of 50 % FA/2× SSC into the petri dish to loosen the coverslip and remove the 12-mm coverslip from the bottom of the petri dish using a fine pair of forceps taking care not to scrape the cells. Continue with **step 8**.
8. Perform three washes, each for 7 min at 42 °C, with pre-warmed 50 % FA/2× SSC pH 7.2–7.4
9. Perform three washes with pre-warmed 2× SSC for 5 min each at 42 °C.
10. Counterstain nuclei by washing in 2× SSC with DAPI for a minimum of 3 min at room temperature.
11. Wash three times with 2× SSC.
12. Mounting of coverslips: Spot an appropriate amount of mounting medium on a slide and place the coverslip on top of the drop with cells facing down (avoid bubbles). Wipe off excess mounting solution and seal the coverslip with a small amount of nail polish.
13. Mounting of petri dishes: Spot an appropriate amount of mounting medium on the coverslip in the bottom of the petri dish. Place an 18-mm glass coverslip on top of the mounting medium (avoid bubbles). Wipe off extensive mounting solution and seal coverslip with a small amount of nail polish.

3.5 Protocol for DNA FISH

The fixation of samples prior to DNA FISH is the same as for RNA FISH. Coverslips or petri dishes with coverslip stored in 70 % ethanol can be used for DNA FISH.

3.5.1 RNAse Treatment

1. After dehydration (through 80, 95, and 100 % ethanol series as described above) of coverslips or petri dishes with glass bottom (*see* Subheading 3.4.4), add 1–2 ml of 2× SSC supplemented with 0.1 mg/ml RNAse A and 10 U/ml RNase H to the coverslip and incubate for 1 h at 37 °C.
2. Wash 3× with 2× SSC.

3.5.2 Denaturation

1. Preheat 50 % FA/2× SSC (pH 7.2–7.4) to 80 °C in a water bath.
2. Preheat a hybridization oven to 80 °C.
3. Add the preheated 50 % FA/2× SSC to the coverslips and incubate in a hybridization oven at 80 °C for 37–40 min (*see* **Note 14**).
4. Quickly remove the denaturation solution from the coverslip and add ice-cold 2× SSC.
5. Wash 2× with ice-cold 2× SSC and keep on ice until hybridization.

3.5.3 Hybridization and Washes

Hybridization and washes are performed as described above (*see* Subheading 3.4.4). Note that for DNA FISH, hybridization and washes are performed at higher temperatures (e.g., 42 °C for the hybridization and 45 °C for the washes).

3.6 Immunofluorescence (RNA or DNA FISH) Protocol

3.6.1 Fixation and Blocking

1. Fixation is performed as described in Subheading 3.4.1 except for the last washing step in 70 % ethanol, instead of washing with 70 % ethanol wash 3× in 1× PBS.
2. Blocking is performed in 1 % BSA in PBS for 15 min at room temperature.

3.6.2 Antibody Incubation

For cells on coverslips:

1. Dilute primary antibody in antibody dilution buffer and apply 50 μl of the antibody solution onto a clean glass slide.
2. Place coverslip cells down onto each of the drops and hybridize for 40 min at room temperature.
3. Wash 3× in 1× PBS for 5 min at room temperature on a rotating shaker.
4. Dilute the secondary antibody in antibody dilution solution and apply 50 μl onto a clean glass slide.
5. Put coverslips cells down onto each of the drops and hybridize for 40 min at room temperature in the dark.
6. Wash 3× in 1× PBS for 5 min at room temperature on a rotating shaker in the dark.
7. Fix for 10 min at room temperature with fixation solution.
8. Wash 3× quickly in 1× PBS.
9. Mount coverslips (IF only) or wash 2× in 2× SSC and proceed with **step 4** of Subheading 3.4.4 for IF-RNA FISH or Subheading 3.5.2 for IF-DNA FISH.

For cells on petri dishes:

1. All solutions are applied directly to the coverslip at the bottom of the petri dish. Incubations and washes are performed as described above.

4 Notes

1. The exact concentration of drug to be used for any particular ESC line should always be verified prior to the experiment, by testing various concentrations and evaluating kinetics and extent of cell death. Different ESC lines can show substantial variations in drug sensitivity.

2. Clones should be selected based on high signal-to-noise ratios, where the two TetO loci are visible in the majority (>90 %) of cells and for which minimal mosaicism is seen.
3. Glass coverslips: Make sure the brand and batch of glass coverslips used (see materials for examples) are suited for culturing ES cells as well as for the live-cell imaging device, the objective and microscope you use to achieve optimal imaging results.
4. Pre-coating of coverslips and dishes: We have noted that certain glass surfaces and dishes are suboptimal for feeder-independent ESC cultures for live-cell imaging. Cells need to grow in relatively thin colonies, with one or at most two cell layers. 0.1 % gelatin is the usual coating agent used—however, if colonies are too round and multilayered for optimal imaging, pre-coating with poly-d-lysine or fibronectin can be used. It should be verified that poly-d-lysine or fibronectin coating does not change the properties of the ES cells (other than cell shape) by checking for pluripotency marker staining; the differentiation kinetics of ES cells should also be compared to those cultured on standard gelatin coating.
5. Cell density: The optimal density of cells seeded at day 0 or at day −1 should be determined in advance, as it is different for every ESC line and differentiation condition used. Titration experiments should be performed on the coverslips or in the petri dishes and the pre-coating used for the actual experiment to determine the initial cell density needed to gain subconfluent and optimal density at the day of imaging.
6. Medium for live-cell imaging: Tolerance to different media needs to be tested for individual ESC lines. Better imaging results can often be obtained using medium without phenol red. Phenol red-containing medium can induce changes in ESC morphology, TetO array signal detection, and cell viability in our experience.
7. Ludin live-cell imaging chamber: Make sure that the coverslip has no scratches or breaks and that the isolation rubber is placed on the coverslip edges and that both are even before carefully screwing the top part on the Ludin chamber. Carefully clean the bottom of the coverslip after assembly of the chamber with a tissue to wipe off remaining medium and cells.
8. Antibiotics: For long-term imaging, we recommend the use of antibiotics in the medium, especially when using a Ludin chamber or an open petri dish.
9. CO_2 perfusion system: Make sure that the gas is pre-warmed and pre-humidified. Dry gas will lead to fast evaporation of medium.

10. 35-mm petri dishes with glass coverslip: The use of petri dishes with a lid limits evaporation as well as the risk of bacterial or fungal contamination of the medium. If a lid on the dish cannot be used, due to a special heating or CO_2 perfusion system, evaporation of the medium can be prevented by covering it with biology grade, sterile-filtered paraffin (or mineral oil). In our experience this does not interfere with ESC growth or morphology.
11. Orientation of the grid: Prior to the live-cell imaging, note down the orientation of the grid (using the numbered fields) on the microscope to find cells more easily after fixation and have them in the same orientation for imaging.
12. UV filter/neutral density filters: Use such filters as much as possible, to decrease phototoxicity.
13. Storage of petri dishes and coverslips: After fixation coverslips can be stored for several months in 70 % ethanol at −20 °C. Make sure that the petri dish or 6-well plate containing the coverslips is thoroughly sealed with parafilm to prevent evaporation of the ethanol.
14. Denaturation for DNA FISH: Denaturation time for optimal DNA FISH results differs greatly between cell types and even for the same ES cell line at different stages of differentiation. The genomic locus being detected can also vary considerably. Test run DNA FISH should be performed to ensure that the conditions are optimal for a given cell type and locus.

Acknowledgements

E. Heard's work is funded by ERC, EU EpiGeneSys Network, EU SYBOSS, and EU MODHEP; and the team is supported by La Ligue contre le Cancer. T. Pollex is funded by the Institut Curie International PhD grant. We would like to thank E. Schulz for critical reading and discussion of the manuscript and O. Masui for his help and insight into some of the materials and methods described here.

References

1. Rastan S (1983) Non-random X-chromosome inactivation in mouse X-autosome translocation embryos–location of the inactivation centre. J Embryol Exp Morphol 78:1–22
2. Borsani G, Tonlorenzi R, Simmler MC, Dandolo L, Arnaud D, Capra V, Grompe M, Pizzuti A, Muzny D, Lawrence C et al (1991) Characterization of a murine gene expressed from the inactive X chromosome. Nature 351:325–329
3. Brockdorff N, Ashworth A, Kay GF, Cooper P, Smith S, McCabe VM, Norris DP, Penny GD, Patel D, Rastan S (1991) Conservation of position and exclusive expression of mouse Xist from the inactive X chromosome. Nature 351:329–331
4. Brown CJ, Ballabio A, Rupert JL, Lafreniere RG, Grompe M, Tonlorenzi R, Willard HF (1991) A gene from the region of the human X inactivation centre is expressed exclusively

from the inactive X chromosome. Nature 349:38–44

5. Lee JT, Davidow LS, Warshawsky D (1999) Tsix, a gene antisense to Xist at the X-inactivation centre. Nat Genet 21:400–404
6. Nora EP, Lajoie BR, Schulz EG, Giorgetti L, Okamoto I, Servant N, Piolot T, van Berkum NL, Meisig J, Sedat J et al (2012) Spatial partitioning of the regulatory landscape of the X-inactivation centre. Nature 485(7398): 381–385
7. Augui S, Filion GJ, Huart S, Nora E, Guggiari M, Maresca M, Stewart AF, Heard E (2007) Sensing X chromosome pairs before X inactivation via a novel X-pairing region of the Xic. Science 318:1632–1636
8. Bacher CP, Guggiari M, Brors B, Augui S, Clerc P, Avner P, Eils R, Heard E (2006) Transient colocalization of X-inactivation centres accompanies the initiation of X inactivation. Nat Cell Biol 8:293–299
9. Xu N, Tsai C-L, Lee JT (2006) Transient homologous chromosome pairing marks the onset of X inactivation. Science 311: 1149–1152
10. Masui O, Bonnet I, Le Baccon P, Brito I, Pollex T, Murphy N, Hupé P, Barillot E, Belmont AS, Heard E (2011) Live-cell chromosome dynamics and outcome of X chromosome pairing events during ES cell differentiation. Cell 145:447–458
11. Chaumeil J, Augui S, Chow JC, Heard E (2008) Combined immunofluorescence, RNA fluorescent in situ hybridization, and DNA fluorescent in situ hybridization to study chromatin changes, transcriptional activity, nuclear organization, and X-chromosome inactivation. Methods Mol Biol 463:297–308

Chapter 3

Single-Molecule Resolution Fluorescent In Situ Hybridization (smFISH) in the Yeast *S. cerevisiae*

Samir Rahman and Daniel Zenklusen

Abstract

Regulating gene expression is a major task for all cellular systems. RNA production and degradation plays a critical role in this process and accurately measuring cellular mRNA levels is essential to understanding gene expression regulation. Classical biochemical assays that study gene expression rely on extracting RNAs from large populations of cells, taking them out of their native context and thereby losing spatial information as well as cell-to-cell variability. In this chapter, we describe a fluorescent in situ hybridization (FISH) technique that circumvents this problem by detecting single RNAs in single cells. The technique employs multiple single-stranded short DNA probes fluorescently labeled with organic dyes that hybridize to target RNAs in fixed cells, allowing quantification and localization of RNAs at the single-cell level and at single-molecule resolution. The protocol described here has been optimized for the yeast *S. cerevisiae*.

Key words Gene expression, Single-cell imaging, Single RNA resolution fluorescent in situ hybridization, Yeast, mRNA detection

1 Introduction

Methods to study gene expression regulation are constantly evolving. Microarray and next generation sequencing technologies have fundamentally changed gene expression analysis by shifting the focus from studying single genes to analyzing expression profiles for entire organisms in a single experiment [1, 2]. However, these new methodologies also come with a number of limitations: the need to isolate mRNA from cells, which is always associated with a potential loss or damage of the material (e.g., degradation), as well as the loss of spatial information. Moreover, individual cells within a population are unlikely to all behave in the same way. As current standard techniques are designed to study only transcriptional changes in whole cell populations, they are unable to detect cell-to-cell differences that can result from genetic variation, biological noise, and/or different characteristics of genes within a population. In particular, the accuracy of kinetic studies performed on cell

Yaron Shav-Tal (ed.), *Imaging Gene Expression: Methods and Protocols*, Methods in Molecular Biology, vol. 1042, DOI 10.1007/978-1-62703-526-2_3, © Springer Science+Business Media, LLC 2013

populations is highly dependent on the synchronicity by which processes occur within individual cells. Therefore, analyzing cellular processes at the single-cell level and understanding cell-to-cell differences has become increasingly important as it allows to explain how changes in the behavior of a whole population of cells can arise through the selection of individuals under specific conditions [3–5].

GFP reporter-based approaches have been extensively used to study gene expression regulation at the single-cell level. However, information obtained from such experiments only provide end-point measurements reflecting the ensemble of the different processes resulting in protein production, including coordination of transcription as well as all the post-transcriptional events such as mRNA transport, mRNA stability and translation efficiency. Therefore, to truly understand the different aspects of gene expression regulation, we have to be able to study mRNA at the single-cell and single RNA level.

The ability to detect single mRNA molecules in single cells by fluorescence microscopy was first demonstrated in normal rat kidney (NRK) cells for the beta-actin mRNA as early as 1998 [6]. However, at the time, access to sensitive, high-resolution cameras and image analysis software was limited to a few (mostly biophysics) laboratories and not widely accessible to most cell biologists. Furthermore, synthesis and labeling of FISH probes either required access to a DNA synthesizer or was hugely expensive when ordered from a company, resulting in only sporadic use of this powerful methodology. However, the recent technology advances have made sensitive high-resolution cameras standard equipment in many laboratories and most imaging facilities, and have thus turned single-molecule resolution FISH (smFISH) into a simple experimental tool for anyone interested in single RNA detection. Indeed, smFISH is about to become a standard tool to study different aspects of gene expression regulation [4, 7–13].

A main feature of smFISH compared to standard RNA FISH protocols is the use of multiple short DNA probes in contrast to a single long antisense DNA or RNA probe [7]. The use of multiple short probes is essential to reduce background, as nonspecific binding of a single short probe will result in a very weak (almost undetectable) signal compared to the signal emitted from binding of the ensemble of all probes. This is different to the nonspecific binding of a single long probe which will lead to an identical (and artifactual) signal to be erroneously considered as a specific signal. Furthermore, using chemically synthesized DNA oligonucleotide probes also allows for very efficient probe labeling, essential for single-molecule detection. The very high labeling efficiency (>90 %) also facilitates the quantification of FISH signal, as the signal intensity will scale linearly with the number of probes hybridizing [6, 14].

In this chapter, we describe a step-by-step smFISH protocol for detecting single *MDN1* mRNA in the yeast *Saccharomyces cerevisiae* using 48 singly labeled 20mer probes. We first describe probe labeling and measuring probe labeling efficiency. We then guide through the different steps: cell fixation, digestion of the cell wall, hybridization, and image acquisition. Finally, we will very briefly describe some aspects of data analysis.

2 Materials

2.1 Probe Design

Any DNA analysis software package or online probe design program. Here we used Stellaris™ Probe Designer version 2.0 (http://www.biosearchtech.com/stellarisdesigner/).

2.2 Probe Labeling

1. Forty-eight 20 nt long DNA oligonucleotides complementary to the yeast *MDN1* mRNA containing mdC(TEG-Amino) for labeling (Biosearch Technologies, Novato, CA).
2. Mono-Reactive CyDye™ Cy3 (GE Healthcare).
3. QIAquick Nucleotide Removal Kit (Qiagen).
4. Nanospectrophotometer.
5. Labeling buffer: 0.1 M sodium bicarbonate, pH 9.0.

2.3 Cell Fixation, Preparation, and Storage

1. Yeast strain (here we use BY4741).
2. YPD growth medium.
3. 125 ml Erlenmeyer flask.
4. Shaking incubator dedicated to 30 °C.
5. 32 % paraformaldehyde solution, EM grade.
6. Lyticase: resuspend in 1× PBS to 25,000 U/ml. Stored at −20 °C.
7. Ribonucleoside–vanadyl complex.
8. β-mercaptoethanol.
9. Sorbitol.
10. 1 M $KHPO_4$, pH 7.5.
11. 70 % ethanol.
12. Noncoated coverslips: No. 1: 0.13–0.17 mm thick; size: 18 mm.
13. Poly-L-lysine.
14. 12-well cell culture plates.
15. Buffer B: 1.2 M sorbitol, 100 mM $KHPO_4$, pH 7.5.
16. Spheroplast buffer: 1.2 M sorbitol, 100 mM $KHPO_4$, pH 7.5, 20 mM ribonucleoside–vanadyl complex, preheated at 65 °C, 20 mM β-mercaptoethanol.
17. 0.1 N HCl.

2.4 Hybridization

1. Deionized formamide (*see* **Note 1**).
2. 20× saline–sodium citrate buffer (SSC): 3 M NaCl, 0.3 M sodium citrate at pH 7.
3. 1 M $NaHPO_4$, pH 7.5.
4. 10 mg/ml BSA (DNase/ RNase free).
5. Ribonucleoside–vanadyl complex.
6. 10 mg/ml *Escherichia coli* tRNA (Roche # 10 109 541 001).
7. 10 mg/ml ssDNA (deoxyribonucleic acid, single stranded from salmon testes, Sigma #D9156).
8. Fluorescently labeled DNA probe.
9. Vacuum concentrator (SpeedVac).
10. 1× PBS.
11. 100 % ethanol.
12. Mounting solution containing DAPI.
13. Glass plate, about 20×20 cm.
14. Parafilm.
15. Cardboard spacers.
16. 12-well cell culture plates.
17. Glass slides.
18. 2× SSC.
19. Wash solution: 10 % formamide/2× SSC.
20. Solution F: 20 % formamide, 10 mM $NaHPO_4$, pH 7.5.
21. Solution H: 4× SSC, 2 mg/ml BSA, 10 mM VRC (preheated at 65 °C).
22. Nail polish (to seal coverslips).

2.5 Image Acquisition

1. Fluorescent wide-field microscope of choice with a 100× high numerical aperture objective (images shown were acquired using a Nikon Eclipse E800 using a 100× 1.4 NA Nikon objective).
2. CCD camera with small pixel size (ideally 6.45-μm or smaller, like Photometrics CoolSNAP HQ2 or Zeiss AxioCam MRm Rev.3).
3. Dichroic filter cubes. For detecting DAPI and Cy3 we use Chroma Filters 31000 (DAPI), SP-102v1 (Cy3), Chroma Technology.
4. Image acquisition software (MetaMorph, ZEN, micro-manager or similar).

3 Methods

3.1 Probe Design

Designing smFISH probes is similar to designing PCR primers, and most PCR primer design software packages can be used (*see* **Note 2**). Here we use the online program Stellaris™ Probe Designer version 2.0 from Biosearch Technologies. Input the *MDN1* coding sequence into the program, which automatically generates a set of probes complementary to the *MDN1* mRNA, optimized for binding to the target RNA sequence. We designed 48 probes against the 5′ region of *MDN1*, each 20 nt long, with a minimum spacer length of 2 nt between the probes and an average GC content of 45 % (*see* **Notes 3** and **4**). The probes were synthesized by Biosearch Technologies (Novato, CA) with a free functional amine at the 3′ end (mdC(TEG-Amino)) to be labeled with an amine-reactive fluorescent dye (*see* **Notes 5–8**). Oligos are resuspended in H_2O at a concentration of 100 μM (do not use buffers that contain primary amines, such as Tris, as these will compete for conjugation with the amine-reactive compound). Unlabeled probes can be stored for many months at −20 °C.

3.2 Probe Labeling

1. Pool 2 μg of each of the 48 unlabeled probes in a single microcentrifuge tube.
2. Transfer 20 μg of probe mix to a new tube.
3. Dry down probe mix using a vacuum concentrator.
4. Suspend probe mix in 20 μl of labeling buffer (0.1 M sodium bicarbonate, pH 9.0).
5. Add suspended probes to a single dye tube and mix vigorously by vortexing. Collect labeling reaction at the bottom of the tube by doing a quick spin using a microcentrifuge.
6. Incubate labeling reaction mix in the dark overnight at room temperature.
7. *Purify labeled probes from unincorporated dye using the QIAquick Nucleotide Removal Kit.* Add 500 μl of buffer PN (from QIAquick Nucleotide Removal Kit) to the labeling reaction and load onto two columns (each column has a binding capacity of 10 μg).
8. Spin through columns according to the protocol.
9. Load the flow-through a second time onto the same column to increase probe recovery.
10. Spin through columns according to the protocol.
11. Wash column twice with buffer PE (from QIAquick Nucleotide Removal Kit) to remove all non-incorporated dye.

12. Spin again to remove residual buffer PE to completely dry down labeled oligonucleotides.
13. Elute the labeled probes using 100 μl of elution buffer (from QIAquick Nucleotide Removal Kit).
14. Measure concentration and labeling efficiency using a nanospectrophotometer.
15. Store probes at −20 °C in the dark.
16. *Calculating labeling efficiency.* The labeling efficiency (LE) is calculated according to Beer–Lambert's law such that LE = [dye]/[probe]*N* where *N* is the number of incorporated fluorophore-binding sites on a probe. First, measure DNA and dye (here Cy3) absorbance at 260 and 552 nm. The extinction coefficient of DNA is dependent on the DNA sequence and can be calculated using various online resources. However, when 48 different probes are used, it is reasonable to approximate an average of 25 % of each base and use this coefficient which is $\varepsilon_{DNA} = 197{,}700\ M^{-1}\ cm^{-1}$. The extinction coefficient of Cy3 is $\varepsilon_{dye} = 150{,}000\ M^{-1}\ cm^{-1}$ at its maximum emission at 552 and $\varepsilon_{dye} = 4{,}930\ M^{-1}\ cm^{-1}$ at 260 nm. 0.1 cm is the light path of the instrument. Labeling efficiency is then calculated using the following formula:

$$[\mathrm{DNA}] = \frac{A_{\mathrm{DNA}} - \varepsilon_{\mathrm{dye}(260)} \times \left(A_{\mathrm{dye}} / \varepsilon_{\mathrm{dye}(\mathrm{max})}\right)}{\varepsilon_{\mathrm{DNA}} \times 0.1\mathrm{cm}}$$

$$[\mathrm{Dye}] = \frac{A_{\mathrm{Dye}(\mathrm{max})}}{\varepsilon_{\mathrm{Dye}} \times 0.1\mathrm{cm}}$$

A_{DNA} = Absorption of DNA at 260 nm.

ε_{DNA} = Extinction coefficient of the DNA.

A_{dye} = Absorption at absorbance max of the dye.

ε_{dye} = Extinction coefficient of the dye.

A labeling efficiency of 1.0 indicates that all modified bases have been labeled. Typically, labeling efficiencies above 0.90 should be achieved. Some spectrophotometers, such as the nanospectrophotometer, can directly output labeling efficiency.

3.3 Cell Fixation, Preparation, and Storage

1. Grow cells in appropriate liquid medium (here in YPD) overnight to mid-log phase (OD_{600} nm = 0.6–0.8).
2. Fix cells by adding 32 % paraformaldehyde directly to culture to a final concentration of 4 %. Fix for 45 min at room temperature by slight rocking.

3. Pellet cells by centrifuging at 2,600×g for 3 min at 4 °C. Discard paraformaldehyde in appropriate container.
4. Wash cells 3× with 10 ml ice-cold Buffer B and collect by spinning at 2,600×g for 3 min at 4 °C.
5. Resuspend cells in 1 ml Buffer B and transfer to a 1.5 ml microcentrifuge tube. Directly continue to **step 6** or store cells at 4 °C overnight.
6. Pellet cells by spinning at 6,000×g for 1min at 4 °C, remove the supernatant and resuspend pellet in 500 μl spheroplast buffer.
7. To digest cell wall, add 5 μl lyticase (25 U per 1OD of cells). Incubate at 30 °C for about 5 min or until cells are digested. The cell wall needs to be digested in order for probes to enter the cell (*see* **Notes 9** and **10**).
8. Collect cells by centrifugation at 1,200×g for 3 min at 4 °C. Do not spin at a higher speed as cells are fragile at this step and can disintegrate after the cell wall is gone.
9. Remove supernatant and carefully wash cells with 1 ml ice-cold Buffer B.
10. Spin at 1,200×g for 3 min at 4 °C and resuspend cells in 1 ml ice-cold Buffer B. Keep on ice.
11. Drop 140 μl of cells on each coverslip (in 12-well cell culture dish, poly-L-lysine treated).
12. Incubate at 4 °C for 30 min.
13. Wash carefully once with 2 ml ice-cold Buffer B in each well (be careful not to wash away cells) (*see* **Note 11**).
14. Discard Buffer B, and carefully add 2 ml 70 % EtOH (−20 °C) in each well. Seal 12-well cell culture plates in parafilm and incubate at −20 °C for at least 3 h (usually overnight). Ethanol helps to permeabilize the cell membrane so that the probes may enter the cell. Cells can be stored at −20 °C for several months.

3.4 Preparation of Coverslips

1. Put one box of 18 mm round coverslips into 500 ml 0.1 N HCl.
2. Boil for 10 min.
3. Rinse extensively with H_2O.
4. Store in 70 % EtOH.
5. Drop 150 μl of 0.01 % poly-L-lysine on coverslip.
6. Leave for 10 min and then remove poly-L-lysine.
7. Let air dry.
8. Wash 2× with H_2O.
9. Air dry. Coverslips can be stored at room temperature for a few weeks.

3.5 Hybridization

1. *Prepare cells for hybridization.* Transfer coverslips to a new 12-well dish and rehydrate cells by adding 2 ml 2× SSC in each well at RT for 5 min (twice).
2. Wash once with 10 % formamide/2× SSC for 10 min at room temperature.
3. *Prepare probes for hybridization.* Mix 20 ng of labeled probe per sample with 40 μg of ssDNA/tRNA mix. The ssDNA/tRNA mix serves as a competitor to prevent nonspecific hybridization.
4. Dry down labeled probe/competitor DNA mix in a SpeedVac.
5. Resuspend probes in 12 μl of solution F.
6. Heat probe mix at 95 °C for 3 min to denature the probes.
7. Add 12 μl of solution H per sample to produce the final hybridization mix.
8. *Hybridization.* Wrap parafilm around the glass plate (*see* **Notes 12** and **13**).
9. Drop 22 μl of hybridization mix per coverslip on the parafilm.
10. Lay each (one) coverslip on each drop with the *cells facing down.* Allow hybridization solution to completely spread on the coverslip, trying to avoid the formation of air bubbles.
11. Place three cardboard spacers on the parafilm (approximately 1 × 1 cm), two on each edge of the glass plate and one in the middle.
12. Place another piece of parafilm on top. The cardboard spacers will prevent the top parafilm layer from touching the coverslips. Seal the two layers of parafilm to make an airtight hybridization chamber to prevent the hybridization solution from evaporating.
13. Cover the hybridization chamber in aluminum foil and incubate for at least 3 h or overnight at 37 °C in the dark.
14. Preheat 10 % formamide/2× SSC at 37 °C (for around 10 min). Add 2 ml preheated 10 % Formamide/2× SSC to each well in a 12-well culture dish.
15. Transfer the hybridized coverslips, *cells facing up*, to the 12-well plate containing the preheated formamide/2× SSC solution and incubate in the dark for 30 min at 37 °C.
16. Wash once more with 10 % formamide/2× SSC at 37 °C for another 30 min.
17. Wash once with 2 ml 1× PBS at RT in the dark for 5 min.
18. Dip coverslips quickly in 100 % EtOH to dry them before mounting.
19. Drop 6 μl of mounting solution containing DAPI for each coverslip on a microscope slide.

20. Place the dried coverslip on the drop of mounting solution *cells facing down.*
21. Keep glass slides at 4 °C in the dark for at least 3 h to allow the mounting solution to harden and seal coverslips with nail polish. After nail polish is completely dry, cells are ready to image. Coverslips can be stored for many weeks at 4 °C and many months at −20 °C.

3.6 Image Acquisition and Analysis

Images are acquired on a wide-field epifluorescence microscope, with an optical sectioning size of 200 nm per *z*-plane and spanning the entire volume of the cell (around 5 μm, depending on your cells). Exposure times will vary depending on your microscope setup (light source, objective, etc.), but typical exposure times are 30 ms for DAPI, 1,000–1,500 ms for Cy3.

After image acquisition, reduce the 3D dataset to a 2D dataset using maximum projection in Image J (*see* **Note 14**). Overlay the maximum projections of the RNA and the DAPI signal to separate nuclear and cytoplasmic signal. This will result in an image with diffraction-limited spots in the cytoplasm corresponding to single RNAs, and more intense spots co-localizing with the nuclear nascent RNAs, corresponding to nascent mRNAs still associated with the corresponding gene (Fig. 1). Intensities of single mRNAs in the cytoplasm should be homogenous.

4 Notes

1. Formamide is a teratogen and easily absorbed through the skin and should therefore be handled with proper safety attire including gloves.
2. *Designing probes.* Designing probes for hybridization follows the same rules as for any oligonucleotide hybridizing to another nucleic acid sequence. Designing smFISH probes can therefore be done using most DNA analysis software packages. As a general rule, probes should have a CG content of about 45 %, not form internal stem loops or dimers with other probes used in the same experiment. Furthermore, probes should be tested "in silico" not to hybridize to other transcripts than the one they are intended to. Biosearch Technologies offers an online tool that also checks for repetitive sequences occurring in different organisms. Probes used in this protocol were designed using this software.
3. *Probe length.* Probes of different lengths can be used. Traditionally, a set of four to five 50 nt long DNA probes were used, each labeled at multiple positions with fluorescent dyes [6, 13]. Alternatively, as done here, using a larger number of shorter probes (~48, 20 nt long) labeled at a single position equally allows single RNA detection [10]. 50 nt probes with

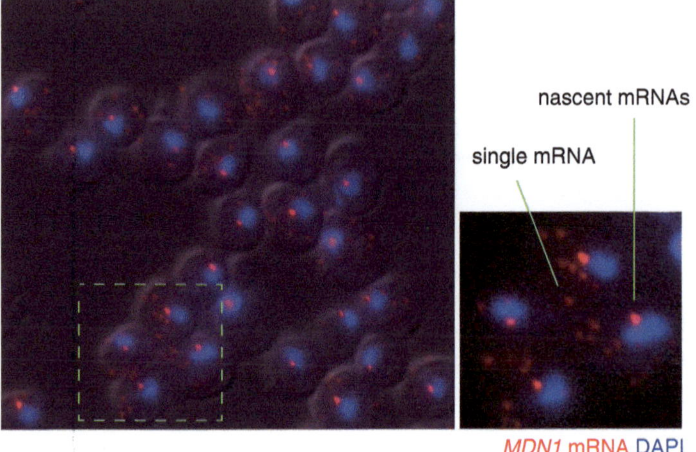

Fig. 1 The image on the *left* is a representative image showing MDN1 mRNA expression in wild type *Saccharomyces cerevisiae* cells, grown in YPD and fixed during mid-log phase (OD 0.6 at 600 nm). The nucleus is stained with DAPI and shown in *blue*. The diffraction-limited spots corresponding to MDN1 mRNAs are indicated in *red*. The spots in the cytoplasm correspond to single MDN1 mRNAs, and the intense spots co-localizing with the DAPI stain correspond to transcription sites with multiple nascent mRNAs. The DIC image has been overlaid to indicate cell boundaries. Images were acquired using a Nikon Eclipse E800 with a 100× 1.4 NA Nikon objective. The image on the *right* is a magnification of the section indicated by the *dotted square*

multiple labels require a smaller region of RNA to allow enough probes for single-molecule detection. However, successful single-molecule detection using 50mers requires high hybridization efficiency for all probes. When using four 50mer probes and only two hybridize due to poor hybridization efficiencies, the observed signal will be of similar intensity than the signal emitted from the nonspecific hybridization of a single probe. Therefore, high hybridization efficiency is essential when using few 50 nt long probes. The use of many 20mer probes containing a single label somehow reduces this problem, as single-labeled probes are barely detectable, and poor hybridization of a small subset of probes will still allow robust single-molecule detection. Generally, *the more probes are used, independent of the size or the labeling density of the probes, the better and more consistent is single mRNA detection.*

4. *Using probes of different lengths.* Probes with different lengths can be used, but the formamide concentration in the hybridization mix and during the washes may have to be adjusted. For 20mer probes, typically 10 % formamide is used, and for 50mer probes it is 40 % formamide. When probes with different lengths are used in the same experiment, the ideal formamide concentration

should be determined empirically. The concentration of the other components in the hybridization and wash buffers can be kept constant.

5. *Ordering probes.* Many companies synthesizing standard DNA oligonucleotides will offer DNA probes suitable for smFISH. Probes can either be purchased already labeled with fluorescent dyes or contain modified nucleotides allowing post synthesis labeling. For post synthesis labeling, a single or multiple modified nucleotides containing a primary amine are incorporated during synthesis. For internal labeling, usually amino-allyl thymidines are incorporated, spaced by at least 8 nt to avoid quenching. When incorporating a single label, usually a 5′- or 3′-modified nucleotide containing a primary amine is added.
6. *Order labeled probes or label post synthesis.* We generally choose post synthesis labeling as it allows that the same set of probes can be labeled with different fluorescent dyes. This allows more flexibility in experimental design, in particular when expression/localization of multiple RNAs is analyzed simultaneously in the same cell. Post synthesis probe labeling is simple and efficient and many different dyes can be used (see below).
7. *Costs.* Probes for smFISH will cost a few hundred US dollars per gene. Even if many more probes are used, synthesizing 20 nt single-modified probes is generally cheaper compared to 50 nt probes with multiple labels, which might cost up to 1,000 US dollars per gene. Generally, end labeling (dye or a primary amine) is cheaper than internal labeling.
8. *Dyes.* Many fluorescent dyes are available suitable for single-molecule RNA detection. Dyes should have high quantum efficiency, be stable at different pH values, and not be sensitive to photobleaching. Generally, the same family of dyes used for immunofluorescence is used for smFISH. As most cells contain more autofluorescent signal in the green than the red and the far red spectrum, it is preferable to use red and far red dyes. However, dyes in the green spectrum have been used successfully. We use cyanine dyes (cy3, cy3.5, and cy5; GE healthcare) as well as DyLight dyes (DyLight 550, 594, and 650; Thermo Scientific); however, many other dyes can be used (Alexa dyes; Invitrogen, Quasar dyes; Biosearch Technologies, and others). When detecting multiple mRNAs simultaneously and using dyes with excitation and emission spectra that are close to each other (e.g., Cy3 and Cy3.5), it is important to use appropriate filter cubes to avoid bleed-through. We use the following filter cubes for the most frequently used dyes: Chroma Filters 31000 (DAPI), 41001 (FITC), SP-102v1 (Cy3), SP-103v1 (Cy3.5), and CP-104 (Cy5) (Chroma Technology, Rockingham, VT). Coupling occurs in a labeling reaction with dyes activated by a NHS-ester in order to react with the free amine of the DNA oligo.

9. *Cell wall digestion.* Time and/or lyticase concentration required for cell wall digestion varies between strain backgrounds. In particular, some mutant strains or cells not grown in standard conditions, can be hard to digest [15]. Generally, with the concentrations indicated in this protocol, most strains will have lost their cell wall after around 10 min. We have also observed differences in lyticase activity for different batches of lyticase obtained from Sigma. Therefore, digestion times might have to be adjusted when using a new batch of lyticase. If the cell wall is not digested, probes will not penetrate the cell and no signal will be observed. Overdigestion on the other hand will lead to disintegration of the cells.
10. *Monitoring cell wall digestion.* Following the digestion process using a microscope is the best way to ensure that cells are optimally digested. The loss of the cell wall leads to a change in light absorption of the cell and can be observed using a standard microscope. When using a bright-field setup, digested cells will become opaque. When phase contrast optics is used, digested cells will appear darker than the nondigested cells. Ideally, keep an undigested sample as a control to monitor the progress of the digestion. To avoid overdigestion, stop the digestion process when around 50 % of the cells are digested and keep cells on ice for the following steps.
11. *No cells on coverslips.* Even though the coverslips are coated with poly-L-lysine, cells do not always efficiently attach to the glass surface until the addition of EtOH. Therefore, all washes should be performed gently in order to not loose cells.
12. *High background signal.* High (probe-specific) background signal is rare but can occasionally be observed. Troubleshooting steps include reducing the amount of probe, shortening hybridization periods down to 3 h, increasing the number of washes, as well as increasing formamide concentration in the hybridization solution. It is important to ensure that probes do not cross hybridize with mRNAs from other genes. Whenever possible, include a deletion strain for your gene of interest or use glucose depletion to deplete your RNA of interest.
13. *No signal.* smFISH is a very robust technique, and it is rare that a set of probes does not emit sufficient signal allowing single-molecule detection. The most frequent reason for the lack of signal is that cells were not sufficiently digested (see above). A useful internal control is to include probes against a second gene that has been previously successfully used and is labeled with a different dye. If a probe set does not lead to sufficient signal, decrease the stringency of the hybridization conditions (lower formamide concentrations) or increase the number of probes. Alternatively, probe labeling efficiency should be re-measured.

14. *Image analysis*. The type of image analysis required depends on the questions asked. When only RNA localization to a specific cellular compartment is studied, often little further analysis is required. When single-cell expression levels are assessed, mRNA levels can either be counted manually, or specific software for spot detection and segmentation can be used [10, 13]. For most RNAs in yeast, image analysis can be performed after datasets are reduced from 3D to 2D, as most genes in yeast are present in low copies. However, when highly expressed RNAs are studied (>30 copies per cell), 2D projections will underestimate RNA numbers.

Acknowledgements

We thank Marlene Oeffinger as well as members of the Zenklusen laboratory for comments and discussions on the manuscript. The laboratory of Daniel Zenklusen is supported by the Canadian Institutes of Health Research (MOP-BMB-232642), the Natural Sciences and Engineering Research Council of Canada, the Fonds de recherche Santé Québec, and the Canada Foundation for Innovation.

References

1. Holstege FC, Jennings EG, Wyrick JJ et al (1998) Dissecting the regulatory circuitry of a eukaryotic genome. Cell 95:717–728
2. Shendure J, Ji H (2008) Next-generation DNA sequencing. Nat Biotechnol 26:1135–1145
3. Elowitz M, Levine A, Siggia E, Swain P (2002) Stochastic gene expression in a single cell. Science 297:1183–1186
4. Larson DR, Singer RH, Zenklusen D (2009) A single molecule view of gene expression. Trends Cell Biol 19:630–637. doi:10.1016/j.tcb.2009.08.008
5. Locke J, Elowitz M (2009) Using movies to analyse gene circuit dynamics in single cells. Nat Rev Microbiol 7:383–392
6. Femino A, Fay F, Fogarty K, Singer R (1998) Visualization of single RNA transcripts in situ. Science 280:585–590
7. Itzkovitz S, van Oudenaarden A (2011) Validating transcripts with probes and imaging technology. Nat Methods 8:S12–S19. doi:10.1038/nmeth.1573
8. Raj A, Peskin CS, Tranchina D et al (2006) Stochastic mRNA synthesis in mammalian cells. PLoS Biol 4:e309. doi:10.1371/journal.pbio.0040309.sv002
9. Raj A, Rifkin SA, Andersen E, van Oudenaarden A (2010) Variability in gene expression underlies incomplete penetrance. Nature 463:913–918. doi:10.1038/nature08781
10. Raj A, van den Bogaard P, Rifkin SA et al (2008) Imaging individual mRNA molecules using multiple singly labeled probes. Nat Methods 5:877–879. doi:10.1038/nmeth.1253
11. Trcek T, Larson DR, Moldón A et al (2011) Single-molecule mRNA decay measurements reveal promoter-regulated mrna stability in yeast. Cell 147:1484–1497. doi:10.1016/j.cell.2011.11.051
12. Vargas DY, Shah K, Batish M et al (2011) Single-molecule imaging of transcriptionally coupled and uncoupled splicing. Cell 147:1054–1065. doi:10.1016/j.cell.2011.10.024

13. Zenklusen D, Larson DR, Singer RH (2008) Single-RNA counting reveals alternative modes of gene expression in yeast. Nat Struct Mol Biol 15:1263–1271. doi:10.1038/nsmb.1514
14. Zenklusen D, Singer RH (2010) Analyzing mRNA expression using single mRNA resolution fluorescent in situ hybridization. Methods Enzymol 470:641–659. doi:10.1016/S0076-6879(10)70026-4
15. Silverman SJ, Petti AA, Slavov N et al (2010) Metabolic cycling in single yeast cells from unsynchronized steady-state populations limited on glucose or phosphate. Proc Natl Acad Sci 107:6946–6951. doi:10.1073/pnas.s1002422107

Chapter 4

Measuring Transcription Dynamics in Living Cells Using Fluctuation Analysis

Matthew L. Ferguson and Daniel R. Larson

Abstract

Single-cell studies of gene regulation suggest that transcription dynamics play a fundamental role in determining expression heterogeneity within a population. In addition, the three-dimensional organization of the nucleus seems to both reflect and influence expression patterns in the cell. Therefore, to gain a holistic understanding of transcriptional regulation, it is necessary to develop methods for studying transcription of single genes in living cells with high spatial and temporal resolution. In this chapter, we describe a recently developed approach for visualizing and quantifying pre-mRNA synthesis at a single active gene in the nucleus. The approach is based on the high-affinity interaction between MS2/PP7 bacteriophage coat proteins and RNA hairpins which are transcribed by the gene of interest. The MS2/PP7 coat protein is fused to a fluorescent protein and binds the nascent mRNA, allowing for detection of single transcription events in the fluorescence microscope. By time-lapse fluorescence imaging and quantitative image analysis, one can generate a time trace of fluorescence intensity at the site of transcription. By temporal autocorrelation analysis, one can determine enzymatic activities of RNAP such as initiation rate and elongation rate. In this protocol, we summarize the experimental concept, design, and execution for real-time observation of transcription in living cells.

Key words Fluorescence fluctuation spectroscopy, Fluorescence correlation spectroscopy, FCS, Transcription, MS2, PP7, Single molecule, Imaging, RNA, Stochastic

1 Introduction

In recent years, live-cell microscopy in the nucleus has revealed previously unanticipated aspects of gene regulation. Chromatin-binding proteins show a surprising degree of dynamic mobility, with dwell times on the order of seconds [1]. Chromatin itself shows conformational fluctuations over several microns [2, 3]. Genes reposition within the nucleus in response to activation and inactivation [4]. These dynamic processes are also reflected in the process of transcription, which can display kinetic behavior ranging from isolated uncorrelated events of RNA synthesis to highly cooperative bursts of transcription [5–10]. One way to integrate these molecular events into a coherent view of gene regulation in a single

Yaron Shav-Tal (ed.), *Imaging Gene Expression: Methods and Protocols*, Methods in Molecular Biology, vol. 1042, DOI 10.1007/978-1-62703-526-2_4, © Springer Science+Business Media, LLC 2013

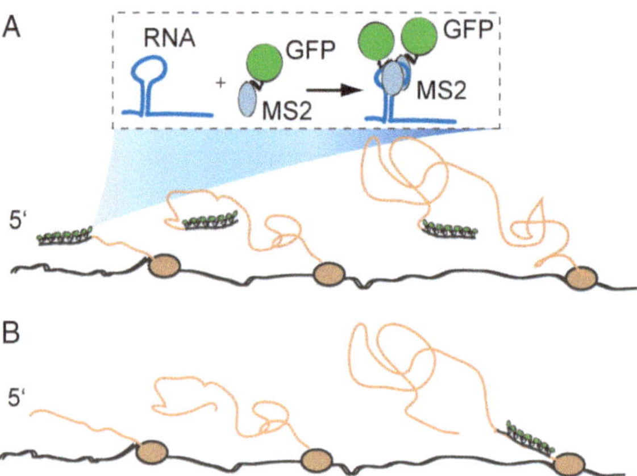

Fig. 1 Scheme for observing nascent RNA in living cells. The approach is based on the high-affinity binding of bacteriophage coat protein (i.e., MS2) to hairpins in the nascent RNA. Each stem loop binds a dimer of the coat protein, and each coat protein is labeled with a fluorescent protein (i.e., GFP). (**a**) 5′ UTR labeling. When stem loops are located in the 5′ UTR, it is possible to visualize the RNA shortly after elongation commences. Here, three nascent RNAs are visible, resulting in a signal which is three times brighter than a single RNA. (**b**) 3′ UTR labeling. When stem loops are located in the 3′ UTR, the nascent RNA is only visible once the polymerase proceeds to the end of the gene. Here, only a single nascent RNA contributes to the fluorescence signal

cell is to visualize the process of transcription directly through the observation of nascent pre-mRNA synthesis at an active locus. Here, we describe the implementation of a newly developed fluctuation analysis approach for quantifying RNA synthesis from a single gene in living cells [7]. The benefit of this method is that it is a direct measure of transcriptional regulation in single cells, independent of upstream steps such as transcription factor binding and downstream steps such as RNA processing and decay.

The approach of observing RNA in living cells was pioneered by Bertrand and Singer and is based on the high-affinity binding of the MS2 bacteriophage coat protein (MCP) to specific RNA hairpins [11] (Fig. 1). Spector and Singer adapted this technique to observe transcription from a reporter gene integrated as a tandem array into the genome [12, 13]. In this implementation, the site of transcription is visible when the fluorescently labeled coat protein binds the newly synthesized RNA at the site of transcription. The active transcription site is visible in the microscope as a punctate fluorescent spot in the nucleus against the background of unbound coat protein (Fig. 2). Recently, another coat protein (from the PP7 phage) was also adapted to RNA imaging, extending the combination of labels that can be utilized in any given experiment [7, 14]. Variations on this basic approach have now been used to observe RNA synthesis in real time in living cells of bacteria, yeast, and mammals [15]. These studies differ in the number of insertions

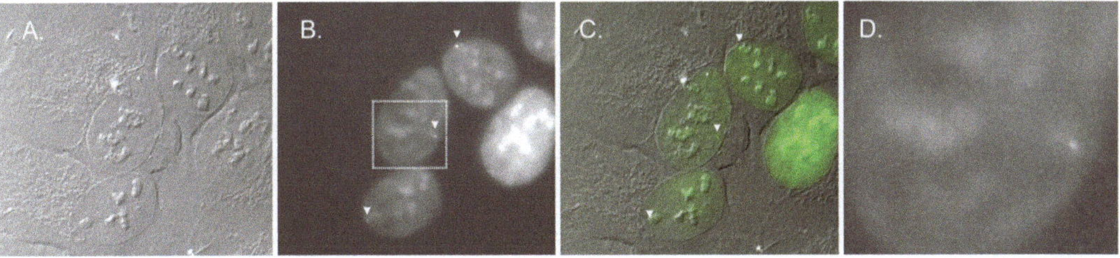

Fig. 2 Visualization of nascent RNA in living cells. (**a**) DIC image of U2-OS cells containing a reporter gene inserted randomly into the genome. Nucleoli are visible as dense bodies within the nucleus. (**b**) Fluorescence image of PP7-mCherry coat protein with a nuclear localization signal. The coat protein accumulates in the nucleus. The nascent transcription site is visible as a punctate spot (*white arrows*). At high coat protein expression levels, one frequently observes nucleolar staining as well. The *box* indicates the region which is magnified in panel **d**. (**c**) Merge of panels **a**, **b**. Transcription sites are indicated with *white arrows*. (**d**) Magnification of the demarcated region in panel **b**. The transcription site is visible near the edge of the nucleus

observed (array vs. single gene), the coat protein used to visualize the nascent RNA (MS2 or PP7), and the type of analysis for extracting data (FRAP, fluctuation analysis, and others). In this chapter, we focus on one specific implementation: using time-lapse imaging and fluctuation analysis to extract transcription data from single genes.

The guiding principle of fluctuation analysis is that observing the fluctuations of a signal around equilibrium can provide information on the underlying non-equilibrium processes. For example, diffusion is a non-equilibrium process which can be measured by recording fluorescence fluctuations in an optical focal volume in solution [16]. Transport mechanisms of ion channels can be determined by recording the fluctuations in current using electrophysiology [17]. The critical feature of fluctuation analysis is the ability to distinguish the fluctuations of interest (say, those due to a biological process) from other fluctuations which are inevitably present in any experiment. In transcription fluctuation analysis, the fluctuations of interest are due to the biochemical events which control RNA synthesis such as initiation and nascent chain elongation. However, numerous other experimental fluctuations are also present, including noise in photon collection, fluctuations in image analysis, fluctuations in transcription site position, variability in photobleaching, and more. The approach described in this protocol is able to separate the fluctuations due to transcription from the fluctuations due to other sources. In fact, in the case we describe in this protocol for analyzing a single gene, the signal fluctuations are dominated by the events of RNA synthesis. In this regime of one or a few genes, fluctuation analysis is the preferred method for analyzing data. However, in the case of a gene array, the signal shows very little fluctuation, and other techniques such as fluorescence recovery after photobleaching (FRAP) are more appropriate [18].

The protocol is divided into four major steps: (1) reporter gene design, stable cell line construction, and sample preparation; (2) live-cell fluorescence time-lapse imaging; (3) image analysis, generation of a transcription time trace, and autocorrelation; and (4) data fitting to extract kinetic parameters.

2 Materials

1. Cell chambers with No. 1.5 coverslip at the bottom (e.g., 35 mm MatTek, Ashland, MA).
2. Tissue culture plates.
3. Tissue culture reagents: serum, phenol-free L15 (used for imaging), DMEM (used for induction and cell culture), trypsin, and 1,000× stock concentration of inducer (e.g., doxycycline).
4. Cells co-expressing fluorescently tagged coat protein (MS2 or PP7) and an RNA hairpin cassette.
5. An inverted fluorescence microscope (e.g., AxioObserver, Zeiss, Thornwood, NY).
6. High numerical aperture objective (e.g., Zeiss 63× C-Apochromat).
7. An autofocusing attachment (e.g., Definite Focus, Zeiss, Thornwood, NY).
8. A microscope stage incubator (e.g., Tokai Hit, INUB-LPS, Shizuoka-ken, Japan).
9. A sensitive EMCCD camera (e.g., Evolve 512, Photometrics, Tucson, AZ).
10. A single-line excitation source (e.g., 488 nm Excelsior, Spectra Physics, Santa Clara, CA).
11. Optics for separation of excitation and emission (e.g., 488 polychroic beam splitter + 535/70 nm emission filter, Chroma, Bellows Falls, VT).
12. Image analysis software (custom programs for transcription fluctuation analysis are available at www.larsonlab.net).

3 Methods

3.1 Reporter Gene Design, Stable Cell Line Construction, and Sample Preparation

1. The first step is preparation of the constructs and stable cell lines. Since these steps are covered extensively in other protocols [10] and vary considerably between applications, we only outline some of the major design considerations in **Note 1** and turn now to sample preparation using a standard human cancer cell line U2-OS as an example.

2. Grow the U2-OS cells. Cells are grown in 10 cm tissue culture plates to a density between 20 and 80 %, taking care to keep cells within proscribed passage numbers.
3. Pass cells to microscopy dishes. Trypsinize the cells and place approximately 5,000–10,000 cells in a 35 mm culture dish with a coverslip affixed to the bottom (*see* Subheading 2). Total volume in the dish is 2 mL. Note that the thickness of the coverslip on the chamber must match the specifications of the objective.
4. Return the cells plated on the microscopy dishes to the incubator under the same conditions as before (media, temperature, CO_2 levels, etc.). Allow the cells to adhere to the bottom of the dish and begin dividing, which usually takes approximately 1 day. Cells typically grow more slowly on glass than plastic, and some cell lines may not grow at all on microscope coverslips (*see* **Note 2**).
5. If the reporter gene is inducible, add the inducer and return cells to the incubator. Inducer incubation times vary considerably depending on the reporter gene and the inducible promoter.
6. Switch the cells to Leibovitz L-15 medium in preparation for live-cell imaging. Leibovitz L-15 is a non-carbonate-buffered medium designed to optimize cell growth under ambient conditions. Using L-15 eliminates the need to maintain CO_2 partial pressure on the microscope stage, thus substantially simplifying the imaging conditions.
7. If continuing induction is necessary, make sure the imaging medium (L-15) contains the inducer (*see* **Note 3**).
8. Place the microscopy dish on the stage and bring the stage incubator to the correct temperature. Allow the microscope, dish, and stage to thermally equilibrate to minimize drift during the experiment (*see* **Note 4**). For mammalian cells, the stage incubator is set according to manufacturer instructions and typically requires at least 15 min to equilibrate.

3.2 Live-Cell Fluorescence Time-Lapse Imaging

1. Imaging single molecules of RNA in vivo is a demanding application, and care must be taken in the design and operation of the microscope to maximize fluorescence collection and minimize autofluorescence. Several aspects of microscope design are discussed in **Note 5**.
2. Establish the illumination conditions for the experiment. Excitation power should be set such that the transcription site is readily visible to the naked eye above background, but each acquisition does not measurably reduce the fluorescence intensity of the cell. Typical illumination conditions are 0.1 mW of collimated 488 nm light after the objective, resulting in an excitation light intensity in the object plane of 10 mW/mm^2.

3. Establish the exposure time for the acquisition. The exposure time should be long enough that the transcription is visible over the background as shown in Fig. 2. However, the exposure must be short enough to enable imaging at moderate frame rates. Typical exposure time values for the Evolve 512 EMCCD camera are 100–200 ms.
4. Establish the *z*-depth acquisition conditions of the experiment. Focus the microscope above and below the transcription site and note the positions. At each time point, the entire *z*-stack will be acquired. For long acquisitions (>1 h), there may be some motion in *z*. Taking a larger range of focal positions will ensure that the site does not go out of focus during the acquisition. Typical values are 10 *z*-slices spaced at 0.5 μm, resulting in a total depth of acquisition of 5 μm, which amounts to ±2.5 μm above and below the in focus plane of the transcription site.
5. Establish the time-lapse acquisition conditions of the experiment. Images should be acquired at regular time intervals between *z*-stacks. These intervals should capture the transcriptional dynamics being measured while not oversampling the transcriptional state of the cell and causing unnecessary photo bleaching or photo toxicity (*see* **Note 6**). Typical values are 10 s between *z*-stacks for 512 stacks. Thus, the total number of frames acquired is 5,120 for a total acquisition time of 85 min. Under these acquisition conditions, it is possible to resolve both initiation and elongation dynamics (see below).
6. Maintaining focus during the experiment. At each time interval, the focus should be adjusted using the autofocus features of the microscope. Specific settings depend on the microscope manufacturer.

3.3 Image Analysis, Generation of a Transcription Time Trace, and Autocorrelation

1. Maximum intensity projection. After images are acquired, only the maximum intensity projection of each *z*-stack will be used for subsequent analysis. Maximum intensity projection on three-dimensional image stacks can be performed using any standard image processing software package, for example, the freely available package ImageJ (http://rsbweb.nih.gov/ij/) or FIJI (http://fiji.sc/).
2. Calculate the brightness of the spot at each time point (Fig. 3a). The brightness of the transcription site reflects the amount of RNA present, and the time dependence of this value is the input into the fluctuation analysis procedure. Therefore, this step is critical to correct analysis of the data, and we have developed and tested an algorithm called Localize© which is described in refs. 19–21 and is available at www.larsonlab.net (*see* **Note 7**). Briefly, once a time series of projections is produced, each transcription site is identified and fit to a 2D Gaussian on top of a tilted planar background as described in reference (Fig. 3b, c) [7].

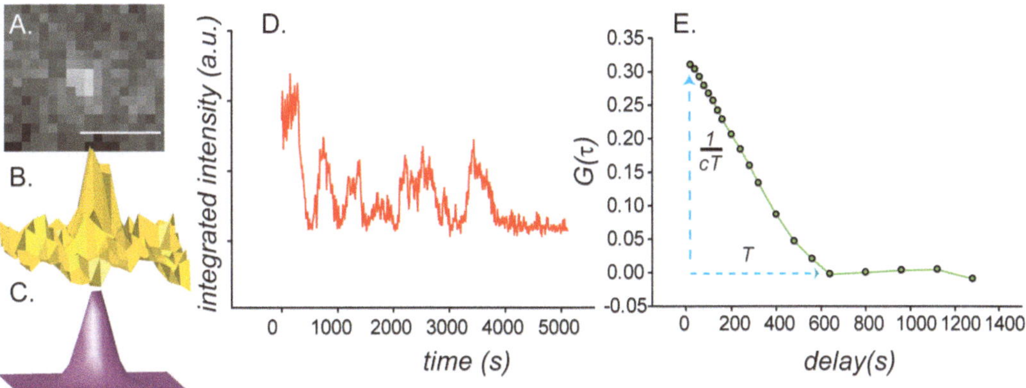

Fig. 3 Fluctuation analysis of transcription activity. (**a**) Representative individual fluorescence image from a single transcription site time series. Scale bar = 1.5 μm. (**b**) Surface plot of panel **a**. (**c**) Two-dimensional Gaussian fit to the data in panel **a**. (**d**) Transcription site intensity trajectory. Each data point in the curve comes from a determination of spot intensity using the Gaussian mask algorithm. The spot is tracked over many frames, and missing frames are interpolated based on the last known position of the spot. The final output of the tracking program is the fluorescence of the transcription site as a function of time. (**e**) Autocorrelation. The time series intensity trajectory is autocorrelated using a multi-tau correlation algorithm. The *x*-axis is the correlation decay; the *y*-axis is the amplitude of the autocorrelation. The fit parameters for the 5′ UTR case are shown graphically: the amplitude of the autocorrelation is related to the number of polymerases (*cT*), and the characteristic decay is related to the dwell time of the polymerase (*T*)

The fluorescence intensity of the transcription site is recorded at each time-step as the integrated intensity of the Gaussian fit above the fluorescence background.

3. Construct a trajectory of the transcription site over time. After the brightness and position of the spot have been measured in each frame, the next step is to track the site over time to generate a single time trace for each transcription site. This step is accomplished using a single-particle tracking algorithm developed by Crocker and Weeks [22] and implemented in Localize© which is available at www.larsonlab.net. Briefly, the tracking algorithm uses the coordinates of the spot in one frame to find the nearest neighbor spot in the subsequent frame. By iterative application of this approach, the full time-dependent intensity of the transcription site is generated (*see* **Note 8**).

4. Filling in the gaps in the trajectory. Often, the transcription site disappears, meaning that no nascent RNA is present. This disappearance is expected in the case of single genes which may only transcribe infrequently. The tracking algorithm handles these events by calculating the intensity at the position where the spot was last located. Once the spot reappears, the spot-finding calculation proceeds normally on the newly visible spot. A full trace is shown in Fig. 3d.

5. The fluorescence intensity time trace is then autocorrelated to a minimum delay time of the imaging interval and a maximum delay time of the total acquisition time. The algorithm used to produce the autocorrelation is based on the multi-tau approach summarized graphically in ref. 23. We have developed an implementation called FCSApp which can be downloaded from www.larsonlab.net. Briefly, autocorrelation with a multi-tau algorithm amounts to multiplying the signal at one point in time by the signal at another point in time at a fixed delay (*see* **Note 9**). That delay can vary from a lower limit of zero to an upper limit of the length of the experiment. The variable τ is this delay, and $G(\tau)$ is the autocorrelation at each delay (Fig. 3e).
6. Mathematically, the process of computing the autocorrelation is equivalent to

$$G(\tau) = \frac{< \delta I(t+\tau) \times \delta I(t) >}{< I(t) >^2} \quad (1)$$

where I is the integrated photon number, t is time (in units of the sampling time), and τ is the delay time. $\delta I = I(t) - <I(t)>$, where $< >$ denotes the ensemble average.

3.4 Data Fitting to Extract Kinetic Parameters

At this point, we emphasize that generation of the autocorrelation curve (Subheading 3.3) from the original raw image data (Subheading 3.2) is primarily a data-processing exercise. No assumptions have been made about the underlying biological processes. In this section, we demonstrate how to fit and interpret transcription autocorrelation curves.

1. Fitting the autocorrelation, $G(\tau)$. It is necessary to first determine which fitting equation applies to the data, based on the experimental design of the reporter construct.
For constructs with stem loops in the 5′ UTR (Fig. 1a), the fitting equation is

$$G(\tau) = \frac{(T-\tau)}{cT^2} H[T-\tau] \quad (2)$$

where c is the initiation rate, T is the total dwell time, and H is the Heaviside function.
For constructs with stem loops in the 3′ UTR (Fig. 1b), the fitting equation is

$$G(\tau) = \frac{k}{c}\left(\frac{2}{3}\right)\frac{1}{(N(N-1))^2} \sum_{n=0}^{N} (N-n)(N-n-1)(2N+n+1)\frac{(k\tau)^n}{n!} \quad (3)$$

where N is the number of stem loops, τ is the delay time, and k is the elongation rate (per stem loop). Note that the number of stem loops N does not appear in the 5′ UTR expression (Eq. 2), because in that approximation, the dwell time is dominated by the post-stem loop region.

These closed form analytical solutions can be used to fit the data $G(\tau)$ using standard approaches such as nonlinear least squares regression [24]. There are multiple commercial software packages for such fitting, and we use SigmaPlot (http://www.sigmaplot.com/).

2. Interpretation of the autocorrelation fit. The common feature of both fitting equations is the presence of a parameter c which is the initiation rate and some time parameter which is either the total dwell time of a nascent transcript (T) or the time it takes a polymerase to move between stem loops ($1/k$). The autocorrelation function therefore gives two physical parameters related to transcriptional activity (Fig. 3e) (*see* **Note 10**).

4 Notes

1. The system for live-cell RNA visualization relies on two different constructs: the reporter gene which contains the stem loops and the fluorescently labeled coat protein which binds the stem loops. We address design considerations for each in turn, starting with the reporter gene.

   One must decide how many stem loops are present and where they are located in the construct. Usually, the reporter contains multiple repeats (e.g., 6, 24, or 96) of the stem loop. This multimerization leads to amplification of the signal because each stem loop binds a dimer of the coat protein, and each monomer is labeled with a fluorescent protein (Fig. 1a). Thus, a 24-repeat sequence (length ~1.3 kb) can be maximally bound by 48 fluorophores [25]. It is the polymeric feature of RNA which makes it amenable to robust single-molecule imaging. However, the use of 24 stem loops to visualize RNA is slightly a historical artifact: smaller constructs may be visible under optimized microscopy conditions.

   The desired position of the stem loops depends on the question of interest. Thus far, the stem loops have been used in untranslated regions of the RNA, for example, the 5′ and 3′ UTRs. Stem loops in the 5′ UTR are transcribed first and are therefore retained longer at the site (Fig. 1a). Stem loops in the 3′ UTR are transcribed last and have a shorter dwell time at the transcription site (Fig. 1b) [7]. The transcription site of a 5′-labeled gene will appear more often than a 3′-labeled gene because there is a greater probability of observing the nascent transcript (Fig. 1a). The residence time of the 5′-labeled transcript is dominated by the time required for transcriptional elongation (*see* Subheading 3.4), while the residence time of the 3′-labeled transcript is determined both from elongation and termination rates. Take note that post-transcriptional control sequences are often located in the 3′ UTR, so stem loops in this region may perturb, for example, RNA half-lives. On the other hand, stem loops in the 5′ UTR may affect translation efficiency.

Transcription site imaging also requires simultaneous expression of the RNA binding protein—either MS2 or PP7 coat protein—under the control of a constitutive promoter and fused to a fluorescent protein of choice (e.g., GFP, YFP, or mCherry). Achieving the appropriate level of coat protein is critical. Too much, the transcription site signal cannot be seen over background fluorescence; too little, the availability of the coat protein may be a limiting factor, leading to dim or invisible transcription sites. The ubiquitin promoter has been used in several studies because it is active in most cell types and expresses at a reasonable level for transcription site imaging [26]. Another variable is the presence or absence of a nuclear localization signal on the coat protein. Many previous studies use coat protein with an NLS, although this signal is not strictly necessary to visualize transcription sites in the nucleus (data not shown).

2. Some cell lines grow poorly or not at all on glass. One method for improving growth on microscopy dishes is to use the following treatment:
   (a) 10 min in 1 N HCl.
   (b) Remove HCl.
   (c) 2 min in 95 % ethanol.
   (d) Rinse three times with sterile PBS.
   (e) Dish is now ready for cell plating.
3. Inducible promoters are often used to control the expression of the reporter gene. It is important to remember, however, that when measuring transcription in single cells by nascent RNA imaging, one is looking at a snapshot of transcription activity. Care must be taken to ensure the reporter is actively transcribed during the measurement period. Some cell types induce better in certain media or under certain growth conditions which are different from the imaging conditions. Switching media immediately before the experiment may disrupt transcription. Therefore, it is important to probe different environmental conditions. For example, the cells may need to be induced for 24 h in the Leibovitz L-15 media. Or, conversely, cells may be best induced in DMEM and then switched to L-15 immediately before imaging.

In general, transcription is very sensitive to environmental parameters. Therefore, it is necessary to control variability and growth conditions such as cell density, growth rate, and induction level.

4. Minimizing experimental drift is a major focus of single-molecule microscopy. In addition to active focus stabilization provided through the microscope, it is important to maintain constant temperature and humidity in the laboratory. In addition, it may be necessary to allow the microscope and the room to thermally equilibrate for many hours under the imaging conditions of the experiment. For example, both the light source and the stage incubator are sources of thermal drift, so it is often beneficial for these instruments to remain on between samples or even imaging sessions.
5. Single-molecule imaging is a demanding application requiring attention to many details of illumination and excitation. A full discussion is beyond the scope of this protocol, but there are nevertheless several criteria which should be kept in mind. Narrowband excitation (e.g., using a laser for epifluorescence imaging) maximizes excitation of the fluorophore and minimizes cellular autofluorescence and phototoxicity. High numerical aperture objectives (NA > 1.2) are necessary for efficient photon collection. Electron-multiplying CCDs (EMCCDs) are essential for single-RNA imaging. Excitation and emission filters must be matched to the fluorophore being used in the experiment. The stage incubator must maintain appropriate temperature and humidity levels during the experiment. The above requirements are the ones that are the most difficult to circumvent in single-molecule experiments.
6. Setting the timescale and imaging parameters is important and also variable for each experiment. Ideally, one should observe a diffraction-limited spot which is visible above the nuclear background (Fig. 2). The fluctuations in brightness reflect the initiation and elongation of RNA, and one should strive to image the spot long enough to observe the spot blinking on and off ~5–15 times during the course of experiment (Fig. 3d). The excitation intensity and the time series duration may need to be adjusted accordingly. Moreover, the total fluorescence intensity of the cell should not decrease by more than 10 % over the course of the experiment.
7. An object in the light microscope which is smaller than the wavelength of light appears as a diffraction-limited spot or point-spread function [27]. Extracting position information from this point-spread function is the subject of numerous papers and is the basis for super-resolution imaging techniques such as PALM [28–30]. In this protocol, however, the spatial information is of incidental importance, while the brightness

of the spot is critical. We use a Gaussian mask algorithm described in [20]. This method for calculating the integrated intensity is a hybrid approach where the actual value of each pixel is weighted by the expected value of the pixel given a specific position of the diffraction-limited spot. Therefore, it is a centroid-type approach but with a point-spread function weighting instead of a top-hat function weighting. This algorithm performs well in the low signal to noise regime of the nucleus.

8. There is often only one or zero visible transcription sites in the nucleus, begging the question why it is necessary to track the site over time. In reality, the spot-finding algorithm will also pick up other features such as nucleoli or single RNAs in the nucleus (Fig. 2c). By constructing a trajectory which connects spots found in individual frames, one is effectively applying an additional filter to the data. In other words, the transcription site is allowed to move a proscribed maximum distance between frames. This eliminates the inclusion of spurious spots from other parts of the nucleus which show up during the time series.
9. The multi-tau algorithm was developed to reduce the error in the correlation at long times. Basically, instead of a single delay bin size (single-tau), the delay bin size width increases with the number of bins. So at long delays (right side of a correlation curve (Fig. 3e)), there are more events in each bin than would have occurred had the bin size remained the same size. The error in multi-tau correlation measurements is discussed in [23].
10. In a traditional fluorescence correlation spectroscopy measurement, the amplitude of the autocorrelation at a delay time of zero ($G(0)$) is equal to $1/N$, where N is the average number of molecules in the focal volume [16]. In the case of transcription, the interpretation is similar: $G(0)$ is related to the number of transcripts present at the site of transcription. For the 5′ UTR, $G(0)=1/cT$, which is exactly the number of nascent transcripts. For the 3′ UTR, $G(0)=(4/3cT)$, which is greater than the number of transcripts. Intuitively, this second relationship derives from the fact that in the case of the 3′ UTR, there is a contribution to the autocorrelation from the increase in fluorescence as the stem loops are transcribed (in addition to the total dwell time) which adds to the amplitude. For the 5′ UTR, this increase in fluorescence is a negligible part of the time-dependent fluorescent signal and therefore does not contribute to the autocorrelation.

   By using these particular models to fit an experimental autocorrelation function, we are making several assumptions about the particular kinetics of transcription initiation and elongation: (1) there are no backward rate constants in initiation or elongation (i.e., once the polymerase starts transcribing, it eventually reaches the end of the gene), (2) the coat protein is not limiting and the on rate is rapid compared to imaging time resolution, and (3) single mRNA diffusion in the nucleus is rapid compared to the imaging time resolution. If these assumptions do not apply, then other models should be used.

Acknowledgments

D.R.L. and M.L.F. are supported by the Intramural Research Program of the NIH, National Cancer Institute, Center for Cancer Research.

References

1. Darzacq X et al (2009) Imaging transcription in living cells. Annu Rev Biophys 38:173–196
2. Sinclair P et al (2010) Dynamic plasticity of large-scale chromatin structure revealed by self-assembly of engineered chromosome regions. J Cell Biol 190(5):761–776
3. Neumann FR et al (2012) Targeted INO80 enhances subnuclear chromatin movement and ectopic homologous recombination. Genes Dev 26(4):369–383
4. Green EM et al (2012) A negative feedback loop at the nuclear periphery regulates GAL gene expression. Mol Biol Cell 23(7):1367–1375
5. Chubb JR et al (2006) Transcriptional pulsing of a developmental gene. Curr Biol 16(10):1018–1025
6. Golding I et al (2005) Real-time kinetics of gene activity in individual bacteria. Cell 123(6):1025–1036
7. Larson DR et al (2011) Real-time observation of transcription initiation and elongation on an endogenous yeast gene. Science 332(6028): 475–478
8. Raj A et al (2006) Stochastic mRNA synthesis in mammalian cells. PLoS Biol 4(10):e309
9. Zenklusen D, Larson DR, Singer RH (2008) Single-RNA counting reveals alternative modes of gene expression in yeast. Nat Struct Mol Biol 15(12):1263–1271
10. Yunger S, Shav-Tal Y (2011) Imaging mRNAs in living mammalian cells. In: Gerst JE (ed) RNA Detection and Visualization, Humana, New York pp 249–263
11. Bertrand E et al (1998) Localization of ASH1 mRNA particles in living yeast. Mol Cell 2(4):437–445
12. Janicki SM et al (2004) From silencing to gene expression: real-time analysis in single cells. Cell 116(5):683–698
13. Darzacq X et al (2007) In vivo dynamics of RNA polymerase II transcription. Nat Struct Mol Biol 14(9):796–806
14. Chao JA et al (2008) Structural basis for the coevolution of a viral RNA-protein complex. Nat Struct Mol Biol 15(1):103–105
15. Larson DR (2011) What do expression dynamics tell us about the mechanism of transcription? Curr Opin Genet Dev 21(5):591–599
16. Elson E, Magde D (1974) Fluorescence correlation spectroscopy. I. Conceptual basis and theory. Biopolymers 13(1):1–27
17. Simonneau M, Tauc L, Baux G (1980) Quantal release of acetylcholine examined by current fluctuation analysis at an identified neuro-neuronal synapse of aplysia. Proc Natl Acad Sci 77(3):1661–1665
18. Lionnet T et al (2010) Nuclear physics: quantitative single-cell approaches to nuclear organization and gene expression. Cold Spring Harb Symp Quant Biol 75:113–126
19. Larson DR et al (2005) Visualization of retrovirus budding with correlated light and electron microscopy. Proc Natl Acad Sci USA 102(43):15453–15458
20. Thompson RE, Larson DR, Webb WW (2002) Precise nanometer localization analysis for individual fluorescent probes. Biophys J 82(5): 2775–2783
21. Trcek T et al (2012) Single-mRNA counting using fluorescent in situ hybridization in budding yeast. Nat Protoc 7(2):408–419
22. Crocker JC, Grier DG (1996) Methods of digital video microscopy for colloidal studies. J Colloid Interface Sci 179(1):298–310
23. Wohland T, Rigler R, Vogel H (2001) The standard deviation in fluorescence correlation spectroscopy. Biophys J 80(6): 2987–2999

24. Bevington PR, Robinson DK (1992) Data reduction and error analysis for the physical sciences. WCB McGraw-Hill. Boston, MA
25. Wu B, Chao JA, Singer RH (2012) Fluorescence fluctuation spectroscopy enables quantitative imaging of single mRNAs in living cells. Biophys J 102(12):2936–2944
26. Lionnet T et al (2011) A transgenic mouse for in vivo detection of endogenous labeled mRNA. Nat Meth 8(2):165–170
27. Larson DR (2010) The economy of photons. Nat Methods 7(5):357–359
28. Betzig E et al (2006) Imaging intracellular fluorescent proteins at nanometer resolution 10.1126/Science.1127344. Science 313(5793): 1642–1645
29. Hess ST, Girirajan TPK, Mason MD (2006) Ultra-high resolution imaging by fluorescence photoactivation localization microscopy. Biophys J 91(11):4258–4272
30. Rust MJ, Bates M, Zhuang XW (2006) Subdiffraction-limit imaging by stochastic optical reconstruction microscopy (STORM). Nat Methods 3(10):793–795

Chapter 5

Tracking Nuclear Poly(A) RNA Movement Within and Among Speckle Nuclear Bodies and the Surrounding Nucleoplasm

Joan C. Ritland Politz and Thoru Pederson

Abstract

The movement of polyadenylated RNA transcripts (poly(A) RNA) through speckles in the nucleus can be detected and studied using fluorescence correlation microscopy (FCM) and photoactivation RNA tracking techniques. Speckles, sometimes called interchromatin granule clusters, are nuclear bodies that contain pre-mRNA splicing factors and poly(A) RNA. In the methods described here, speckles are marked in live cells using monomeric red fluorescent protein fused to SC35, a splicing protein that is a common speckle component. Endogenous poly(A) RNAs are tagged by in vivo hybridization with fluorescein-labeled oligo(dT) and FCM is performed at the marked speckles and in the nucleoplasm to measure the mobility of the tagged poly(A) RNA. The majority of the nuclear poly(A) RNA population diffuses rapidly throughout the nucleoplasm, and thus this method allows one to ask whether poly(A) RNA that is located in speckles at a given time is undergoing a dynamic transit or is, in contrast, a more immobile, perhaps structural, component. To visualize the movement of poly(A) RNA away from speckles, poly(A) RNA is tagged with caged-fluorescein-labeled oligo(dT) and speckle-associated poly(A) RNAs are specifically photoactivated using a laser beam directed through a pinhole in a rapid digital imaging microscopy system. The spatial distribution of the now-fluorescent RNA as it moves from the speckle photoactivation site is then recorded over time. Temperature and/or ATP levels can also be varied to test whether movement or localization of the poly(A) RNA is dependent on metabolic energy.

Key words Nuclear poly(A) RNA, Interchromatin granule clusters, Nuclear speckles, RNA mobility, Photoactivation, Fluorescence correlation microscopy, Fluorescence correlation spectroscopy, Live cell imaging, RNA tracking, Intranuclear transport, Nuclear diffusion

Abbreviations

FCM	Fluorescence correlation microscopy
mRFP-SC35	Monomeric red fluorescent protein fused to SC35 protein
oligo	Oligodeoxynucleotide
poly(A) RNA	Polyadenylated RNA transcripts

Yaron Shav-Tal (ed.), *Imaging Gene Expression: Methods and Protocols*, Methods in Molecular Biology, vol. 1042, DOI 10.1007/978-1-62703-526-2_5, © Springer Science+Business Media, LLC 2013

1 Introduction

Nuclear poly(A) RNA consists of both mRNA (unspliced and spliced) and noncoding RNAs (the latter also including primary transcripts and processed descendants). When the movement of poly(A) RNA molecules is measured throughout the nucleoplasm of live rat myoblasts using FCM [1–5], the majority of the population is observed to move rather rapidly, but there is also a significant fraction that moves very slowly [6]. This latter fraction could represent molecules that are constrained in motion due to confinement within nuclear bodies such as speckles. It is possible that such RNAs could even serve as a "scaffolding" to organize and stabilize the overall molecular components and architecture of nuclear bodies. Thus, it was of interest to determine the mobility of poly(A) RNA within these nuclear speckles.

Here we describe our established methods for the use of FCM to measure and compare the mobility of poly(A) RNA inside speckles and in the nucleoplasm and the use of photoactivation-tracking techniques to visually track the movement of poly(A) RNA as it moves between speckles and the nucleoplasm. Standard cloning techniques are employed to construct mRFP-SC35, a chimeric fluorescent protein that is a marker for speckles [7]. A stable HeLa cell line is generated that expresses mRFP-SC35 at levels that do not affect cell growth or appear to alter cell behavior. Endogenous nuclear poly(A) RNA is then tagged in these cells using fluorescently labeled oligo(dT) as a hybridization probe, and FCM is used to monitor mobility of speckle-associated poly(A) RNA [8]. FCM allows the measurement of very rapid fluctuations of fluorescent signal intensity over time within a very small confocal volume—about a femtoliter—which is similar or smaller in size than a speckle. The more rapid the fluctuations, the faster the molecules are moving in and out of the interrogated volume and the greater the rate of decay in the data's autocorrelation function. Curve-fitting algorithms can then be used to model the number of differently diffusing components and to estimate their diffusion coefficients [2, 6, 8]. Photoactivation tracking is next employed to allow analysis of poly(A) RNA movement away from speckles. Poly(A) RNA is tagged with caged-fluorescein-labeled oligo(dT) and this tag is specifically photoactivated on speckle-associated poly(A) RNA using a laser beam directed through a pinhole in the light path of a rapid digital imaging microscopy system. The distribution of the newly fluorescent signal is then digitally recorded over time as it moves away from the photoactivation site [8–11].

2 Materials

2.1 Oligo Labeling

1. Eppendorf 1.5 ml tubes.
2. 1 mM sodium bicarbonate buffer, pH 8.
3. Amine-reactive fluor, i.e., 6-(fluorescein-5-carboxamido)-hexanoic acid succinimidyl ester (5-SFX, Invitrogen Life Technologies) or a "caged" carboxyfluorescein succinimidyl ester [9, 12].
4. Dimethyl sulfoxide (DMSO), highly pure, anhydrous.
5. 30–60mer antisense oligodeoxynucleotide of choice synthesized with internal C6-aminohexyl groups present approximately every 10 bases (i.e., 43mer oligo(dT) with 5 thymidine C6-aminohexyl groups at positions 2, 12, 22, 32, and 42, custom ordered from Integrated DNA Technologies, Coralville, IA) (*see* **Note 1**).
6. Sephadex G-50 columns pre-equilibrated in 10 mM TEAB (triethylene ammonium bicarbonate, pH 8.5) or water (*see* **Note 2**).
7. Lyophilizer.
8. Sterile deionized water.
9. UV spectrophotometer.

2.2 Cell Culture and Oligo Uptake

1. Sterile Dulbecco Modified Eagle's Minimal Essential Medium (DMEM) without phenol red.
2. Sterile fetal bovine serum (FBS).
3. Sterile trypsin (0.5–2.5 %) for detaching cells from culture plates (Gibco/BRL).
4. Sterile PBS solution for washing cells.
5. Sterile 35 mm dishes with #1.5 glass bottoms (for photoactivation experiments, e.g., #P35G-1.5-20-C from MatTek Corp.)
6. Sterile 8-well Lab-Tek™ II chambered cover glass dishes (for FCM, Nalge Nunc International).
7. Sterile cationic lipid suited for transfection of cell type of interest (e.g., Lipofectamine 2000).
8. Sterile OptiMEM for transfection, if cells transfect better without serum.
9. Sterile 1.5 ml Eppendorf and/or15 ml conical tubes for mixing transfection reagents.
10. Cells containing a stably expressing SC35-mRFP (to obtain this cell line, contact the laboratory of David Spector:

http://spectorlab.cshl.edu/) or another cell line expressing marked nuclear body of interest.

11. Antisense oligos labeled with fluorescein or caged fluorescein (*see* Subheading 3.1).

2.3 Fluorescence Correlation Microscopy (FCM)

1. 20–50 nM Alexa 488 (Molecular Probes, Eugene, Oregon) in sterile deionized water for standardization.
2. 20–50 nM fluorescein-labeled oligo(dT) in sterile deionized water for standardization (*see* Subheading 3.1 for labeling procedure).
3. Confocal microscope equipped with FCM module, avalanche diode detector, and lasers for red and green excitation with appropriate filters. A commercial example of such a system is shown at http://microscopy.zeiss.com/microscopy/en_us/products/confocal-microscopes/single-molecule-imaging.html#inpagetabs-1, but the system can also be homemade [2, 8].
4. Stage chamber and dish holder that can be heated and reliably maintained at 37 °C.

2.4 Photoactivation-Tracking Experiments

1. High end inverted wide field microscope equipped with lasers for long UV (e.g., 351 and 364 nm lines of an argon laser (Coherent)) plus red and green excitation; automated filter wheels for both excitation and emission; back-thinned, high-resolution, low-noise CCD camera; dish heater; and rapid image acquisition and analysis software. A commercial example is Nikon's Eclipse Ti (modified for UV excitation), but a system also can be homemade [11].
2. Metal plate with pinhole (i.e., Edmund Scientific, 35 μm diameter for 40× objective, 100 μm diameter for 100× objective) aligned in the light path between the UV laser and the cell specimen for directed photoactivation (*see* ref. 11 for diagram).
3. Phenol red-free medium and either a CO_2 source to create a 5 % concentration in the cell chamber or HEPES-buffered medium.
4. Power meter (1815-C power meter, 818-UV detector; Newport Corp.) to measure uncaging power directed inside cells.

3 Methods

3.1 Oligodeoxynucleotide Labeling

3.1.1 Fluorescently Label Probes: Sample Reaction

1. 40 μl of a 1 M solution of 6-(fluorescein-5-carboxamido)-hexanoic acid succinimidyl ester [5-SFX, Invitrogen Life Technologies #F-6106] in anhydrous DMSO (*see* **Note 3**).
2. 40 μl sterile 1 M sodium bicarbonate, pH 8.
3. 80 μl of a 3.26 μg/μl (0.25 nm/μl) solution of amino-modified oligo(dT) (containing a C6-aminohexyl group at positions 2,

12, 22, 32, and 42) in sterile water (the equivalent of 100 nmol of amine-reactive group).

4. 40 μl sterile ddH_2O. Set up reaction and incubate overnight at room temperature in the dark.

3.1.2 Removal of Unreacted Fluorescein and Measurement of Labeling Efficiency

1. Columns can be prepared in 25 ml pipets in a darkened room (*see* **Notes 4** and **5**).
2. Collect the void volume in 1 ml aliquots in Eppendorf tubes. The usual elution peaks of the dye-coupled oligo and free dye are separated by about 10 ml of eluate.
3. Close and pierce the covers of the Eppendorf tubes containing the oligo-dye fractions (*see* **Note 6**) with a sharp forceps tip, lyophilize and resuspend in 50 μl sterile water.
4. Alternatively, free dye can be removed using a spin column such as #28304, Qiagen.
5. Measure the concentration and labeling efficiency using a spectrophotometer (*see* **Note 7**). Most dyes show some fluorescence at 260 nm; therefore, the absorbance needs to be corrected before calculating concentration. To do this, record the absorbance (A) at both 260 nm for DNA and 494 nm for fluorescein. Remember to re-zero the spectrophotometer when you switch wavelengths. Calculate the corrected base concentration ($A_{base} = (A_{260}) - (A_{494} \times 0.32)$) and then number of dye molecules/base using $(A_{dye})(\varepsilon_{base})/(A_{base})(\varepsilon_{fluorescein})$, where $\varepsilon_{fluorescein} = 30{,}000\ M^{-1}\ cm^{-1}$, $\varepsilon_{base} = 10{,}000\ M^{-1}\ cm^{-1}$, and 0.32 in the first equation is the resultant correction factor for fluorescein. A convenient website that does this for you can be found at http://probes.invitrogen.com/resources/calc/basedyeratio.html.

3.2 Stable Cell Lines and Oligo Uptake

3.2.1 mRFP-SC35 Stable Cell Line

A human SC35 cDNA [7] was cloned into a vector encoding monomeric RFP [13]. A HeLa stable cell line containing the mRFP-SC35 was generated using standard techniques [8] and maintained in DMEM with 10 % fetal bovine serum and 0.5 mg/ml G418. Other marker proteins with various fluorescent protein tags can be used so long as their emission wavelength can be separated from the fluor used to label the oligo tags (usually fluorescein or Alexa 488).

3.2.2 Transfection of Oligo

1. Cells are plated into 2-, 4-, or 8-well Lab-Tek II dishes for FCM or onto 25 mm cover slips for photoactivation experiments at ~40 % confluency and allowed to grow overnight in medium without G418 and phenol red.
2. Standard cultured cell lines such as HeLa and U2OS can then be transfected with oligo(dT) or oligo(dA) (as a control) at a final medium concentration of 0.125 μM with, for example, Lipofectamine 2000. Oligos and cationic lipid are separately

diluted in OptiMEM, mixed, incubated for 15 min at room temperature, and added directly to cells followed by an equal volume of medium. Labeled oligos (both fluorescent and "caged") should always be protected from light. Appropriate oligo and transfection reagent concentrations must be determined empirically for different cell lines. Oligo uptake should be seen in approximately 50 % of the cells and should not affect cell viability as judged in control experiments to measure cell doubling over the next 24 h.

3. Cells are incubated at 37 °C with oligo for 2–3 h in the cell culture incubator (in the dark) followed by a 1 h incubation without oligo in DMEM (with serum) and then imaged immediately. Cells containing oligo should be protected from light until imaged. Hybridization can be assayed in parallel sample cells at this point using in situ reverse transcription in which the hybridized oligo serves as the primer for reverse transcription to extend a DNA product at these sites. This procedure has been described in detail before [14], so is not described here.

3.3 FCM

1. FCM can be carried out using a homemade instrument [2] or using a confocal microscope with an FCM module (*see* Subheading 2). The dish holder (and stage chamber, depending on the configuration of the microscope) must be pre-warmed to 37 °C for 2–3 h. It is very important to pre-equilibrate the microscope to prevent drifting during data acquisition [5].
2. Transfected cells growing at 37 °C in 6 of the 8 wells in Lab-Tek II dish are mounted on the pre-warmed stage of the FCM microscope.
3. A solution of 50 nM Alexa 488 in sterile deionized water is placed in 1 or 2 wells of an 8-well dish, and FCM readings are taken (commonly 10 readings of 30 s each). The resulting autocorrelation curves are fit to a one component model using curve-fitting software such as Quickfit [15]. Using the known diffusion coefficient of Alexa 488 (425 $\mu m^2/s$ at 25 °C, [16]), the actual detection volume can be determined. This measurement must be performed each time an experiment is started and is used to help fit the autocorrelation curves obtained in the experiment (see below). The confocal volume is in the femtoliter range; a typical measurement is a radial diameter of 0.44 μm and a *z*-axis height of 1.6 μm when exciting Alexa 488 (and 0.5 μm and 2 μm, respectively, for mRFP).
4. A 50 nM solution of the fluorescein-labeled oligo is placed in a well of the dish the cells are growing in and allowed to equilibrate to 37 °C for at least half an hour on the stage. This solution is then subjected to FCM and the autocorrelation curves fit as above. This gives the mobility of the free oligo in solution at 37 °C.

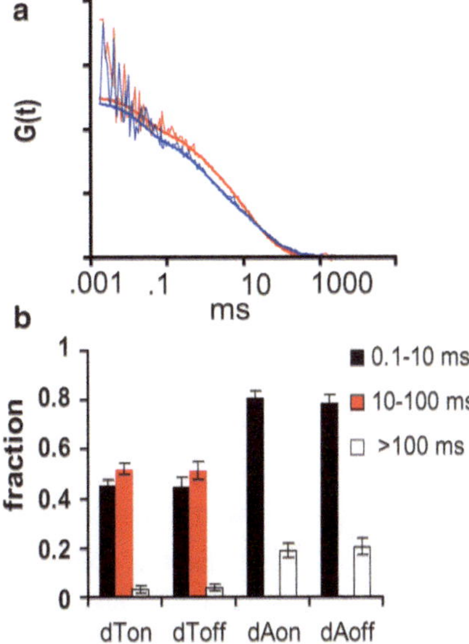

Fig. 1 Mobility of oligo(dT) and oligo(dA) on and off speckles in mRFP-SC35 HeLa cells measured using FCM. (**a**) Autocorrelation curves of cells containing oligo(dT), based on FCM measured either within a speckle (*blue*) or in the nucleoplasm (off a speckle, *red*). Best fit curves are shown as *smooth lines*. (**b**) Fraction of oligo(dT) and oligo(dA) present in different mobility classes measured within speckles (on) and in the nucleoplasm (off). Each kinetic component is indicated by *black*, *red*, or *open bars* in the histogram. The 10–100 ms component (*red*) was undetectable in oligo(dA)-containing cells (From ref. 8)

5. Inside the cells, a speckle, identified by excitation with the mRFP laser, or a site in the nucleoplasm, is aligned in the laser path using a galvanometer scanner in a point-addressable mode (*see* **Note 8**). The *x*, *y*, *z* coordinates of the site are recorded and images of the region of interest are taken after excitation for first mRFP and then the fluorescein. FCM measurements are recorded and ten readings of 30 s each are typically averaged for each interrogated spatial site. Readings that show bleaching (signal intensity decreases during the measurement time) should be discarded. Laser intensity should be in the range of 1–5 kW/cm^2 (0.1–0.5 mW laser power at the focus). Excitation filters (i.e., 485 nm DF 22, Omega Optical) can be placed in the laser light path to reduce the intensity of the mRFP if it is bright enough to bleed into the fluorescein channel. This should be monitored using control cells that have not taken up oligo.
6. Best fits for the FCM autocorrelation curves (Fig. 1a) are obtained using software such as Quickfit [15] and MaxEnt [17] or various software packages that accompany commercial FCM modules. Commonly, best fits to one, two, or three component models are evaluated. Figure 1b compares the fraction

of molecules in different mobility classes in cells containing oligo(dT) versus oligo(dA) controls. The unhybridized fast-moving component is present in both, but the somewhat slower-moving hybridized fraction is present only in oligo(dT)-containing cells. *See* also refs. 6, 8.

3.4 Photoactivation Tracking

3.4.1 Photoactivation and image acquisition

A standard experiment is as follows:

1. A 25 mm glass-bottomed dish containing growing cells transfected with caged-fluorescein-labeled oligo(dT) is mounted in the temperature-controlled chamber on an inverted microscope (*see* Subheading 2) in the dark.
2. A speckle, visualized by exciting mRFP, is aligned in the center of the microscope field.
3. The "caged" oligo present in this speckle is photoactivated using a UV laser directed through a pinhole set in the laser path in front of the microscope [10, 11]. Using a 100 μm pinhole and a 100× water immersion objective lens, the uncaging site within the nucleus can be focused to a spot 1.0–1.5 μm in diameter, which is similar to or smaller than speckles. Uncaging time is typically 65 ms and beam power is measured and adjusted to 90–120 μW at the beginning of each experiment. This results in ~1.6 kW/cm^2 in the focused spot on the sample and both HeLa cells and rat L6 myoblasts survive this treatment with no obvious ill effects. However, care must be taken not to substantially exceed these values, since even long wavelength UV light can damage cells at higher intensities. Other cell lines, especially primary cells, will need to be tested for their viability after exposure to the photoactivation light.
4. The movement of the photoactivated signal away from the speckle is immediately tracked over time by capturing rapid sequential digital images. It is usually sufficient to capture 2D images, although 3D images can be successfully obtained if image size, resolution, exposure times, and laser intensities are adjusted appropriately. A typical image size is 512 × 512 pixels and typical exposure times are 5–10 ms every 500 ms for 30 s for the photoactivated fluorescein. Signal spreads throughout the nucleus during this assay time [8]. An image of the mRFP-labeled speckle distribution is captured before and after tracking the photoactivated fluorescein signal.

3.4.2 Control experiment

A control experiment using cells containing a non-hybridizing "Caged" Oligo(dA) must also be performed, exactly as described above. Most of the oligo(dA) will leave the uncaging site very rapidly (by 2 s), since it is not hybridized, and the behavior of this fraction can be used to model the behavior of unhybridized oligo(dT) during image analysis.

3.4.3 Image Analysis

1. Deconvolution. 3D image stacks are deconvolved using exhaustive photon reassignment or other comparable algorithms to remove out of focus light.
2. Estimate of diffusion coefficient. When analyzed in 2D, the signal moving away from the photoactivation site will generally spread relatively uniformly and follow a Gaussian distribution with small fluctuations. Observed deviations from a pure Gaussian distribution may reflect the presence of obstacles the RNA molecules encounter in the intranuclear landscape. The mean square displacement can be plotted as a function of the width (at e^{-2}) of the Gaussian distribution at increasing times and, in the case of poly(A) RNA, is linear indicating the process is diffusive (10, 18 and *see* **Note 9**). The slope of this line gives an estimate of the diffusion coefficient, but this estimate depends on where the Gaussian width is measured, does not account for optical blurring, and is not sensitive to small fluctuations in the distribution. More refined algorithms (i.e., 8, 19) can be used to fit the entirety of the Gaussian distributions over multiple time points to better estimate a diffusion coefficient when necessary.
3. Movement to other speckles. The distribution of signal throughout the nucleus can be analyzed to determine whether poly(A) RNA leaving one speckle can enter another. In our HeLa cell studies, the signal photoactivated in one speckle was found in other speckles after signal dispersion [8].
4. Mobility changes with respect to temperature or various drug treatments. The incubation temperature can be changed and mobility remeasured after equilibration and/or drugs can be added to the medium to test their effects on poly(A) RNA movement in and out of the speckle.

4 Notes

1. Both the aminohexyl linker arms and the spacing of the dye along the oligo(dT) are designed to help prevent RNase H degradation of the poly(A) tail after hybridization (*see* refs. 9, 20).
2. TEAB is a volatile buffer removed during lyophilization.
3. A previously unopened ampule of DMSO should be used to dissolve each new batch of succinimidyl ester to ensure it is anhydrous. Succinimidyl esters do not survive well in solution; if necessary to store, freeze at –80°C.
4. It is sufficient to close blinds and turn off fluorescent lights for the column separation.
5. Score and snap off the top inch of a sterile and disposable 25 ml pipet, advance the cotton plug to plug the pipet tip

(using a 1 ml pipet), fill with room temperature Sephadex G-50 suspension (pre-swelled in 4XSSC), and equilibrate with 5 mM TEAB by allowing several column bed volumes of buffer to pass through. Then clamp the outflow and add the oligo-dye coupling reaction sample.

6. If caged oligo is being fractionated, place 50 µl aliquots from each fraction into small Eppendorfs and organize these into open-bottomed tube holder. Expose this to UV light using long wavelength UV light on a UV box and identify the aliquots containing the labeled oligo once the appropriate samples are photoactivated and fluoresce. *See* also ref. 11.
7. Caged oligos must be photoactivated before this step—a dilute sample can be uncaged by laying the Eppendorf on a UV box producing long wavelength light for about 30 min—this results in almost complete uncaging.
8. The confocal volume is close to or smaller than that of a speckle; deconvolved images of speckles in the stable SC35 cell line showed radial diameters of about 0.5–2.5 µm and *z*-axis heights of about 1.5–2.5 µm [8].
9. A good first description of diffusive behavior is given by Howard Berg in Random Walks in Biology, a wonderful book [21].

Acknowledgments

We thank Richard Tuft for expert assistance with the uncaging experiments, Jörg Langowski and Nina Baudendistel for expert help with the FCM, and David Spector and Kannanganattu Prasanth for construction and characterization of the SC35-mRFP plasmid.

References

1. Magde D, Elson E, Webb W (1972) Thermodynamic fluctuations in a reacting system. Measurement by fluorescence correlation spectroscopy. Phys Rev Lett 29:705–708
2. Wachsmuth M, Waldeck W, Langowski J (2000) Anomalous diffusion of fluorescent probes inside living cell nuclei investigated by spatially-resolved fluorescence correlation spectroscopy. J Mol Biol 298:677–689
3. Bacia K, Kim SA, Schwille P (2006) Fluorescence cross-correlation spectroscopy in living cells. Nat Methods 3:83–89
4. Digman MA, Gratton E (2009) Fluorescence Correlation Spectroscopy and Fluorescence Cross-correlation Spectroscopy. Wiley Interdiscip Rev Syst Biol Med 1:273–282
5. Dross N, Spriet C, Zwerger M et al (2009) Mapping eGFP oligomer mobility in living cell nuclei. PLoS One 4:e5041
6. Politz JC, Browne ES, Wolf DE et al (1998) Intranuclear diffusion and hybridization state of oligonucleotides measured by fluorescence correlation spectroscopy in living cells. Proc Natl Acad Sci USA 95:6043–6048
7. Prasanth KV, Sacco-Bubulya PA, Prasanth SG et al (2003) Sequential entry of components of the gene expression machinery into daughter nuclei. Mol Biol Cell 14:1043–1057
8. Politz JCR, Tuft RA, Prasanth KV et al (2006) Rapid, diffusional shuttling of poly(A) RNA between nuclear speckles and the nucleoplasm. Mol Biol Cell 17:1239–1249

9. Politz J (1999) Use of caged fluorochromes to track macromolecular movement in living cells. Trends Cell Biol 9:284–287
10. Politz JC, Tuft RA, Pederson T et al (1999) Movement of nuclear poly(A) RNA throughout the interchromatin space in living cells. Curr Biol 9:285–291
11. Politz JCR, Tuft RA, Pederson T (2004) Photoactivation-based labeling and in vivo tracking of RNA molecules in the nucleus. In: Spector D, Goldman R (eds) Live cell imaging: a laboratory manual. Cold Spring Harbor Laboratory Press, Cold Spring Harbor, NY, pp 177–185
12. Mitchison TJ, Sawin KE, Theriot JA (1994) Caged fluorescent probes for monitoring cytoskeleton dynamics. In: Celis JE (ed) Cell biology: a laboratory handbook. Academic, New York, NY, pp 65–74
13. Campbell RE, Tour O, Palmer AE et al (2002) A monomeric red fluorescent protein. Proc Natl Acad Sci USA 99:7877–7882
14. Politz JC, Singer RH (1999) In situ reverse transcription for detection of hybridization between oligonucleotides and their intracellular targets. Methods 18:281–285
15. Press WH, Teukolsky SA, Vetterling WT et al (1992) numerical recipes in C: the art of scientific computing. Cambridge University Press, Cambridge
16. Petrasek Z, Schwille P (2008) Precise measurement of diffusion coefficients using scanning fluorescence correlation spectroscopy. Biophys J 94:1437–1448
17. Modos K, Galantai R, Bardos-Nagy I et al (2004) Maximum-entropy decomposition of fluorescence correlation spectroscopy data: application to liposome-human serum albumin association. Eur Biophys J 33:59–67
18. Cardullo RA, Mungovan RM, Wolf DE (1991) Imaging membrane organization and dynamics. In: Dewey TG (ed). Biophysical and biochemical aspects of fluorescence spectroscopy. Plenum Pub. Corp. New York, NY, pp 231–260
19. Fogarty KE, Kidd JF, Tuft RA et al (2000) Mechanisms underlying InsP3-evoked global Ca2+ signals in mouse pancreatic acinar cells. J Physiol 526:515–526
20. Ueno Y, Kumagai I, Haginoya N et al (1997) Effects of 5-(N-aminohexyl)carbamoyl-2′-deoxyuridine on endonuclease stability and the ability of oligodeoxynucleotide to activate RNase H. Nucleic Acids Res 25:3777–3782
21. Berg H (1993) Random walks in biology. Princeton University Press, Princeton, NJ

Chapter 6

Nuclear Trafficking and Export of Single, Native mRNPs in *Chironomus tentans* Salivary Gland Cells

Tim P. Kaminski, Jan-Hendrik Spille, Claudio Nietzel, Jan Peter Siebrasse, and Ulrich Kubitscheck

Abstract

Real-time observation of single molecules or biological nanoparticles with high spatial resolution in living cells provides detailed insights into the dynamics of cellular processes. The salivary gland cells of *Chironomus tentans* are a well-established model system to study the processing of RNA and the formation and fate of messenger ribonucleoprotein particles (mRNPs). For a long time, challenging imaging conditions limited the access to this system for in vivo fluorescence microscopy. Recent technical and methodical advantages now allow observing even single molecules in these cells. We describe here the experimental approach and the optical techniques required to analyze intranuclear trafficking and export of single native mRNPs across the nuclear envelope.

Key words mRNP export, Light sheet fluorescence microscopy, Single-molecule microscopy, Single-molecule tracking, Nucleocytoplasmic trafficking, Fluorescence labeling

1 Introduction

The salivary gland cells of *C. tentans* undergo multiple DNA replications without cell division. The chromatids are organized in giant polytene chromosomes exhibiting a characteristic banding pattern due to the parallel alignment of the sister chromatids, while the remaining volume of the nucleus is devoid of chromatin. Transcriptionally active regions on chromosomes are expanded or "puffed." On chromosome IV there are two giant puffs, the Balbiani rings (BR) 1 and 2, which encode for large saliva proteins and display an especially high transcriptional activity. BR2 actually contains two BR genes, which encode for long mRNA transcripts (~37 kb). These transcripts are packed into large mRNPs with a diameter of 50 nm [1]. These particles represent an ideal system to study mRNP trafficking with single-molecule microscopy [2, 3], because their size leads to a relatively low mobility. These mRNPs

Yaron Shav-Tal (ed.), *Imaging Gene Expression: Methods and Protocols*, Methods in Molecular Biology, vol. 1042, DOI 10.1007/978-1-62703-526-2_6, © Springer Science+Business Media, LLC 2013

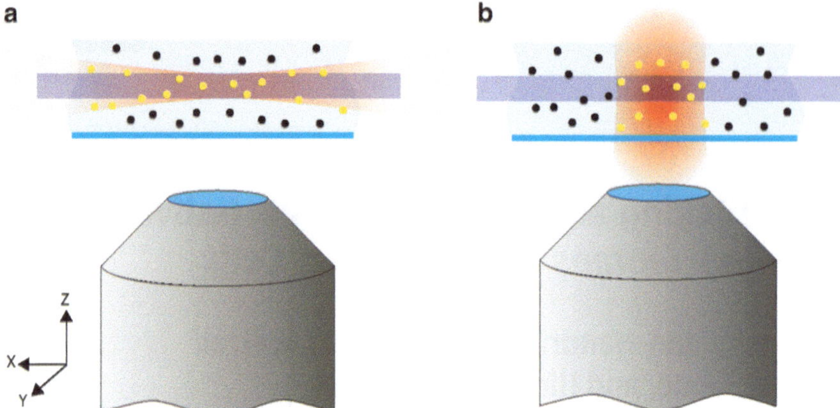

Fig. 1 Comparison of light sheet fluorescence microscopy and epi-illumination. (**a**) Light sheet microscopy uses an illumination pathway (*red*), whose optical axis lies in the focal plane of the detection objective (*dark blue*). Therefore, only fluorophores (*black* and *yellow dots*) in the detection focal plane are excited (*yellow dots*). (**b**) Epi-illumination excites not only the fluorophores in the focal plane but also above and below it. The emission of fluorophores outside the focal plane contributes to the background and decreases the SNR

have to be remodeled to traverse the NPC, which should slow down nuclear export, too.

However, the salivary gland cells present a difficult imaging situation, even after explantation out of the larva. The cell nuclei are located deep inside the glands. To study mRNP export, the nuclear envelope (NE) has to be marked. A sharp image of the NE can only be obtained at the equator of a nucleus, which is usually in a distance of 50–80 μm from the cover glass. Even worse, to maintain ongoing mRNP export, the glands require incubation in larval hemolymph, which has a dark red color due to the insect hemoglobin. Furthermore, the salivary gland cell cytoplasm contains diffracting and autofluorescent structures. Therefore, a robust microscopy technique is needed, which reduces the background to yield a sufficient signal-to-noise ratio (SNR), can function with low illumination intensity to reduce photobleaching, and allows a high temporal resolution. All these requirements are fulfilled by light sheet fluorescence microscopy (LSFM).

The basic concept of LSFM is the separation of illumination and detection pathway. The sample is illuminated perpendicular to the detection pathway (Fig. 1). In this configuration the focal plane of the detection objective can selectively be illuminated. This principle eliminates most background fluorescence, because fluorophores outside the illumination plane are not excited. Beyond that, photobleaching is minimized and, since no dichroic mirror is needed, signal intensity is increased in comparison to epi-illumination fluorescence microscopy. With LSFM the whole field of view is imaged simultaneously, and not sequentially scanned like in classical confocal imaging. This allows a high temporal resolution. LSFM allows to image single molecules in vivo with frame rates up to 500 Hz. To sum up, LSFM allows fast and sensitive imaging while avoiding out-of-focus fluorescence.

Aside from the microscope setup, fluorescent labeling is a critical step in single-molecule microscopy. The fluorescent molecules or particles in question have to be bright enough to produce a satisfactory SNR. In addition, the fluorescence label must not interfere with any biological process. Messenger RNPs diffuse through the nucleus and are exported through an NPC accompanied by a multitude of export factors. Therefore, the tag should not change the size of the mRNP or interact with the mRNP export machinery. The attachment of chemical dyes to the tag is preferable to using a fluorescent protein, due to their superior photo-physical properties.

The *C. tentans* hrp36, a homologue of mammalian hnRNPA1, is incorporated in multiple copies into nascent mRNPs during transcription and remains incorporated in mRNPs during intranuclear trafficking and nuclear export [4]. Hrp36 contains a C-terminal M9 domain, which mediates the transportin-dependent shuttling of the protein between the nucleus and the cytoplasm. The protein can easily be expressed in bacteria and covalently labeled with cysteine-reactive dyes after purification. To improve the labeling ratio and to avoid modification of internal cysteines, we introduced an N-terminal tetra-cysteine motif. When fluorescently labeled hrp36 was microinjected into the cytoplasm of salivary gland cells, it quickly translocated to the nucleus, where it was stably incorporated into nascent mRNPs thus marking the transcription sites. Messenger RNPs labeled in this manner were only modified by a few dye molecules and the 10 amino acid-long tetracysteine (TC) motif attached to the hrp36. The salivary gland cells contain a large number of endogenous hrp36. The injection of appropriately diluted solutions of labeled hrp36 therefore allowed controlling the extent of mRNP labeling. Final concentrations in the picomolar range yielded the defined labeling of single mRNPs. We focus here on the analysis of the trafficking and nuclear export of single mRNPs.

2 Materials

2.1 Microscope Setup

Our light sheet microscopes are based on commercial inverse microscopes (Fig. 2). Currently we use a Ti-U (Nikon, Düsseldorf, Germany) and an Axiovert 200 (Carl Zeiss, Jena, Germany). We use three lasers in the illumination beam path: a Sapphire 488LP (488 nm, 100 mW, Coherent, Santa Clara, CA, USA), a PL532.400 (532 nm, 400 mW, Pegasus GmbH, Wallenhorst, Germany), and a Cube 640-40C (640 nm, 40 mW, Coherent, Santa Clara, CA, USA). The lasers are coupled via an acousto-optical tunable filter

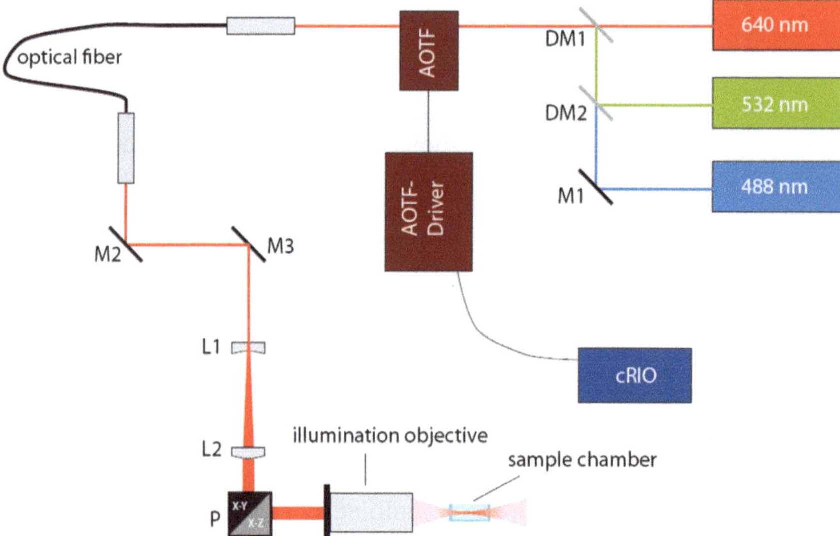

Fig. 2 Schematic drawing of the optical setup. *DM* dichroic mirror, *M* mirror, *L* cylindrical lens, *P* periscope. The *upper part* of the sketch shows a top view (*X–Y*) of the instrument, and the *lower part* gives a side view (*X–Z*)

(AOTF; AA Optoelectronics, Orsay Cedex, France) into an optical multispectral mono-mode fiber (kineFlex, PointSource, Hamble, UK). The collimated fiber output is directed via mirrors to a cylindrical beam expander comprising two cylindrical lenses (CKX540-C and CKV522-C, Newport, Darmstadt, Germany). The resulting collimated elliptical beam is directed to the illumination objective via a periscope with two adjustable mirrors. The elliptical illumination spot is focused into the sample by a 10× objective lens (plan apochromat 106, NA 0.28, Mitutoyo, Kawasaki, Japan). A micrometer screw allows adjusting the position of the illumination objective on the illumination axis. The specimen is mounted in a special glass chamber (105-044-V2-40, Hellma GmbH & Co. KG, Müllheim, Germany). A 40× water immersion objective lens (NA 1.2, C-Apochromat, Carl Zeiss, Jena, Germany) projects the image onto an EMCCD camera with 128 × 128 pixels (iXon BI DV-860, pixel size 24 μm, Andor Technologies, Belfast, Ireland). An additional fourfold magnifier yields an object field pixel size of 150 nm. A double band-pass filter (z532/633m, Chroma Technology, Bellow Falls, USA) and notch filters remove residual scattered excitation light. The sample can be positioned in X and Y by micrometer screws and in Z by a linear stepper motor (M-232 DC Mike, Physik Instrumente, Karlsruhe, Germany). Image acquisition is controlled by Solis software (Andor Technologies, Belfast, Northern Ireland). A cRIO-9076 module containing an 8-channel digital input/output module (NI 9401) and a 4-channel analog output module (NI 9263) (both from National Instruments, Munich, Germany), which is triggered by the EMCCD and governed by a custom-programmed LabVIEW (National Instruments,

Munich, Germany) virtual instrument, controls the AOTF. FPGA programming was done by SET GmbH (Smart embedded Technologies Wangen/Allgaeu, Germany).

2.2 Protein Labeling

1. AlexaFluor647 maleimide (Life Technologies).
2. Bio-Gel P6, superfine (Bio-Rad).
3. Glass Econo-Column with 10 cm length and 1 cm inner diameter (Bio-Rad).

2.3 Protein Expression and Purification

1. Isopropyl β-D-1-thiuogalactopyraoside (IPTG).
2. Glutathione Sepharose 4B (GE Healthcare).
3. Thrombin protease (GE Healthcare).
4. Disposable PD-10 column (GE Healthcare).
5. Physiological buffer containing a reducing agent, for example, phosphate-buffered saline (PBS) pH 7.2–7.4 containing 1 mM TCEP.
6. Amicon Ultra centrifugal filter units with MWCO 10 and 30 kDa (Millipore).
7. Equipment for separation of proteins by SDS-PAGE.

2.4 Microinjection Instrumentation

1. FemtoJet (Eppendorf).
2. InjectMan NI2 (Eppendorf).
3. Femtotips II (Eppendorf).
4. Microloader (Eppendorf).

2.5 C. tentans Dissection Instrumentation

1. Nettle powder (Alfred Galke GmbH).
2. 60 mm petri dish.
3. Scalpel No. 10.
4. Forceps: Dumont 3c, Dumont 5, Dumont 5, Live Insect Forceps blunt.
5. Vannas Spring Scissors—2.5 mm blades.
6. Transfer tip, created by cutting the first 3 mm of a blue (1 mL) pipette tip. Incubate the front end for 120 min in 4 % BSA. The transfer tip can be used until the glands stick to the tip.
7. Stereomicroscope.
8. KIMTECH SCIENCE Precision Wipes.
9. Mineral oil M5904 (Sigma).
10. Polytetrafluoroethylene (PTFE) building blocks (2 mm × 4 mm × 2 mm solid block) (custom-built).

2.6 Calibration Solution

1. AlexaFluor647 maleimide (Life Technologies).
2. Phosphate-buffered saline (PBS).

3 Methods

3.1 Protein Expression and Purification

Heterologous expression of recombinant protein takes place in an *E. coli* host system followed by affinity-chromatography purification via glutathione-S-transferase (GST) and subsequent thrombin cleavage of the target protein. This protocol can be applied for both proteins, TC-hrp36 and TC-NTF2, with small modifications.

1. Inoculate 100 mL LB medium containing 50 μg/mL ampicillin and 17 μg/mL chloramphenicol from a frozen BL21[DE3] pLys stock and grow overnight at 37 °C and 180 rpm in an orbital shaker.
2. Transfer 10 mL starter culture to 1 L of 2YT medium with appropriate antibiotics and grow cells at 37 °C and 180 rpm in an incubated orbital shaker to OD_{600} ~0.65.
3. When the OD is reached, cool the culture to room temperature and induce protein expression using 0.5 mM IPTG.
4. Express the protein overnight at room temperature with vigorous shaking.
5. Collect cells at 6,000 × *g* for 15 min, resuspend pellets in a small amount of buffer, and pool in a high-performance centrifuge tube.
6. Lyse cells by sonication at 14 kHz in ice water for 2 min, in 5 cycles with 2 min cooling time between each cycle.
7. Spin down cell debris at 40,000 × *g* for 30 min. Transfer cleared cell lysate to 1 mL well-washed and equilibrated Glutathione Sepharose resin and incubate for at least 30 min at room temperature with gentle shaking.
8. Wash away unbound protein with 3× 50 mL buffer by centrifuging the resin at 500 × *g* for 5 min and aspirating the supernatant.
9. Transfer the resin to a single-use column, mix it with 50 U thrombin protease, incubate for at least 1 h at room temperature, and elute the cleaved protein in 1 mL fractions.
10. Test fractions for protein content by a qualitative Bradford test and pool them in an Amicon centrifuge filter with 30 kDa MWCO (10 kDa MWCO for cleaved NTF2).
11. Bring the eluate to a volume between 250 and 1,000 μL and aliquot to 250 μL. The purified protein can now be used for subsequent labeling or be stored at −80 °C. In case of storage, the protein should be shock-frozen in liquid nitrogen.
12. SDS-PAGE analysis should be performed with 12 % polyacrylamide (PAA) gel for TC-hrp36 and with a 15 % PAA gel for TC-NTF2, respectively. Check whether proteins with the expected molecular weight are obtained, 36 kDa for hrp36, and 16 kDa for TC-NTF2.

3.2 Fluorescent Labeling of Proteins

The fluorescent dye is coupled to the recombinant protein via thiol-reactive groups (*see* **Note 1**).

1. Determine volume and molar concentration of the protein and calculate the volume of dye needed for labeling with respect to a fourfold molar excess of dye to protein.
2. Mix the dye into the protein by pipetting up and down and incubate for at least 2 h at room temperature in the dark with gentle shaking.
3. Prepare a gel filtration column by mixing ~0.5 g of Bio-Rad P6 gel into 25 mL of pre-warmed buffer and let the mixture swell for 30 min at room temperature with casual slewing.
4. Pour the swollen gel in a single step into a small desalting column and let the gel bed settle down.
5. Let the buffer drain from the column and apply your labeling reaction carefully to the surface of the gel bed. Once the labeled protein has completely entered the geld bed, top it with a small amount of PBS/TCEP and apply gentle pressure to the column.
6. The protein of interest is located in the first visible band. The second band contains only unbound dye and is discarded.
7. The labeled protein can now be concentrated in a centrifuge filter if necessary or directly be stored at −80 °C in 10 μL aliquots after nitrogen shock freezing.

3.3 Preparation of Salivary Glands

The *C. tentans* larvae are raised in water-filled aerated plastic dishes. They are fed with nettle plant powder that has been fermented. For dissection, the fourth instar larvae are collected and transferred into a beaker with water of the growth dishes. During dissection the larvae are kept in PBS.

1. Fill a petri dish with PBS and transfer previously collected larva into it.
2. Place the petri dish under a stereomicroscope.
3. Grab the larva with a Dumont 3c forceps at the abdominal third.
4. Stretch the larva by scrubbing along with a No. 10 scalpel.
5. Place scalpel blade between the first and second segment and decapitate the larva. During cutting, try to keep the larva strained by pressing the scalpel slightly cranial.
6. Make a second cut 3–4 segments abdominal from the first cut.
7. If the salivary glands are already visible, dissect them carefully. The glands are innervated and fixed by several ligaments. Cut the nerve and the ligaments carefully away. If the glands are not visible, try to grab the gut and pull it slightly out, which will push the glands out.

8. Transfer the salivary gland with a transfer tip to a sample chamber, which was prior to that coated with poly-L-lysine and filled with PBS.
9. Position the gland in the sample chamber in such a way that several nuclei are close (~0.5 mm) and almost parallel to the long side of the sample chamber.
10. Remove the PBS from the sample chamber to attach the gland to the bottom glass.
11. Refill the sample chamber with PBS.

The salivary glands are dissected carefully and transferred into a poly-L-lysine-coated specimen chamber. To reduce the required volume of hemolymph, two PTFE spacers were placed at the ends of the sample chamber.

3.4 Hemolymph Collection

1. Collect as many larvae as required (to fill one sample chamber with hemolymph, about 40 μl with usage of PTFE spacers) and transfer them into a beaker containing water of the growth dishes.
2. Take a larva and clean it with a dry precision wipe.
3. Put it on a microscope slide, grab it at the abdominal end, and cut it lengthwise with a spring scissor.
4. Collect hemolymph with a gel loader tip and transfer it to a 0.2 ml reaction tube.
5. Repeat **steps 2–4** until the required volume of hemolymph is collected.
6. Centrifuge the hemolymph and flash freeze it in liquid nitrogen.
7. Store at −80 °C.

3.5 Alignment of the Optical Setup

The alignment of the light sheet microscope has a tremendous impact on the data quality. A bad alignment decreases the SNR significantly. The illumination focus has to be positioned in the focal plane of the detection objective and in the field of view (FOV) of the EMCCD. We developed a fast and robust method to align the light sheet microscope:

1. Fill a sample chamber with 160 μl of calibration solution and place it in the holder.
2. Illuminate with the red laser with 100 % illumination power.
3. Adjust the z-position of the sample chamber and the axial position of the illumination objective, so that illumination focus is visible within the sample chamber.
4. Move the sample chamber up until you see that the illumination beam enters the sample chamber slightly above lower edge.
5. Use transmitted light to focus the detection objective to the inner edge of the sample chamber.

6. Align the light sheet beam by use of the two periscope mirrors into the camera FOV. Now move the light sheet focus into the camera FOV. The focus of the light sheet appears as a darker area in the sheet because at the illumination focus there is practically no background fluorescence. You may move the illumination objective slightly back and forth. This makes it easier to perceive the illumination focus.
7. To verify the perpendicular orientation of illumination and detection pathway to each other, move the detection objective slightly up and down. If they are perpendicular, the defocusing pattern of the light sheet will be mirror symmetrical about the *y*-axis (see Fig. 1). If illumination and detection pathway are not perpendicular, the defocusing pattern is cone-shaped. Furthermore, if you move the detection objective up and down, it will look like the illumination focus is moving from the left to the right side or vice versa. This "movement" gives you information of how the light sheet is tilted. Adjust the angle and position of the illumination beam until the illumination and detection pathway are perpendicular.
8. The sheet may be rotated along the illumination axis. To correct this, the cylindrical beam expander has to be readjusted.
9. Check again with the camera if the illumination focus is in the EMCCD FOV.

3.6 Microinjection of Salivary Glands

To deliver the labeled proteins to the salivary gland cells, we use microinjection (*see* **Note 2**).

1. Before microinjection the injection solution should be centrifuged for 20–30 min at 22,000 × *g*.
2. Start the FemtoJet and InjectMan device.
3. Load 3–5 μl of the microinjection solution into a Femtotip II with a microloader. Avoid touching the base of your vessel with the microloader.
4. Watch the salivary gland with transmitted light and move the top of the salivary gland into focus.
5. Move the gland out of your FOV.
6. Mount the injection needle according to manufacturer's advice.
7. By pressing the "clean button," get rid of the remaining air of the microinjection needle tip. To get rid of remaining air bubbles, you may gently knock on the needle.
8. Adjust the compensation pressure to 75–100 hPa to ensure a slow constant efflux of the needle. The injection pressure should be adjusted to 500 hPa. Vary the injection time from 0.1 to 0.5 s according to the required injection volume.

Increasing the injection pressure may damage the cell. Therefore, it is safer to modify the injection duration to regulate the injected volume.

9. It requires some training to find the needle with the 40× objective. Use transmitted light to find the needle. Move the needle so that it just touches the liquid surface above the objective. Then move the needle back and forth and from the left to the right. If the needle is above your FOV, you will see a weak shadowing. Slowly move the needle down until the tip is focused. Move the needle to an edge of your field of view and move it back by pressing the "home button."
10. Move a salivary gland cell into your FOV and move the equatorial plane of the nucleus into focus. By pressing "home" again, the needle is positioned back into your FOV. Move the needle manually into the cytoplasm and press the joystick button once or more until the intended injection volume is reached. Remove the needle gently. Performing this step in the "axial" mode can help to reduce the damage. Inject as many cells as needed. Microinjection of 5–10 cells of each gland takes ~10 min.
11. Replace the PBS by 40 μL hemolymph with a gel loading tip. Cover the hemolymph with mineral oil to avoid evaporation.

3.7 Imaging mRNPs by LSFM

For observing the mRNP during export (*see* **Note 3**), we stain the NE by co-injection of fluorescent NTF2. Usually, we employ Alexa Fluor647-labeled hrp36 and Alexa Fluor546-labeled NTF2. The camera is operated in frame transfer mode with 20 ms integration time, 1 MHz horizontal readout rate, and 0.9 μs vertical shift time. Thus, the frame rate is approximately 50 Hz.

We image the stained NPCs and labeled Hrp36 sequentially by the following recording scheme:

1. 950 frames excitation of mRNP fluorescence.
2. 50 frames excitation of NTF2.
3. Repeat **steps 1** and **2**.
4. Pause for 5 s.
5. Start again at the first step.

Recording is only stopped to readjust the FOV. The maximal duration of the measurements was 90 min (*see* **Note 4**). The image data must be saved in a file format that allows to recover all data acquisition parameters.

3.8 Data Evaluation

3.8.1 Single-Particle Tracking

For earlier single-particle tracking studies, we used "DiaTrack 3" for Windows (Semasopht, company closed) and gave a detailed description of the procedure [5]. Meanwhile, several software packages and custom-built plug-ins for ImageJ allow reliable single-particle tracking.

3.8.2 mRNP Export Analysis

For quantification of mRNP export events, we used a kymograph approach based on three custom-programmed ImageJ plug-ins (GaussProfiler, Kymograph_mt2, and KymoReader). First, we extracted the exact membrane position with the GaussProfiler plug-in by finding the brightest pixel in an NE reference image and fitting a Gaussian vertically to the membrane line. The respective pixels of the hrp36 image sequence were then used to construct a kymograph of the NE region (Kymograph_mt2 plug-in). The kymographs were finally screened for mRNP-NE interaction events using the KymoReader plug-in to extract dwell times. For a detailed mRNP export trajectory analysis, the positions of hrp36 molecules were determined by 2D Gaussians and their movement relative to the nuclear envelope registered with sub pixel localization accuracy. The ImageJ plug-ins and a detailed user guide are available from the authors' homepage (http://www.chemie.uni-bonn.de/pctc/kubitscheck).

4 Notes

1. Protein labeling: To increase labeling efficiency and specificity and to improve signal-to-noise ratio in single-molecule microscopy, the TC motive was introduced to the N-terminus of the protein. The reduction of disulfide bonds before labeling reaction and a pH close to 7 are crucial parameters for a good labeling efficiency.
2. Microinjecting salivary gland cells: Microinjection is the most direct method for protein delivery to cells, and if it is carried out carefully, then cell viability is not impaired. Furthermore, it enables the use of organic fluorescent dyes, which we prefer for single-molecule microscopy because of their brightness, photostability, and small size. During microinjection the salivary glands are kept in PBS, and after injection PBS is replaced by hemolymph. It is possible to inject the gland cells in hemolymph, but the microinjection needle is prone to clogging.
3. Intracellular single-molecule tracking: Successful single-molecule imaging of single molecules has several preconditions. The concentration of labeled molecules, here hrp36-labeled mRNPs, has to be very low so that a perception of single diffraction-limited signals is possible. If tracking of the single particles is performed, then the maximal concentration of fluorescent particles is limited by the particle mobility and imaging frame rate to ensure that the tracking algorithm can reliably discriminate single particle trajectories without mixing them up. On the other hand, a low concentration of particles increases the measuring time, which is disadvantageous. We aim at a final concentration in the picomolar range (40–200 pM).

4. Using an EMCCD: A key requirement for successful single-molecule imaging is a sensitive camera and its efficient use. The user may adjust several important parameters. To reduce the dark current caused by thermal fluctuation in the CCD, it is necessary to cool the camera down [6]. Many EMCCDs offer forced air and water cooling. Water cooling is indispensable to avoid undesirable vibrations of the air fan. Choose the slowest horizontal readout speed, which is compatible with the desired frame rate. High horizontal readout speeds increase the readout noise. Another source of noise is clock-induced charge or "spurious noise." This is affected by the "vertical readout speed" and the "vertical clock voltage." A faster "vertical readout speed" and a low "vertical clock voltage" decrease the spurious noise. Very low vertical clock voltages or/and fast "vertical readout speed" may reduce the charge transfer efficiency.

   According to the manufacturer's advice, the preamplifier gain should be as high as possible for low light applications like single-molecule microscopy. Excessive EM gain decreases the dynamic range and does not improve the signal-to-noise ratio. As a rule of thumb, one input electron should not be multiplied to more than 4×–5× the root mean square number of readout noise electrons.

   The magnification of the optical system has a significant impact on the localization precision. There is a trade-off between the higher number of pixels available for the position determination at higher magnifications and the reduced number of photons per pixel. A pixel size of 80–100 nm in the object plane should ideally be used [7, 8].

Acknowledgments

U.K. gratefully acknowledges financial support by the DFG Grant Ku 2474/7-1. T.K. and J.H.S acknowledge support by the German National Academic Foundation.

References

1. Skoglund U, Andersson K, Strandberg B, Daneholt B (1986) Three-dimensional structure of a specific pre-messenger RNP particle established by electron microscope tomography. Nature 319:560–564
2. Veith R, Sorkalla T, Baumgart E, Anzt J, Häberlein H, Tyagi S, Siebrasse JP, Kubitscheck U (2010) Balbiani ring mRNPs diffuse through and bind to clusters of large intranuclear molecular structures. Biophys J 99:2676–2685
3. Siebrasse JP, Veith R, Dobay A, Leonhardt H, Daneholt B, Kubitscheck U (2008) Discontinuous movement of mRNP particles in nucleoplasmic regions devoid of chromatin. Proc Natl Acad Sci USA 105:20291–20296
4. Siebrasse JP, Kaminski T, Kubitscheck U (2012) Nuclear export of single native mRNA molecules observed by light sheet fluorescence microscopy. Proc Natl Acad Sci USA 109: 9426–9431
5. Zhang WW, Chen Q (2009) Optimum signal-to-noise ratio performance of electron multiplying charge coupled devices. World Acad Sci 30, Eng Tech, 264–269.
6. Quan T, Zeng S, Huang Z-L (2010) Localization capability and limitation of

electron-multiplying charge-coupled, scientific complementary metal-oxide semiconductor, and charge-coupled devices for superresolution imaging. J Biomed Opt 15:066005

7. Siebrasse JP, Kubitscheck U (2009) Single molecule tracking for studying nucleocytoplasmic transport and intranuclear dynamics. Methods Mol Biol 464:343–361
8. Pei-Hsun W, Nathaniel N, Yiider T (2010) A general method for improving spatial resolution by optimization of electron multiplication in CCD imaging. Opt Express 18:5199–5212

Chapter 7

Single mRNP Tracking in Living Mammalian Cells

Alon Kalo, Pinhas Kafri, and Yaron Shav-Tal

Abstract

The translocation of single mRNPs (mRNA–protein complexes) from the nucleus to the cytoplasm through the nuclear pore complex (NPC) is an important basic cellular process. Originally, in order to visualize this process, single mRNP export was examined using electron microscopy (EM) in fixed *Chironomus tentans* specimens. These studies described the nucleocytoplasmic translocation of huge mRNPs (~30 kb) transcribed from the Balbiani-ring genes. However, knowledge of the in vivo mRNP kinetics in cell compartments remained poor up until recently. The current use of unique fluorescent protein tags, which are able to bind to mRNA transcripts, has allowed the detection and measurements of single mRNP kinetics in living cells. This has demonstrated that mRNP movement is affected by the size of the transcript and the splicing process. It was found that mRNP rates of translocation are slower in the nucleus compared to the cytoplasm and that the cell nucleus contains interchromatin tracks in which mRNPs diffuse. In order to track single mRNP movement in living cells, it is important to be able to identify single mRNP molecules transcribed from a certain gene, at the single-cell level. Single-molecule analysis of gene expression requires advanced imaging systems and analytical software in order to detect and follow the movement of single mRNPs. In this chapter we describe the methods required for the detection and tracking of single mRNP movement in living mammalian cells.

Key words mRNP, mRNA export, Single-particle tracking

1 Introduction

Analysis of gene expression at the single-cell and single-molecule levels can be performed in fixed and living cells. In eukaryotic cells, the transcription site in the nucleus and the translation site within the cytoplasm are separated by the nuclear envelope. Therefore, the transition of proteins and RNA molecules between the nucleus and cytoplasm through the nuclear pore complex (NPC) is an essential process. Generally, this process occurs by soluble transport proteins which form a transport receptor–cargo complex [1]. Two groups of proteins have been characterized and

Alon Kalo and Pinhas Kafri have contributed equally to this work.

Yaron Shav-Tal (ed.), *Imaging Gene Expression: Methods and Protocols*, Methods in Molecular Biology, vol. 1042,
DOI 10.1007/978-1-62703-526-2_7, © Springer Science+Business Media, LLC 2013

suggested to participate in the transport of proteins and RNA molecules between the cell compartments. The first group of transport proteins belongs to the Karyopherin-B family, which are responsible for the transfer of proteins, tRNA, siRNA, miRNA, and rRNA. The second group are the nuclear export factors (NXF) and these participate mainly in mRNA export [2]. The first study that visually demonstrated the transition of a single mRNP through the NPC used EM microscopy in fixed samples of *C. tentans* [3]. EM was used to visualize specific mRNPs expressed from the Balbiani-ring gene, as they depart from the gene and move through the nucleus towards the nuclear envelope. This study showed that the average translocation time of Balbiani-ring mRNPs in the nucleus can be measured [4]. Such studies have shown that the export of an mRNP involves the restructuring of the transcript while crossing the NPC. However, static images of fixed cells are not capable of providing sufficient information about mRNP dynamics. In order to obtain reliable information of single-molecule movement, live-cell imaging combined with visual detection of single mRNA molecules is an essential tool for the study of single-molecule kinetics in living cells.

A method commonly used for the detection of single mRNA transcripts in living cells utilizes the fluorescent tagging of a unique repeated sequence found within the mRNA of interest. This is based on the integration of 24 MS2 sequence repeats into a gene of interest. These 24 MS2 sequence repeats are transcribed as part of the 3′UTR of the gene of interest to form step-loop structures in the mRNA [5]. The specific binding of an RNA binding protein to these stem loops, namely, the MS2 coat protein fused to a fluorescent protein (e.g., YFP–MS2), renders the mRNA fluorescent. 24 stem-loop repeats are sufficient for the identification and tracking of single mRNPs in mammalian cells as observed with GFP–MS2 tagging [6–10]. Since the MS2 sequence repeats originate from bacteriophage, they do not interfere or interact with any mammalian components.

Direct single mRNP tracking (single-particle tracking, SPT) and data analysis [11–13] allow the extraction of several basic parameters, such as movement characterization and the calculation of diffusion coefficients. In SPT analysis, the square of the mean distances (termed the mean square displacement, MSD) that a single particle moves during a particular time period is calculated. Several MSD values are obtained over different time periods within a sequence of images and are typically plotted on an MSD versus time plot to obtain information about the movement characteristics and diffusion coefficients.

SPT has been used to examine the nucleoplasmic travels of mRNPs and for following the export process, in real-time. mRNP movement in the cell has been characterized by SPT with several types of movements being described: diffusive, directed, confined, and, rarely, lack of movement [13]. These findings support the theory that mRNPs move in between dense chromatin domains [8, 11]. mRNP export rates had been measured in several biological systems such as radioactive pulse-chase experiments and direct microinjection into the cell nucleus and fixation at different times [14–19]. These studies have demonstrated the wide range of export times ranging from several minutes to 2 h. Recent studies have demonstrated the transient interactions occurring between single mRNAs and nuclear pores and have shown that nuclear mRNPs diffuse in between dense chromatin regions [11, 20].

In this chapter we explain the basic requirements and the experimental steps required for the detection and tracking of single mRNPs in living mammalian cells. The experimental approach discussed here is based on the tagging of mRNA transcripts of a certain gene containing the MS2 sequence repeats and the detection of the transcribed mRNPs using a specific YFP–MS2 coat protein targeted to the MS2 sequence. This chapter will describe (a) the construction of a gene construct containing the MS2 sequence repeats, (b) the generation of stable mammalian cell lines that contain the gene of interest, and (c) the acquisition of time-lapse movies for the detection and tracking of single mRNPs in single mammalian cells.

2 Materials

2.1 Cloning of GOI Containing a Fluorescent Fusion Protein and 24 MS2 Sequence Repeats

1. pSL-24MS2 vector: Plasmid containing the 24 MS2 sequence repeats.
   Can be obtained from http://www.addgene.org/27120.
2. Cyan fluorescent protein (CFP) expression vector.
3. YFP–MS2 plasmid for the expression of the tagged protein.
4. Restriction enzymes.
5. 5 μ/ml T4 DNA ligase.
6. Competent Escherichia coli bacteria for transformation (*see* **Note 1**).

2.2 Electroporation for Stable Integration of the GOI–MS2

1. Dulbecco's Modified Eagle's Medium (DMEM).
2. Fetal bovine serum (FBS).
3. Trypsin for detaching cells from tissue culture plate.
4. PBS solution for washing cells.
5. 1–4 μg GOI–MS2 plasmid.
6. Electroporator, e.g., Bio-Rad Gene Pulser Xcell (Bio-Rad).
7. Gene pulser cuvette 0.4 cm.
8. Salmon sperm DNA.
9. Adherent cell line of choice. Human U2OS and HeLa cell lines are recommended.

10. Antibiotics for stable selection depending on the selection marker in the gene construct.
11. Cloning cylinders.
12. Cryotubes.
13. DMSO.

2.3 Time-Lapse Live-Cell Imaging of mRNP Dynamics in Living Cells

1. Fluorescent microscope of choice (confocal or wide field) used at a 100× magnification (oil objective). A wide-field microscope should be equipped with a CCD camera for obtaining best resolution for the detection of single mRNPs.
2. Microscope incubation chamber including temperature and CO_2 control for growing cells during imaging.
3. Glass-bottomed tissue culture plates: 35 mm Petri dishes with 14 mm glass-bottomed microwell (0.16–0.19 mm thickness; MatTek).
4. Image analysis software such as Imaris (Bitplane), MetaMorph (Molecular Devices), or ImageJ (NIH, http://rsb.info.nih.gov/ij/).
5. Deconvolution software such as Huygens Deconvolution Software (Scientific Volume Imaging), AutoQuant (Media Cybernetics), or DeltaVision (Applied Precision).
6. "Ptrack" script written in Matlab [8]. Will be provided upon request from the Shav-Tal lab.

3 Methods

3.1 Construction of an Expression Vector Containing the GOI, a Fluorescent Fusion Protein, and the MS2 Sequence Repeats

1. In order to obtain a cell line stably expressing your GOI fused to a fluorescent protein, insert the GOI into a fluorescent protein expression vector containing an antibiotic resistance gene (*see* **Note 2**). For this protocol, it is preferable to use the CFP vector (e.g., pECFP-C1) since there is good spectral separation of CFP from the YFP channel that is used to image the YFP–MS2 coat protein-tagged mRNPs.
2. Insert the MS2 sequence repeats from the pSL24MS2 plasmid into the 3′UTR of your GOI using the multiple cloning site downstream to the GOI. Preferable restriction sites will be the BamHI at the 5′ of the MS2 sequence repeats and BglII at the 3′ of the MS2 sequence repeats. BamHI/BglII digestion will result in a ~1.3 kp fragment containing the 24 MS2 sequence repeats (*see* **Note 1**). An adaptor with suitable restriction sites can be added to the 3′UTR sequence of the GOI if there are no suitable sites for the insertion of the MS2 sequence repeats fragment. Figure 1 depicts the cloning steps to be taken for generating a plasmid expressing a GOI containing 24 MS2 sequence repeats in the 3′UTR (*see* **Note 3**).

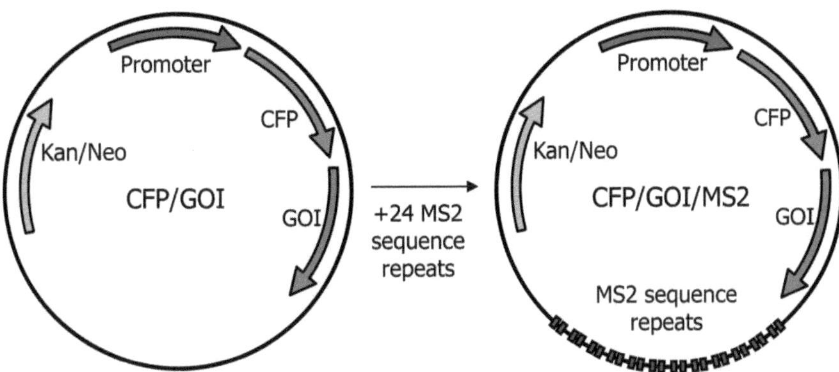

Fig. 1 Structure of the gene construct with the 24 MS2 sequence repeats. The gene of interest (GOI) is fused to a fluorescent protein, in this case cyan fluorescent protein (CFP). The 24 MS2 sequence repeats are cloned into the 3′UTR of the GOI. Antibiotic resistance is conveyed by a resistance gene, here the kanamycin/neomycin gene, for bacterial and mammalian selection, respectively

3.2 Generation of a Cell Line Stably Expressing the GOI–MS2 Construct

1. Split the cells 1 day prior to electroporation. Make a single-cell suspension and plate in a 10 cm tissue culture dish in fresh medium. On the next day, confluence should reach 50–80 %. Transfection efficiency is reduced if cells are too aggregated.
2. Day of transfection: Wash the cells with 1× PBS, trypsinize gently by adding 1–1.5 ml of trypsin to the cells, and incubate for 1–5 min at 37 °C. Add medium containing 10 % FBS and transfer the cells to a 15 ml tube. Centrifuge for 5 min at 200 × *g* and aspirate the medium.
3. Suspend in 1 ml cold medium plus serum.
4. Apply 200–250 μl of the cells (containing approx. 200,000 cells) to a sterile cuvette and add the GOI–MS2 plasmid (2–10 μg of DNA per transfection) to the cells.
5. Tap gently to mix and wait for 10 min at room temperature.
6. Electroporate using either preset protocols or your own settings. The following electroporation conditions have been successfully used with the Bio-Rad Gene Pulser Xcell when transfecting these human cell lines: U2OS: 170 V, 950 μF; HeLa: 150 V, 500 μF; HEK-293: 300 V, 500 μF. Different protocols should be tried to check for best transfection efficiency (*see* **Note 4**).
7. Plate electroporated cells in a 10 cm plate with 10 ml fresh medium plus serum. Gently mix cells and medium and incubate at 37 °C.
8. Next day: Add appropriate antibiotics to the medium to begin selection. Change the medium and antibiotics every 3 days. Continue with the selection for about 2–3 weeks until single colonies develop.

9. Use cloning cylinders to collect colonies. Place the cylinders on well-separated colonies. Pipette 100 μl of trypsin into each cylinder, incubate for 1–5 min, and then add 100 μl of medium to each cylinder. Gently pipette and suspend the colony and transfer to a 24-well plate. Add 0.5 ml fresh medium plus antibiotics to each well.
10. After the colonies expand in the wells, screen for positive colonies which integrated the GOI. A fluorescent microscope is used for detection of the GOI–MS2 protein product that will fluoresce in the CFP channel.
11. Freeze positive colonies for further use. Prepare in advance cold freezing medium (90 % FBS/10 % DMSO) according to the number of vials required and keep on ice. Trypsinize the cells and resuspend by gentle pipetting. Collect the cell suspension into 15 ml tubes and spin down at 200 × *g* for 5 min. Remove the supernatant and gently mix the pellet with 1 ml of the pre-made freezing medium. Transfer the suspended cells to a cryotube and immediately put on ice. Store at −70 °C for 24 h, before transferring to a liquid nitrogen container for long-term storage.

3.3 Live-Cell Imaging and Tracking of Cellular mRNPs

1. Split the cells so as to reach 50–80 % confluence 1 day prior to transfection.
2. Day of transfection: Trypsinize the cells, wash in PBS, and transfer ~2×10^5 cells suspended in 200 μl of cold medium into a 0.4 cm cuvette (as explained above in Subheading 3.2).
3. Add 1–4 μg of YFP–MS2 DNA plasmid and 40 μg salmon sperm DNA to the cells. Tap gently to mix and wait for 10 min at RT (*see* **Notes 4** and **6**).
4. Electroporate the cells.
5. Transfer the cells to the glass-bottomed tissue culture plates and plate in the center of the glass. After first attachment of the cells to the glass (~2–5 h), add 1 ml of fresh medium to the plate.
6. Once the cells have spread properly onto the glass (several hours), transfer the plate to a fluorescence microscope equipped with an incubator that provides conditions of 37 °C and 5 % CO_2. Search for transfected cells containing a strong transcription site and expressing labeled mRNPs. Positive cells can also be detected using the CFP channel for the fused CFP–GOI protein. The mRNPs are seen as many small round dots moving in the cells (*see* **Note 5**). Adjust the imaging conditions according to the level of the expressed signal versus the diffusive level of background. Avoid exposing the cells to high-intensity illumination during the search in order to reduce bleaching of the fluorescent signal.

7. Determine the experimental parameters required for time-lapse imaging such as exposure time and the number of frames. We recommend performing 1 min movies with a time interval in the range of 0.5–1 s with a 100× magnification using 2×2 binning. The light intensity and exposure time need to be high enough to observe the signal over the background.
8. If required, enhance the mRNP signal using a deconvolution algorithm. We find that 5–20 iterations with the Huygens Essential Deconvolution Software efficiently enhances mRNP signal in the time-lapse movies.
9. Analyze the acquired movies using image analysis software. We detect and track mRNAs using the Imaris software according to the following steps:
   (a) Load the movie into the Imaris software, select "ortho slice" as the initial scene, and go to image processing in order to create a time-lapse movie by pressing the "swap time and z" button (*see* **Note 7**). For single-particle tracking, press on "Add new Spots" button and check "Track Spots (over time)."
   (b) Estimate the diameter of a single mRNA signal by measuring the length of a single spot in the movie. Find the diameter distance by pressing on the "Slice" button in order to view the movie in slice mode. In this mode estimate the diameter by marking two points across a chosen spot. The estimated diameter will appear on the right side of the screen under the "Distance" rubric. We recommend choosing the particle diameter in the range of 200–400 nm.
   (c) Choose the intensity value of a single mRNA by changing the threshold. This allows determining the number of spots to be considered as a real mRNAs in the next steps of analysis.
   (d) Determine the "MaxDistance" and "MaxGapSize" parameters. "MaxDistance" regulates the maximum distance a spot moves between each frame in μm. Given a certain value, a spot that appears above the maximum distance determined between two consequent frames will be considered as a new track. The track length for each spot will be calculated by this parameter. In order to determine the max distance value, check the box labeled "Frame" under "Surpass Scene" options. The "Frame" will create a grid dividing the movie into pixels including a scale bar. Using the grid lines along with the scale bar helps to determine the distance a certain spot moves between consequent frames. "MaxGapSize" value regulates the maximum frames that a certain spot cannot be detected by the program, but still remains part of the same track. We recommend setting this

value to zero in order to extract coordinates for continuous tracks only.

(e) Under "Filter type" choose "Track duration" and use the histogram to erase tracks with less than minimal frames for each track. We recommend using 7 as the threshold as we usually consider tracks smaller than 7 frames as irrelevant. In addition, we usually do not consider tracks that are more than 30 frames long (*see* **Notes 8** and **9**).

(f) Press the button labeled in green >> to end the analysis process. Under the "Statistics" tab, press the "Settings" button and under "Spots" check the rubrics "Position X" and "Position Y." In order to extract the data to an Excel sheet, press the "Export Excel" button and save it in the appropriate location.

The Excel sheet contains several parameters: The "Position X" and "Position Y" titles represent the coordinates for each frame of every track. The "Parent" title represents a single track that allows distinguishing between every single track to its own coordinates.

10. Analyze the extracted data from the Excel sheet using the "ptrack script" [8, 13] written for MATLAB. Copy the X and Y coordinates of each track according to the "Parent" number. Open MATLAB software and under the "Edit" Tab press "Paste special," then "Next" and "Finish." Run the "ptrack" script and press the "Update list" tab in order to upload the copied coordinates as a new list in the "Data list" box. In the "Scale" box, insert the pixel size (μm) and time interval (ms) according to the parameters used to make the movie. At the bottom of the screen, press the "Curve Fit" button and then press "Analyze."

After the analysis process is complete, the "ptrack" screen will present plots and kinetic parameters related to the current track. Figure 2 displays a plot representing the MSD values with respect to time. According to this plot, the script will extract the fit for three types of movements, so the type of the particle movement can be characterized along with its diffusion coefficient and rate.

The parameters that are extracted from "ptrack" script are the following:

1. Curve fitting represents the score for three types of movements:
   (a) Diffusive
   (b) Drift
   (c) Corralled

   Determine the type of the particle movement according to the highest score given for the three types of movements.
2. Average velocity of the particle in case of drift movement (μm/s).

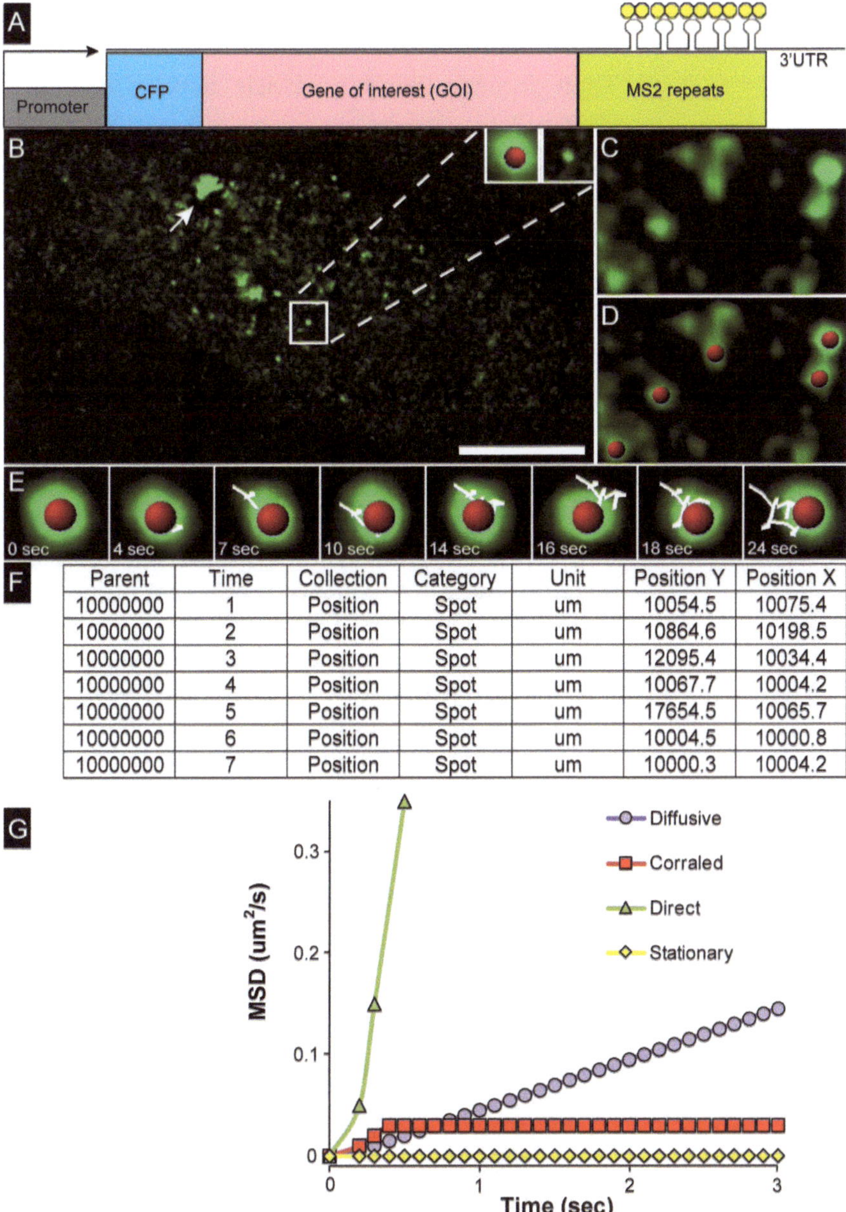

Parent	Time	Collection	Category	Unit	Position Y	Position X
10000000	1	Position	Spot	um	10054.5	10075.4
10000000	2	Position	Spot	um	10864.6	10198.5
10000000	3	Position	Spot	um	12095.4	10034.4
10000000	4	Position	Spot	um	10067.7	10004.2
10000000	5	Position	Spot	um	17654.5	10065.7
10000000	6	Position	Spot	um	10004.5	10000.8
10000000	7	Position	Spot	um	10000.3	10004.2

Fig. 2 Establishment of a biological system for the detection and tracking of single mRNPs in living mammalian cells. (**a**) Schematic description describing the biological system for marking individual transcripts of a particular gene. This construct is a part of a plasmid DNA containing a promoter of choice, the gene of interest fused to CFP and containing 24 MS2 repeats in the 3′UTR region. *Yellow circles* represent the YFP–MS2 protein which binds to the MS2 loops that are transcribed as a part of the transcript. (**b**–**d**) Cell expressing YFP–MS2 showing the site of transcription (large *green dot*, *white arrow*) and single YFP–MS2-tagged mRNPs. Single mRNPs (*red dots*) are detected by the Imaris software. The enlarged *box* shows a single mRNP marked as a *red dot*. (**e**) Single-particle tracking over time. *White track* depicts the direction of the particle movement during the time-lapse movie. (**f**) Excel table extracted from the Imaris software containing tracking parameters for the use in "ptrack" script. The example given is for one individual track. *Parent column* represents a separate track for each particle. *Time column* represents the number of frames taken to each individual track. *Unit column* represents the distance unit from each *X*, *Y* coordinates. *Position X,Y columns* represents the coordinates at any time point for all detected particles. (**g**) Plot representing the MSD (μm^2/s) values with respect to time (s). The coordinates extracted from the tracking analysis are fitted to three types of plots describing three types of particle movement. The highest score for the three fitting curves is used to determine the type of particle movement. Bar = 5 μm

3. Total distance that the particle traveled during the whole track (μm).
4. Instantaneous diffusion coefficient of the particle calculated by the shortest time frames ($\mu m^2/s$).
5. Diffusion coefficient (D) for diffusive movement ($\mu m^2/s$).
6. Drift v for drift movement (μm/s).
7. Radius corralled (Rc) for corralled movement is the calculated radius volume of the space in which the particle travels in a diffusive movement.

4 Notes

1. MS2 repeats: Bacteria tend to discard repeated sequences and therefore the MS2 sequence repeats are prone to shortening during transformation. One possibility is to use special commercially available competent bacteria, e.g., Stbl2 competent cells that grow at 30 °C, that have less recombination events that dispose of the repeats. However, we have found that careful monitoring of the number of repeats following transformation and single colony picking with regular component bacteria allows the detection of bacterial clones that harbor the 24-sequence repeats in a constant manner. This requires the presence of flanking restriction sites on both sides of the MS2 sequence repeats in order to accurately assess the number of repeats within the vector. Due to these limitations, it is useful that the MS2 sequence repeats be cloned last into the expression vector.
2. Antibiotics resistance: In order to generate a cell line stably expressing the GOI–MS2 construct, it is preferable to insert the GOI into an expression vector already containing an antibiotic resistance gene. However, if the target plasmid, harboring the GOI, does not contain an antibiotic resistance gene, it is possible to perform a stable co-transfection by adding a second plasmid carrying a resistance gene. To avoid false-positive clones, we suggest to co-transfect the cells using 1:5, 1:7, and 1:10 ratios in favor of the GOI plasmid.
3. The location of MS2 sequence repeats in the gene construct: A typical gene construct contains the cDNA and an upstream promoter. The MS2 sequence repeats are normally cloned after the coding region, within the 3′UTR (before the poly(A) signal), thereby not interfering with the translation of the gene. While another option is to place these repeats upstream of the coding region within the 5′UTR, this seems less preferable since the presence of the stem-loop secondary structures could interfere with ribosomal scanning prior to translation.

4. Transfection efficiency: The YFP–MS2 protein is usually expressed by transient transfection and therefore it is important to use adherent cells that easily transfect. Also we suggest using stable cell lines in which the GOI + MS2 sequence repeats are stably integrated. We find the U2OS and HeLa cell lines useful for this line of work.
5. Detection of mRNPs and active transcription sites: CFP-positive clones transfected with the YFP–MS2 protein should show a YFP-labeled active nuclear transcription site and cellular mRNPs.
6. High levels of the YFP–MS2 protein might interfere with the detection of the single mRNPs due to masking of the specific signal by the high background. In order to avoid high background, use lower amounts of the YFP–MS2 plasmid or examine the cells at earlier times post-transfection. Note that the YFP–MS2 protein comes in different versions: Either it has a nuclear localization sequence (NLS) only and then the protein is mostly nuclear, or it can contain an NES sequence as well and then the protein distributes between the cytoplasm and nucleoplasm [11].
7. Verification of movie parameters: Once the movies are deconvolved and uploaded into the Imaris software for tracking analysis, verification of correct movie parameters must be done in order to work with original image properties. First, turn the uploaded file into an X, Y, and T movie by pressing on the "Image Processing" button and "Swap Time and Z." This will create a time series according to the total length and time interval taken, instead of a static X, Y, and Z image. For the first step before tracking analysis, press the "Edit" tab and "Image Properties" and check if the time series is properly set according to the microscope parameters, such as voxel size and number of time frames.
8. Filtering of irrelevant tracks: While reviewing each movie in the Imaris software, it is important to detect the time point when the cell begins to move. Due to cell movement, it is hard to determine mRNA motion and therefore we usually do not track from this point. This kind of filtering can be performed during analysis by deletion of these track coordinates. Verify that the coordinate number of each track is the same, especially when comparing between different types of movements for various mRNA lengths.
9. Minimum frames of a relevant track: Determining the type of movement based on single-particle tracking requires a minimal number of movie frames in order to create a plot representing the mean square displacement (MSD) values with respect to time. The type of movement can be determined according to the trend line connecting the points (Fig. 2). In order to calculate and

draw a graph that provides reliable information, it is essential to use at least seven time frames for each track during analysis. A minimum of seven frames will result in six different displacements for each time interval throughout the movie. These will be used to calculate six different MSDs for generating the plot and will allow evaluating the type of the particle movement according to the shape of the trend line.

Acknowledgements

The work in the Shav-Tal lab is supported by the European Research Council (ERC).

References

1. Erkmann JA, Kutay U (2004) Nuclear export of mRNA: from the site of transcription to the cytoplasm. Exp Cell Res 296:12–20
2. Kohler A, Hurt E (2007) Exporting RNA from the nucleus to the cytoplasm. Nat Rev Mol Cell Biol 8:761–773
3. Mehlin H, Daneholt B, Skoglund U (1992) Translocation of a specific premessenger ribonucleoprotein particle through the nuclear pore studied with electron microscope tomography. Cell 69:605–613
4. Singh OP, Bjorkroth B, Masich S, Wieslander L, Daneholt B (1999) The intranuclear movement of Balbiani ring premessenger ribonucleoprotein particles. Exp Cell Res 251:135–146
5. Bertrand E, Chartrand P, Schaefer M, Shenoy SM, Singer RH, Long RM (1998) Localization of ASH1 mRNA particles in living yeast. Mol Cell 2:437–445
6. Fusco D, Accornero N, Lavoie B, Shenoy SM, Blanchard JM, Singer RH, Bertrand E (2003) Single mRNA molecules demonstrate probabilistic movement in living mammalian cells. Curr Biol 13:161–167
7. Lionnet T, Czaplinski K, Darzacq X, Shav-Tal Y, Wells AL, Chao JA, Park HY, de Turris V, Lopez-Jones M, Singer RH (2011) A transgenic mouse for in vivo detection of endogenous labeled mRNA. Nat Methods 8: 165–170
8. Shav-Tal Y, Darzacq X, Shenoy SM, Fusco D, Janicki SM, Spector DL, Singer RH (2004) Dynamics of single mRNPs in nuclei of living cells. Science 304:1797–1800
9. Brody Y, Neufeld N, Bieberstein N, Causse SZ, Bohnlein EM, Neugebauer KM, Darzacq X, Shav-Tal Y (2011) The in vivo kinetics of RNA polymerase II elongation during co-transcriptional splicing. PLoS Biol 9:e1000573
10. Ben-Ari Y, Brody Y, Kinor N, Mor A, Tsukamoto T, Spector DL, Singer RH, Shav-Tal Y (2010) The life of an mRNA in space and time. J Cell Sci 123:1761–1774
11. Mor A, Suliman S, Ben-Yishay R, Yunger S, Brody Y, Shav-Tal Y (2010) Dynamics of single mRNP nucleocytoplasmic transport and export through the nuclear pore in living cells. Nat Cell Biol 12:543–552
12. Aizer A, Brody Y, Ler LW, Sonenberg N, Singer RH, Shav-Tal Y (2008) The dynamics of mammalian P body transport, assembly, and disassembly in vivo. Mol Biol Cell 19: 4154–4166
13. Saxton MJ, Jacobson K (1997) Single-particle tracking: applications to membrane dynamics. Annu Rev Biophys Biomol Struct 26:373–399
14. Bastos RN, Volloch Z, Aviv H (1977) Messenger RNA population analysis during erythroid differentiation: a kinetical approach. J Mol Biol 110:191–203
15. Mariman E, Hagebols AM, van Venrooij W (1982) On the localization and transport of specific adenoviral mRNA-sequences in the late infected HeLa cell. Nucleic Acids Res 10:6131–6145
16. Luo MJ, Reed R (1999) Splicing is required for rapid and efficient mRNA export in metazoans. Proc Natl Acad Sci USA 96: 14937–14942
17. Jarmolowski A, Boelens WC, Izaurralde E, Mattaj IW (1994) Nuclear export of different classes of RNA is mediated by specific factors. J Cell Biol 124:627–635

18. Dargemont C, Kuhn LC (1992) Export of mRNA from microinjected nuclei of Xenopus laevis oocytes. J Cell Biol 118:1–9
19. Riedel N, Bachmann M, Prochnow D, Richter HP, Fasold H (1987) Permeability measurements with closed vesicles from rat liver nuclear envelopes. Proc Natl Acad Sci USA 84:3540–3544
20. Grunwald D, Singer RH (2010) In vivo imaging of labelled endogenous beta-actin mRNA during nucleocytoplasmic transport. Nature 467:604–607

Chapter 8

Imaging Nascent RNA Dynamics in *Dictyostelium*

Jonathan R. Chubb, Michelle Stevense, Danielle Cannon, Tetsuya Muramoto, and Adam M. Corrigan

Abstract

Dictyostelium cells have great utility for live imaging of single gene transcriptional dynamics. The cells allow efficient molecular genetics, for targeting of RNA reporters and fluorescent proteins to individual, defined loci. *Dictyostelium* cells share many signalling, chromatin and nuclear characteristics of larger eukaryotes, yet the cells have a relatively simple scattered differentiation programme, allowing imaging of transcriptional events in the context of stochastic developmental choices. This review will detail the methods and considerations for imaging nascent RNA dynamics at single genes in living *Dictyostelium* cells.

Key words Chromatin, *Dictyostelium*, Morphogen, Live-cell imaging, Transcriptional bursting, Pulsing, Stochastic gene expression, Transcriptional noise, Transcriptional pausing

1 Introduction

Although several techniques have been used to measure gene expression at the single-cell level in *Dictyostelium*, our emphasis is to measure nascent RNA production from single alleles using live-cell RNA detection. Other techniques can be used and include RNA FISH, to describe the instantaneous transcript content at transcription sites in fixed cells, and targeting specific genes with GFP expression, allowing transcriptional behaviour to be measured via total cell fluorescence (albeit subject to influences in RNA processing, transport, translation and decay, in addition to GFP protein folding, fluorophore maturation and protein turnover). These two technologies are extremely useful, but we sought to combine the instantaneous read-out of transcription, with the dynamics of the live-cell environment, without complications of protein reporters.

All authors have contributed equally to this work.

Yaron Shav-Tal (ed.), *Imaging Gene Expression: Methods and Protocols*, Methods in Molecular Biology, vol. 1042, DOI 10.1007/978-1-62703-526-2_8, © Springer Science+Business Media, LLC 2013

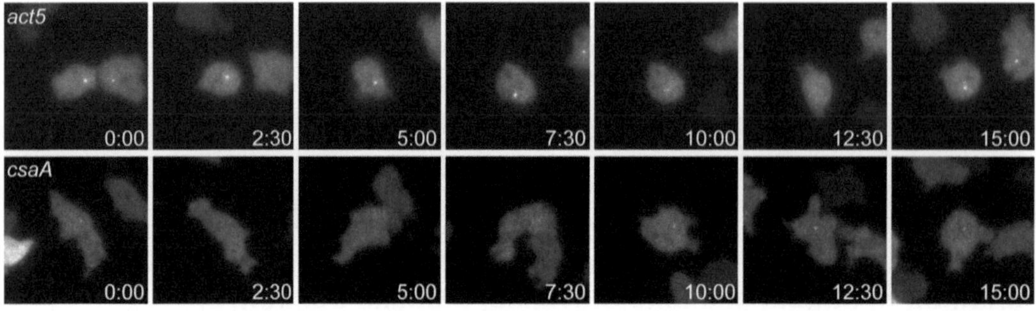

Fig. 1 Transcriptional pulsing of the *act5* and *csaA* genes in 3 h developed *Dictyostelium* cells. 24 MS2 repeats were targeted into the coding sequence of single genes. Targeted clones were then transformed with an MS2-GFP expression vector. MS2-GFP is recruited to the newly synthesised MS2 stem loops at the site of transcription and can be detected as a fluorescent spot, which appears and disappears (pulses) at irregular intervals. Images are maximal projections of 3D stacks captured at 2.5 min intervals

This review concentrates on a method using the specific high affinity interaction between the RNA stem loop from the genome of the MS2 RNA bacteriophage and the MS2 bacteriophage coat protein [1]. 24 MS2 loops are targeted into endogenous gene loci. Upon transcription, the loops are incorporated into the nascent RNA and detected by a fusion of the coat protein and GFP. Recruitment of GFP to nascent RNA results in a fluorescent spot at the site of transcription [2]. Using fluorescence microscopy, we monitor the dynamics of these spots, which appear and disappear at irregular intervals (Fig. 1). This phenomenon was termed "pulsing", although it is also referred to as "bursting" [3] and may relate to observations of high levels of heterogeneity in transcript number observed in fixed cells by FISH [4]. One model invoked, which may unite these observations, describes genes fluctuating between "off" and "on" states at irregular intervals, with only the "on" state competent to make RNA [5–7]. Since a consequence of this bursting model is considerable transcript heterogeneity between cells, basic transcriptional mechanism has been proposed as a driver of symmetry breaking in cell populations, a critical process in differentiation from microbes to mammalian stem cells [8]. Other intrinsic (not from external signals or stresses) sources of heterogeneity have also been proposed [9]. In some contexts, the words "pulsing" and "bursting" are likely to reflect distinct phenomena [10, 11].

2 Materials

1. HL5 medium: 14 g/L Bacto proteose peptone, 7 g/L yeast extract, 13.5 g/L glucose, 0.5 g/L KH_2PO_4, 0.5 g/L Na_2HPO_4. Autoclave and store at room temperature. Antibiotics penicillin and streptomycin are added to HL5 (*see* **Note 1**).

2. LF medium: 11 g/L glucose, 0.68 g/L KH_2PO_4, 5 g/L casein peptone, 26.8 mg/L NH_4Cl_2, 37.1 mg/L $MgCl_2$, 1.1 mg/L $CaCl_2$, 8.11 mg/L $FeCl_3$, 4.84 mg/L Na_2-EDTA·$2H_2O$, 2.30 mg/L $ZnSO_4 \cdot 5H_2O$, 1.11 mg/L H_3BO_3, 0.51 mg/L $MnCl_2 \cdot 4H_2O$, 0.17 mg/L $CoCl_2 \cdot 6H_2O$, 0.15 mg/L $CuSO_5 \cdot 5H_2O$, 0.1 mg/L $(NH_4)6Mo_7O_{24} \cdot 4H_2O$. Stericup-GV filter unit can be used for filter sterilisation of the LF medium. Antibiotics penicillin and streptomycin are also added.
3. Fetal bovine serum (FBS): Thaw serum in a water bath at room temperature. Swirl the bottle every 5–10 min. Place the serum bottle into a 56 °C water bath and leave for 30 min. Gently swirl the bottle every 5 min. Remove the heat inactivated serum after 30 min and allow the bottle to reach room temperature. Aliquot and store at −20 °C.
4. Imaging medium: 20 % HL5/10 % FBS/70 % LF (*see* **Note 2**).
5. Imaging chambers: Bioptechs Delta TPG dishes (0.17 mm thick), Lab-Tek chambered cover glass dishes.
6. G418.
7. Tissue culture dishes.
8. KK2: 20 mM potassium phosphate, pH 6.2.

3 Methods

3.1 General Considerations

3.1.1 *Dictyostelium* as an Experimental System for Transcription Imaging Studies

Dictyostelium has emerged as a major system for imaging single gene transcriptional dynamics in living cells. In addition to haploid molecular genetics, allowing rapid targeting of transcriptional reporters to single endogenous genes, the cells undergo a simple and well-defined differentiation programme, allowing observation of transcriptional changes that drive and respond to development, crucial for furthering our understanding of how cells perceive external signals and make appropriate decisions. The cells have components and organisation more reminiscent of larger eukaryotes than budding yeast [12, 13], often the instinctive choice of simple genetic model. It is possible to image complete cell cycles (average 8 h) whilst resolving subnuclear events, and these events can then be visualised in multicellular structures as the cells develop. The nascent RNA dynamics of 12 endogenous *Dictyostelium* genes have been described [2, 14–16] and all but one of these at multiple stages of development. This, at the time of writing, compares favourably with the numbers of native genes tagged in yeast [17] and mouse [18].

3.1.2 Imaging Conditions

The main concerns for imaging *Dictyostelium* cells are autofluorescence and phototoxicity. The standard liquid growth medium, HL5 [19], is autofluorescent, masking all but the strongest expression of GFP. This is not a problem when imaging developing cells, as media removal (starvation) is the driver for differentiation. However, for imaging undifferentiated cells, the HL5 autofluorescence must be dealt with. The original solution was LF (low fluorescence) medium, developed by Harry MacWilliams. LF is suboptimal for cell growth, causing cells to starve and initiate development, a response alleviated by using a mix of 75 % and 25 % HL5, which is suitable for imaging complete cell cycles [13]. A further richer improvement uses 70 % LF/20 % HL5/10 % heat-inactivated fetal bovine serum [15].

The *Dictyostelium* response to intense fluorescent light can be observed within a few tens of seconds. The cells lose polarity, becoming pancake-shaped, followed by lysis if light is maintained. This rapid behaviour has led to suggestions that the cells are unusually photosensitive—perhaps as they are normally soil dwelling. It may be more likely that as *Dictyostelium* have a highly dynamic cytoskeleton, they reveal stress more rapidly than would a comparatively immobile mammalian fibroblast. Indeed our own studies on mammalian fibroblast cell lines (using cell division as a proxy for cell health) suggest that they have similar photosensitivity [20, 21]. The motility of *Dictyostelium* itself is an issue restricting long-term cell tracking, but this can be mitigated by use of slower moving strains, AX2G and AX3.

Although transcriptional events can be detected as low- to medium-intensity objects on basic fluorescent microscopes, to countering the phototoxic effect requires management of light levels. Although different experiments, with their own signal-to-noise requirements and imaging periods, use different capture conditions, general suggestions for wide-field imaging are to use both UV and narrowband GFP filters (therefore minimising lower wavelength light). An empirical rule to reduce photodamage is to decrease the intensity of excitation, by either reducing light source power or inserting a neutral density filter, whilst increasing the exposure duration. The image quality (signal-to-noise ratio) is not unduly affected by this change, provided the required frame rate incorporates increased exposure times. A major initial advance was the introduction of EM-CCD cameras. Although these have lower resolution than standard CCD cameras, the payoff is much greater detection sensitivity. Our general live imaging station comprises an inverted Axiovert 200 microscope (Zeiss) with light provided by a DG4 lamp (Sutter). For single-channel GFP imaging, we use a narrowband GFP filter set (#41020, Chroma). For two-colour imaging, we use a GFP/mCherry filter set (#59022, Chroma). To reduce illumination we use different strength ND filters (Chroma) and/or reduced lamp power. Cells are imaged with an ImagEM EM-CCD camera (C9100-13, Hamamatsu). The system is managed by Volocity Acquisition software. 3D stacks are captured using a controllable XY stage (ASI) with a piezo attachment for rapid 3D capture at multiple *xy* positions. Spot detection is unreliable with low-NA objectives. Our standard objective is 63×/1.4NA (Zeiss). We also use confocal and spinning disc microscopes. A specific protocol for photobleaching on a confocal microscope is detailed below.

3.1.3 Molecular Biology

Our approach is to integrate a 1.3 kb array containing 24 MS2 repeats just downstream of the ATG [16]. The repeats are followed by the selectable marker for gene targeting, which is a cassette providing resistance to blasticidin S. The transcription of MS2 runs directly into the terminator of the resistance cassette, and based upon the sizes of transcripts detected by Northern blots, this terminator is used effectively. One initial aim was to compare the transcription dynamics of different promoters, so using the same stem loop-terminator combination insert allowed gene length to be constant in comparisons between genes, simplifying analysis of pulsing patterns [16]. However, modern blasticidin resistance cassettes are flanked by loxP sites [22], which allows rapid generation of loci generating longer transcripts running into the gene body and native terminator. This has the additional advantage of freeing the marker for subsequent mutagenesis, to define pulsing regulators (*see* **Note 3**). For expressing the MS2-GFP fusion protein, we transform a plasmid using G418 selection into recombinants which already have targeted MS2 loops, then select phenotypically normal clones with comparable expression levels. Subsequent transformations (e.g. to mutate specific regulators) usually use the hygromycin selectable marker, although knockout vectors can also be co-transformed with the MS2-GFP plasmid using G418 selection. Attempts to target the MS2-GFP coding sequence to a specific driver locus have to date yielded insufficient fluorescence for effective transcript detection. Our MS2-GFP expression vector expresses the fusion protein more homogenously in cells grown attached to tissue culture plastic, rather than in suspension culture.

As the cells are haploid, we have tended to select genes present on a duplicated region of chromosome 2. This region is duplicated in strains AX3 and AX4, but not in AX2-based strains, and allows the targeting of MS2 repeats into essential genes. We have selected other, non-duplicated genes, which give minimal phenotypes when mutated.

3.2 Imaging Transcriptional Pulses During Growth and Development

1. During development, *Dictyostelium* cells are typically imaged under agar. To initiate development, 2.5×10^6 cells are plated in 1 ml KK2 buffer onto 35 mm Petri dishes containing KK2/2 % agar. After 10 min for cells to adhere, buffer is gently removed and the plate kept flat and topside up in a humidified chamber at 22 °C (a clear plastic box with wet paper towels is sufficient). Prior to imaging, a 1 cm^2 piece of agar is cut out

and inverted onto an imaging dish, and a critical step is to cover the agar block with mineral oil (we use the same oil for PCR) to prevent drying. To mimic growth phase conditions using agar overlay, agar is made up with LF agar and the cells plated in imaging media instead of KK2 buffer (*see* **Notes 4–6**).

2. For imaging under agar, typical imaging conditions, using the wide-field imaging station described above, are 50 ms exposures with a 75 % ND filter, capturing two channels (GFP/mCherry) and 19 slices per stack, using full DG4 power. With these conditions, *Dictyostelium* cells have been imaged every 8 s for a period of 5 min without morphological problems or every 30 s for 40 min. When *Dictyostelium* cells undergo differentiation upon starvation, they tend to become progressively more resilient to photodamage and the duration of imaging can be increased accordingly.
3. Although automated analysis is replacing manual spot identification and cell tracking, the following guidelines form the basis of both manual and developing automatic routines. When analysing data, ambiguous cell tracks are discarded from further analysis. For spots, a straightforward threshold is two or more bright adjacent voxels (in *xy* or *z*), with both brighter than all other voxels in the *z*-plane. Intensity values are measured from $5 \times 5 \times 5$ voxel cubes centred on the brightest pixel. Intensity is defined as mean intensity per voxel within cubes, averaged over each time point within a pulse, after background subtraction. Background intensities are calculated from similar cubes applied to cell bodies, averaged over two measurements [14, 16]. We have expressed pulse intensity as the mean value for many pulses in all comparative studies.

3.3 Long-Term Imaging of Nascent RNA Dynamics

For long-term imaging of growing and dividing *Dictyostelium* cells, the balance between sensitivity and viability becomes critical. Fluctuations between different heterogeneous cell states occur over timescales of 1–2 cell cycles [23, 24], so to image the transcriptional basis of these fluctuations requires imaging of transcriptional events over these periods. This limits analysis to genes with very high activity, to allow detection of transcription at low light levels [15]. To date, we have only imaged complete cell cycles (mean 8–12 h for *Dictyostelium*) for two genes, *act5* (β-actin) and *scd* (delta-9 fatty acid desaturase). Measured pulse frequencies and durations are lower under these imaging conditions than under agar, reflecting the lower signal-to-noise ratio. Long-term imaging does not use agar overlay, as we presume this to be restrictive for division, so the cells project more height in the *z*-axis, necessitating more *z*-depth (and therefore more light) during capture.

1. MS2 cell lines are grown in HL5 medium (with 20 μg/ml G418) on 10 cm tissue culture dishes at 22 °C. Cell density: $0.5–1.5 \times 10^7$ cells/10 cm dish. Cells are only used 4–8 days

after defrosting to restrict clonal diversification. To minimise autofluorescence, HL5 is replaced with imaging medium 2 h prior to imaging.

2. After incubation for 2 h, cells are collected from the 10 cm dish and plated on microscope observation chamber in imaging media at a density of 1×10^5 cells/cm^2. After 1 h, the cells have adhered and ready for long-term imaging.

3.4 Stimulation of Transcription by Exogenous Factors During Development

The stimulation of transcription by extracellular signals underlies so many biological phenomena yet has been relatively impervious to dynamic single-cell analysis. This is a disappointing neglect of basic biology, as the quantitative relationships between stimuli and the transcriptional response are at the core of so many models of signalling and developmental phenomena. We have carried out such an experiment on developing *Dictyostelium* cells, using the *ecmA* gene. This gene responds to DIF-1 and cAMP during development, where it switches on around 10 h after onset of starvation [25, 26]. Before this time the gene will not respond to external stimuli. After this, it is already strongly activated by endogenous signalling. To reproducibly capture the responsive phase, developmental timing (which can fluctuate by an hour or two) must be as predictable as possible. Therefore, consistent cell handling and use of methods to enhance synchrony are essential.

1. *Dictyostelium* cells are cultured in HL5 with selection (G418) removed 1 day prior to initiation of development. Reproducibility is improved if cell density is maintained in the exponential phase in culture and cells not maintained in culture >14 days.
2. Cells are developed on fresh KK2 2 % agar plates. KK2 is added to the agar surface once solidified, to prevent dehydration, then removed prior to plating cells. The time for cells to settle on the agar is a consistent 10 min, and the initiation of development is synchronised, by reducing the temperature to 6 °C for 16 h [27]. This greatly increases reproducibility in developmental timing, as cells build up at the onset of aggregation under these conditions. Aggregation (or streaming) occurs rapidly after elevation to 22 °C. After 5 h at 22 °C, cells are responsive to stimulation, with transcription detectable in only a few cells in the absence of stimulus. The disaggregation of cells is essential to allow single-cell imaging. Disaggregation is achieved by passing aggregates ten times through a needle with a gauge of 0.9 mm (BD Microlance3;304827) in KK2 with 10 mM EDTA (*see* **Note 7**).
3. To ensure proper mixing of external signals, equal volumes are used (200 μl of double dose of stimulus in KK2 added to existing 200 μl KK2 in the well containing the cells). Imaging fields are selected in the middle of the wells to reduce effects of

unequal mixing. Eight well dishes allow multiple signal concentrations to be tested on the same starting cell population. Care is required to avoid the imaging dish from moving during signal addition. For *ecmA*, 3.5 min capture intervals allow many *xy* data points to be captured whilst best representing the pulse durations (mean 8–11 min [14]). Without the use of agar overlay, the cells attain a greater height in the *z*-axis, so 10 μm *z*-stacks are captured.

3.5 Photobleaching Transcription Sites in Dictyostelium Cells

1. To estimate absolute and relative transcription rates, recovery of spot fluorescence after bleaching can be used [28–30]. The turnover of MS2 protein on MS2 stem loops is negligible [31], so recovery of spot fluorescence depends upon new transcription only. Cells are prepared under agar.
2. FRAP experiments are performed using a 488 nm laser on a Zeiss LSM 710 confocal microscope with a 63×/1.4 NA oil-immersion objective. The resolution of the confocal is greater than our normal live-cell station (with an EM-CCD) so transcription sites are much clearer. A square area of 1.8 μm^2 is used for bleaching experiments. A pre-bleach stack is acquired, after which the square area is centred on the spot and bleached immediately with the 488 nm laser at 40 % power with 50 iterations. Recovery images are acquired every 5 s for 120 s, with each stack containing 13 slices with intervals of 0.779 μm. As cells and spots are moving, around 50 % of spots move before they are bleached. These cells are not considered further. We concentrate on spots where the bleaching was not complete (0.2 of the initial fluorescence) to allow accurate tracking of recovery.
3. Images are analysed manually using Volocity 6.1.1 and GraphPad Prism 5.0 software. Mean intensity of the transcription site is recorded at every time point, using a 10 × 10 pixel square over 4 *z* planes (*xy* pixel size 0.13 μm). Similar volumes are used to measure cell and nuclear backgrounds. Nuclear intensity is subtracted from spot intensity at each time point [32]. Values are normalised by pre-bleach spot intensity minus pre-bleach nuclear intensity. To account for photobleaching, intensities are scaled to normalised cell background:

$$I_{\text{normalised}} = \left(\frac{I_{\text{spot}_{(t)}} - I_{\text{nuclear}_{(t)}}}{I_{\text{spot}_{(t=0)}} - I_{\text{nuclear}_{(t=0)}}} \times \frac{I_{\text{background}_{(t=0)}}}{I_{\text{background}_{(t)}}} \right)$$

Cells are collected into groups with similar background intensities and similar levels of bleach for analysis purposes to minimise sources of heterogeneity. Examples FRAP time series and pooled recovery curve are shown in Fig. 2.

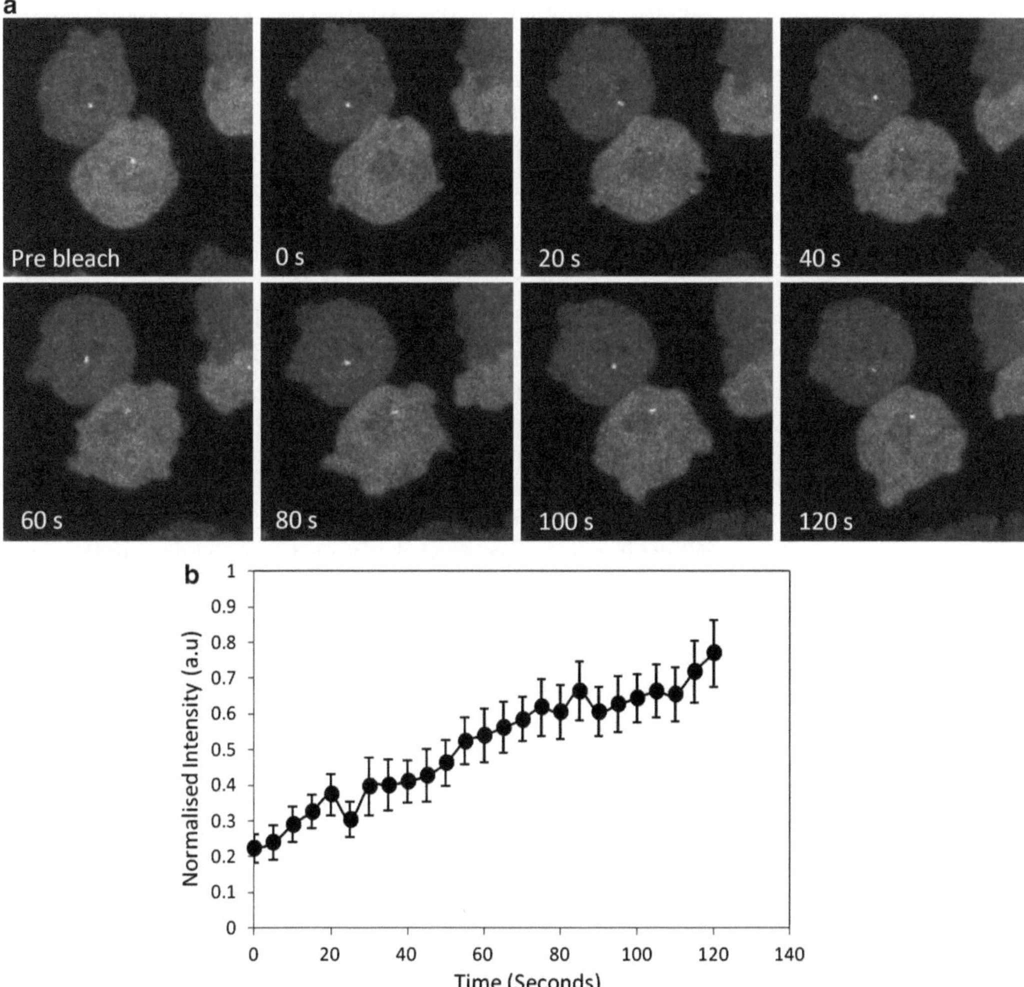

Fig. 2 (**a**) FRAP on nascent RNA spots at the *act5* locus in *Dictyostelium* cells. The nascent RNA spot in the lower cell was bleached by repeat scanning of the region around the spot with a 488 nm laser on a LSM710 confocal. Spot recovery is shown in maximal projections of 3D stacks, displayed at 20 s intervals post-bleach. (**b**) Typical fluorescence recovery curve after bleaching of *act5* transcription spots in undifferentiated cells. Data reproduced from ref. 16

4 Notes

1. The traditional HL5 recipe tends to contain precipitated particles affecting image quality, although ForMedium HL5 contains far fewer particles. These can be removed by filtration.
2. LF medium alone is not suitable for long-term cell growth. Supplying 10 % or more HL5 and 10 % FBS enhances the growth at least to confluence. Filter-sterilised HL5 reduces background fluorescence.

3. We are testing alternatives to the original MS2 loops, such as PP7 loops [17]. We have tried to use the lambda N-Box B detection system [33] but have not yet observed any RNA spots using this method.

4. Imaging under agar has a number of benefits, particularly in terms of reducing potential for photodamage. Firstly, the thickness of the cells is reduced, meaning fewer *z*-slices are needed to span the depth of the cells (this has the added advantage of being faster, allowing a greater number of *xy* stage positions to be visited in the time interval, useful for high-throughput experiments). Secondly, the optical thickness is decreased so less out-of-focus light interferes with the image, improving quality. Finally, from an automated image-processing viewpoint, cell geometry is essentially restricted to two dimensions; cells cannot overlap or move on top of one another, greatly speeding up image segmentation and improving tracking of cells over time. Whilst it is to be expected that long-term imaging under agar is likely to impede cell aggregation and morphogenesis during development, cells taken after 3 h starvation continue to differentiate under agar for several hours, forming actively migrating streams at similar times after starvation as cells not transferred to the imaging dish. Cells on agar blocks inverted earlier in starvation progress to streaming less reliably.

5. When imaging nascent transcription, a number of strategies can be employed to improve the accuracy of identifying genuine transcription spots and reducing the rate of false positives. Firstly, spacing of *z*-slices in stacks is chosen so spots will span more than one slice. The presence of a spot-like object in two or three adjacent slices improves the certainty with which the spot can be assigned. However, other objects within the cell can be of a roughly similar size and exhibit autofluorescence such that they may confuse spot assignment. To overcome this issue, which can be problematic in *Dictyostelium*, we use a marker such as H2B-RFP to identify nuclei or use a GFP filter which greatly reduces autofluorescence (#41020, Chroma). Autofluorescent objects tend to be located outside the nucleus, so using a second channel to limit our search to within nuclei, we can assign spots with more certainty. An additional benefit of using a nuclear marker is the addition of spatial information; the position of the transcription spot within the nucleus is also of interest. The drawbacks of having to image an additional channel are that the problem of photodamage is exacerbated, and the increased time taken to image a stack may limit the frame rate or the number of points that can be visited in a single experiment. Also, a channel is taken up which could potentially be used to identify some other cellular property. MS2-GFP with a nuclear localisation signal could overcome these problems. One point of practical note when using a nuclear marker

in motile cells is that both channels should be imaged directly within each *z*-slice, to ensure that capture of spot position and nuclear boundary are as simultaneous as possible.

6. An important parameter for live imaging of nascent transcription is the frame rate used for capture. The rate chosen is usually a balance between a number of considerations. As described above, the frame rate, in combination with the number of *z*-slices per stack and the duration of capture, should not be so high as to damage the cells. Furthermore, if the cell line is reasonably motile, then the frame rate must be fast enough that cells can be tracked from frame to frame. In practice this means ensuring that the typical distance moved between frames is less than the average separation between cells. Variegation in the level of MS2-GFP can help to distinguish and track adjacent cells at higher densities; however, overconfidence in assigning matches from frame to frame will result in inaccuracies in the pulsing dynamics measured. Finally and most importantly, the frame rate should reflect the timescale of the processes of interest. For live-cell transcription, there are a number of timescales which may not all be captured in a single experiment. The initiation and elongation of individual polymerases may influence the intensity over timescales of the order of tens of seconds, depending on the length of the gene. Transcription pulses typically persist for several minutes. Variation in gene activity due to changes in the microenvironment may occur over longer timescales, up to hours.

7. Disaggregated *Dictyostelium* cells continue to move by chemotaxis and in response to cell contact when plated in microscope dishes. These problems are avoided by plating cells at low density. However, this reduces the number of data points; therefore, a compromised density is used. The loss of 3D tissue context will likely have a large influence on cell behaviour and development; however, when kept overnight at high doses of DIF-1 and cAMP, a fully differentiated state is reached, indicating the developmental path can be travelled, despite lack of cell contacts. A further benefit of disaggregation is the likelihood that each cell receives the same signal concentration, difficult to assess with penetration into tissue. A further limitation of using native tissue context is background fluorescence. In single cells, this is not a problem as spots are only obscured at low levels of transcription; however, even with deconvolution, background tissue fluorescence obscures all but the brightest spots. This might be overcome by increasing the number of MS2 repeats, although this might distort gene dynamics. Another possibility is using squashed 3D tissue, under agar, although tracking becomes extremely difficult and controlling the amount of stimulus each cell is exposed to would be challenging.

Acknowledgements

Work in the lab is supported by a Wellcome Trust Senior Research Fellowship to J.R.C.

References

1. Bertrand E et al (1998) Localization of ASH1 mRNA particles in living yeast. Mol Cell 2(4):437–445
2. Chubb JR, Trcek T, Shenoy SM, Singer RH (2006) Transcriptional pulsing of a developmental gene. Curr Biol 16(10):1018–1025
3. Golding I, Paulsson J, Zawilski SM, Cox EC (2005) Real-time kinetics of gene activity in individual bacteria. Cell 123(6):1025–1036
4. Raj A, Peskin CS, Tranchina D, Vargas DY, Tyagi S (2006) Stochastic mRNA synthesis in mammalian cells. PLoS Biol 4(10):e309
5. Raj A, van Oudenaarden A (2008) Nature, nurture, or chance: stochastic gene expression and its consequences. Cell 135(2):216–226
6. Chubb JR, Liverpool TB (2010) Bursts and pulses: insights from single cell studies into transcriptional mechanisms. Curr Opin Genet Dev 20(5):478–484
7. Lionnet T, Singer RH (2012) Transcription goes digital. EMBO Rep 13(4):313–321
8. Balazsi G, van Oudenaarden A, Collins JJ (2011) Cellular decision making and biological noise: from microbes to mammals. Cell 144(6):910–925
9. Huh D, Paulsson J (2011) Non-genetic heterogeneity from stochastic partitioning at cell division. Nat Genet 43(2):95–100
10. Stavreva DA et al (2009) Ultradian hormone stimulation induces glucocorticoid receptor-mediated pulses of gene transcription. Nat Cell Biol 11(9):1093–1102
11. Harper CV et al (2011) Dynamic analysis of stochastic transcription cycles. PLoS Biol 9(4):e1000607
12. Stevense M, Chubb JR, Muramoto T (2011) Nuclear organization and transcriptional dynamics in Dictyostelium. Dev Growth Differ 53(4):576–586
13. Muramoto T, Chubb JR (2008) Live imaging of the Dictyostelium cell cycle reveals widespread S phase during development, a G2 bias in spore differentiation and a premitotic checkpoint. Development 135(9): 1647–1657
14. Stevense M, Muramoto T, Muller I, Chubb JR (2010) Digital nature of the immediate-early transcriptional response. Development 137(4):579–584
15. Muramoto T, Muller I, Thomas G, Melvin A, Chubb JR (2010) Methylation of H3K4 Is required for inheritance of active transcriptional states. Curr Biol 20(5):397–406
16. Muramoto T et al (2012) Live imaging of nascent RNA dynamics reveals distinct types of transcriptional pulse regulation. Proc Natl Acad Sci USA 109(19):7350–7355
17. Larson DR, Zenklusen D, Wu B, Chao JA, Singer RH (2011) Real-time observation of transcription initiation and elongation on an endogenous yeast gene. Science 332(6028): 475–478
18. Lionnet T et al (2011) A transgenic mouse for in vivo detection of endogenous labeled mRNA. Nat Methods 8(2):165–170
19. Sussman R, Sussman M (1967) Cultivation of Dictyostelium discoideum in axenic medium. Biochem Biophys Res Commun 29(1):53–55
20. Thomson I, Gilchrist S, Bickmore WA, Chubb JR (2004) The radial positioning of chromatin is not inherited through mitosis but is established de novo in early G1. Curr Biol 14(2):166–172
21. Muller I, Boyle S, Singer RH, Bickmore WA, Chubb JR (2010) Stable morphology, but dynamic internal reorganisation, of interphase human chromosomes in living cells. PLoS One 5(7):e11560
22. Faix J, Kreppel L, Shaulsky G, Schleicher M, Kimmel AR (2004) A rapid and efficient method to generate multiple gene disruptions in Dictyostelium discoideum using a single selectable marker and the Cre-loxP system. Nucleic Acids Res 32(19):e143
23. Spencer SL, Gaudet S, Albeck JG, Burke JM, Sorger PK (2009) Non-genetic origins of cell-to-cell variability in TRAIL-induced apoptosis. Nature 459(7245):428–432
24. Sigal A et al (2006) Variability and memory of protein levels in human cells. Nature 444(7119):643–646

25. Williams JG (2006) Transcriptional regulation of Dictyostelium pattern formation. EMBO Rep 7(7):694–698
26. Verkerke-van Wijk I, Brandt R, Bosman L, Schaap P (1998) Two distinct signaling pathways mediate DIF induction of prestalk gene expression in Dictyostelium. Exp Cell Res 245(1):179–185
27. Alcantara F, Monk M (1974) Signal propagation during aggregation in the slime mould Dictyostelium discoideum. J Gen Microbiol 85(2):321–334
28. Maiuri P et al (2011) Fast transcription rates of RNA polymerase II in human cells. EMBO Rep 12(12):1280–1285
29. Darzacq X et al (2007) In vivo dynamics of RNA polymerase II transcription. Nat Struct Mol Biol 14(9):796–806
30. Yunger S, Rosenfeld L, Garini Y, Shav-Tal Y (2010) Single-allele analysis of transcription kinetics in living mammalian cells. Nat Methods 7(8):631–633
31. Boireau S et al (2007) The transcriptional cycle of HIV-1 in real-time and live cells. J Cell Biol 179(2):291–304
32. Phair RD, Misteli T (2000) High mobility of proteins in the mammalian cell nucleus. Nature 404(6778):604–609
33. Daigle N, Ellenberg J (2007) LambdaN-GFP: an RNA reporter system for live-cell imaging. Nat Methods 4(8):633–636

Part II

Imaging the Genome and Chromatin Dynamics

Chapter 9

Monitoring Dynamic Binding of Chromatin Proteins In Vivo by Single-Molecule Tracking

Davide Mazza, Sourav Ganguly, and James G. McNally

Abstract

Single-molecule fluorescence microscopy has been used for decades to quantify macromolecular dynamics occurring in specimens that are in direct contact with a coverslip. This has permitted in vitro analysis of single-molecule motion in various biochemically reconstituted systems as well as in vivo studies of single-molecule motion on cell membranes. More recently, thanks to improvements in fluorescent tags and microscopes, it has been possible to follow individual molecules inside thicker specimens such as the nucleus of living cells. This has enabled a detailed description of the live-cell binding of nuclear proteins to DNA. In this protocol we describe a method to quantify intranuclear binding using single-molecule tracking (SMT).

Key words Single-molecule tracking, Microscopy, Transcription factor, DNA binding

1 Introduction

Many cellular processes are controlled by the binding of relatively small and mobile molecules to larger and stable scaffolds. For example, intracellular signaling is generally mediated by the binding of a small and mobile ligand to an often bulky, slow-moving receptor in the plasma membrane. Similarly, the processes carried out in the nucleus represent a striking example of "regulation by immobilization": Transcription, DNA replication, and DNA repair are all controlled by the interaction of soluble factors with specific targets on the slowly diffusing, chromatinized DNA. If we want to fully understand these processes, it is necessary to quantitatively describe the interactions between the regulatory factors and the DNA. The inherent heterogeneity of the intracellular milieu makes it important to measure these interactions within a living cell, because the intracellular complexity can hardly be reproduced in vitro.

Fluorescence microscopy, mostly because of its high specificity and low invasiveness, has been a common choice to probe live-cell

Yaron Shav-Tal (ed.), *Imaging Gene Expression: Methods and Protocols*, Methods in Molecular Biology, vol. 1042, DOI 10.1007/978-1-62703-526-2_9, © Springer Science+Business Media, LLC 2013

binding events [1–3]. In this context, over the last decade, traditional bulk techniques such as Fluorescence Recovery After Photobleaching (FRAP) and Fluorescence Correlation Spectroscopy (FCS) have been modified and adapted to quantify binding to immobile scaffolds [4–9]. These techniques have proven to be powerful in the qualitative and quantitative analysis of interactions, for example, allowing the measurement of the fraction of chromatin-associated molecules and the average time spent by a factor on DNA.

While informative, FRAP and FCS measurements typically provide only indirect information about a molecule's behavior, since multiple factors can contribute to the molecule's mobility [10]. In contrast, SMT allows direct observation of a molecule's movement, making it easier to infer the processes that govern such movement. For example, transient binding to an immobile substrate will be visualized by SMT as a transient cessation of motion [11, 12], whereas in either FRAP or FCS, this transient binding must be indirectly inferred by fitting the data with a mathematical model that incorporates a term describing such binding.

Another advantage of single-molecule techniques is that they reveal the dynamic properties of each molecule rather than just the population average as measured by FRAP or FCS [13]. Thus by SMT it is possible to measure not just the average residence time of a molecule on chromatin, but rather the entire distribution of chromatin residence times [11]. For this reason single-molecule methods are invaluable for identifying processes unique to only a subfraction of molecules, which would otherwise be obscured in an ensemble average.

A final advantage of single-molecule tracking (SMT) is that it is not constrained by the resolution limit imposed by diffraction, as in classical fluorescence microscopy. When single emitters are observed, the true position of the molecules can be localized with an accuracy that depends on the amount of light (the number of photons) recorded over the background [14]. For bright single molecules, individual molecules can be localized with a precision of 20 nm or below, ten times better than the resolution limit.

SMT has been widely used in membranes and in in vitro two-dimensional systems, but its application to three-dimensional environments such as the cell nucleus has been somewhat limited, mostly due to the deterioration of the signal-to-background ratio caused by out-of-focus molecules. In the last few years however, there has been an increasing interest in monitoring nuclear protein dynamics using single-molecule approaches [12, 15–17], and a number of groups have begun to make such measurements using several different strategies to improve the visualization of single molecules in 3D environments [18–20].

In this protocol, we will describe in detail our approach to quantify binding of nuclear proteins to DNA using SMT.

To provide an overview of the workflow, we begin with an outline of the basic steps of our approach. First, we create a HaloTag [21] fusion to the protein of interest. As described in Subheading 2.1 this enables subsequent in vivo labeling of the protein with a derivative of tetramethylrhodamine, which is a much more photostable fluorophore than fluorescent proteins, such as GFP. Next, we detect the single molecules using a microscope that minimizes out-of-focus light by means of a slanted excitation light beam known as HILO illumination [20] (Subheading 2.2). Then we collect images of single molecules in the cell nucleus as a function of time (Subheading 3.2) to produce movies of single-molecule motion. We then track the motion of every molecule in the movie to generate trajectories (Subheading 3.3). Finally, we analyze each trajectory to identify the segments that correspond to chromatin-binding events. As described in Subheading 3.4, we identify binding events by characterizing the behavior of a control protein that is known to be tightly bound to chromatin (histone H2B in our case). By analyzing many trajectories of the protein of interest, we can determine what fraction of molecules are bound and for how long. This produces an estimate of the total bound fraction and the distribution of residence times on chromatin for the protein of interest.

2 Materials

2.1 Fluorescent Labeling for SMT

The fluorescence labeling (as outlined in Subheading 3.1) involves construction of a HaloTag fusion protein (Promega Corp.), transfection of the fusion protein into cells, and then labeling of the fusion protein with a cell-permeable organic dye tetramethylrhodamine (TMR). An alternative is to label with fluorophores emitting at shorter wavelengths (Oregon Green, diAcFAM, or Coumarin) which will also bind to the HaloTag fusion protein, but we found that TMR was more photostable than the other fluorescent tags. The reagents required to prepare the samples for SMT are listed below:

(a) Expression plasmid containing your protein of interest fused with the HaloTag.

(b) Lipofectamine LTX for transient transfections.

(c) No. 1 Lab-Tek chambers (Nalge Nunc Intl.).

(d) DMEM without phenol red, complemented with 10 % fetal bovine serum, 1 % L-glutamine, and 0.5 % penicillin/streptomycin.

(e) Phosphate-buffered saline.

(f) A HaloTag fluorescent ligand (HaloTag-TMR in our case), diluted in DMSO to a concentration of 5 μM.

2.2 Instrumentation for SMT

SMT on the surface of a specimen is typically done with a wide-field microscope set up in total internal reflection (TIRF) mode with lasers as excitation sources and an EM-CCD as detector [13, 22, 23]. TIRF imaging greatly reduces background light, but only permits single-molecule detection in the thin sheet of the specimen that is in contact with the coverslip. Thus TIRF cannot be used to image single molecules inside a thicker specimen.

Several different approaches are currently available to reduce background for SMT inside a thicker specimen (*see* **Note 1**), but the simplest is highly inclined laminated optical sheet (HILO) microscopy [20]. This is achieved with a TIRF illuminator adjusted such that a shaft of light is now directed into the specimen instead of internally reflected (Fig 1a).

In this protocol we will describe how to collect SMT data using HILO illumination, but the subsequent procedures for data collection and analysis can be applied to any microscope that can acquire single-molecule images. The key components of our custom-built HILO microscope are:

(a) A 25 mW 561 nm DPSS laser (Kineflex Mustang, Qioptiq Photonics Ltd.) for the excitation of HaloTag-TMR labeled molecules.

(b) A 50 mW 488 nm diode laser (Obis, Coherent Corp.) to collect reference images with GFP fusions.

(c) Optical components to expand and align the beam to the microscope port.

(d) A movable mirror to achieve HILO or TIRF illumination.

(e) A microscope frame (Olympus IX-81).

(f) A high numerical aperture, high-magnification objective (150× NA 1.45, Olympus).

(g) A back-illuminated EM-CCD (Evolve 512, Photometrics) camera coupled to the side port of the microscope. Each of the 512×512 pixels of the camera has a physical size of 16 μm, corresponding to a pixel size in the image of 106 nm, approximately half the size of the diffraction-limited signal produced by a single molecule.

(h) Filter sets to select the imaged wavelengths.

(i) An environmental chamber mounted on the stage to maintain cells at 37 °C and 5 % CO_2.

2.3 Software for Tracking and Data Analysis

Software is needed to analyze the time-lapse single-molecule data and then generate trajectories for each molecule. Software is also needed to analyze the trajectories to identify binding events and quantify them. While different commercially available and custom-made solutions are available for tracking single particles (e.g., Diatrack (Semasopth Corp.) or Imaris (Bitplane)), here we outline

Labellng methods and microscopy approaches to single molecule tracking in the nucleus of living cells.

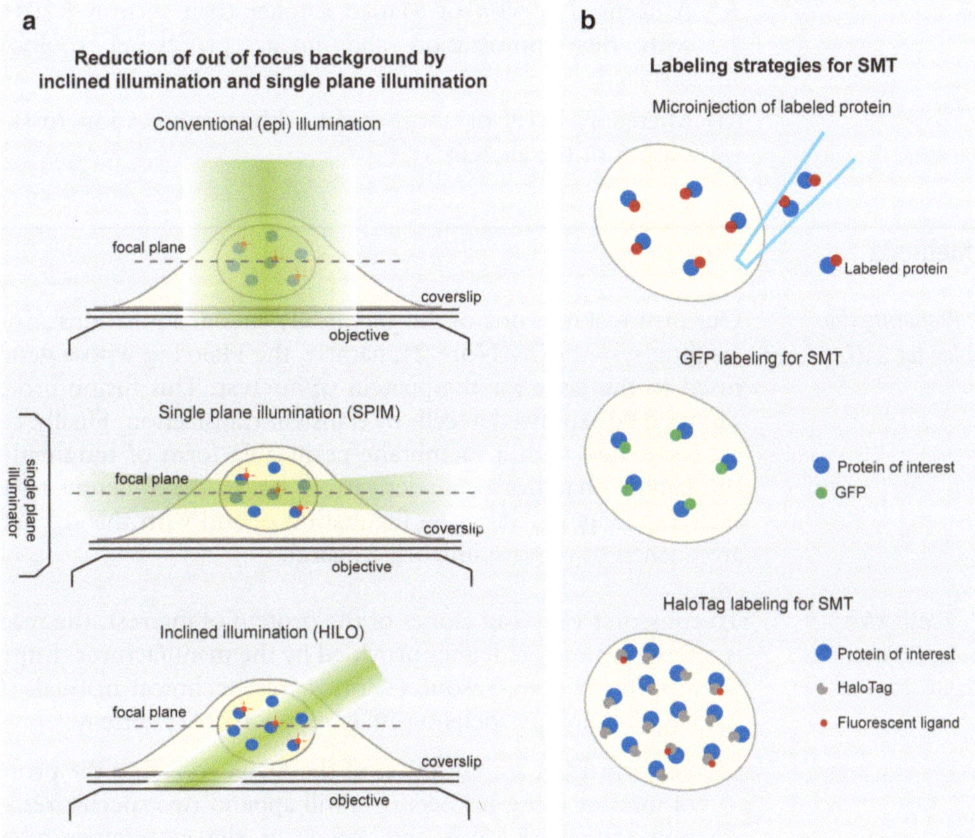

Fig. 1 (**a**) Illumination modes for SMT. In epifluorescence, the background signal due to out-of-focus fluorescence hampers the localization of single molecules. Alternative illumination schemes have been designed to improve the optical sectioning in wide-field microscopy. Single plane illumination microscopy (SPIM) offers the highest optical sectioning power, but requires the use of custom-built microscopes using two orthogonally arranged objectives. A simpler approach is provided by highly inclined laminated optical sheet (HILO) microscopy, using a tilted beam to reduce the excitation of out-of-focus fluorescence. (**b**) Fluorescence labeling strategies for SMT. SMT in living cells is commonly performed by microinjecting the purified protein of interest previously labeled with a photostable organic dye. A less laborious procedure is to use fluorescent proteins to tag the molecule of interest, but the rapid photobleaching of GFP prevents tracking molecules for longer than a handful of frames. With posttranslational labeling systems such as HaloTag, it is possible to easily label the protein of interest with a photostable organic dye

how to use the software "TrackRecord" that we have developed in Matlab (the MathWorks Inc.), which combines a package for single-particle/single-molecule tracking (based on the Crocker and Grier algorithm [24]) with the routines to isolate and analyze the behavior of chromatin-bound molecules:

(a) A downloaded version of the tracking software "TrackRecord" available at http://code.google.com/p/single-molecule-tracking/.

(b) A PC or Macintosh with at least 2 GB of RAM.

(c) A licensed version of Matlab (higher than version R2010a) with the optimization and image processing toolboxes installed.

(d) Microsoft Excel or another spreadsheet application to store output of the analysis.

3 Methods

3.1 Preparing the Samples for SMT

Our protocol uses one of the genetically encoded posttranslational labeling systems (*see* **Note 2**), namely, the HaloTag whose gene is fused to the gene for the protein of interest. This fusion protein can then be expressed in cells by transient transfection. Finally, cells are incubated with a membrane-permeable form of tetramethylrhodamine that has been derivatized to bind covalently to the HaloTag (Fig 1b). After an incubation period with this dye, cells are washed to remove unlabeled ligand.

3.1.1 Construction of HaloTag Fusion Protein and Transfection into Cells

To construct HaloTag clones of the protein of interest, the reader is referred to the guidelines provided by the manufacturer (http://www.promega.com/resources/protocols/technical-manuals/0/halotag-technology-focus-on-imaging-protocol/). Briefly:

1. Perform PCR amplification of the DNA encoding the protein of interest using primers that will append two specific restriction sequences (SgfI and PmeI) at the extremities of the fragment.
2. Digest the extracted DNA and the HaloTag expression vector, and ligate together to produce the fusion protein (*see* **Note 3**).
3. The plasmid encoding the fusion protein must then be either stably or transiently transfected into the cell line(s) of interest. Plate the transfected cells in Lab-Tek chambers to prepare them for microscopy.

   Perform transient transfections using Lipofectamine LTX 1 day prior to single-molecule experiments according to the manufacturer instructions.

3.1.2 Labeling Protocol with TMR Ligand

Once a functional HaloTag fusion protein is expressed in cells, it needs to be labeled with the tetramethylrhodamine (TMR) fluorescent ligand. To load TMR into cells and allow it to bind to the HaloTag fusion protein, we made several modifications to the manufacturer's loading protocol because we found that a) the recommended ligand concentration was at least 1,000× higher than what was needed for single-molecule experiments and b) the recommended rinsing protocol did not eliminate all the unbound fluorescent ligand inside the cells (*see* **Note 4**). We have found that the following labeling protocol is effective for our single-molecule experiments:

1. For each of the samples, dilute the 5 μM HaloTag-TMR ligand to 1:1,000 in phenol red-free DMEM to a final concentration of 5 nM.
2. Remove the medium from the wells of the Lab-Tek chambers containing the HaloTag-expressing cells.
3. Add 1 ml of the diluted HaloTag solution to the well.
4. Incubate at 37 °C for 25 min.
5. Wash three times with phenol red-free DMEM.
6. Incubate at 37 °C for 15 min.
7. Repeat **steps 5** and **6** once.
8. Replace the cell medium once more with fresh phenol red-free DMEM.

3.2 Acquiring SMT Data

Once the cells contain a low concentration of TMR-labeled HaloTag fusion proteins, the next step is to acquire time-lapse movies to record how the single molecules move inside the cell nucleus.

3.2.1 Setting Up the Microscope for Single-Molecule Detection

1. Wear protective goggles and remove jewelry to avoid accidental reflection of the laser beam.
2. Turn on the microscope, the lasers, the stage incubator, and the EM-CCD camera.
3. Tune the laser power to yield approximately 10 mW on the back aperture of the objective.
4. Set the exposure time of the camera and the imaging rate. We usually select an exposure time of 10 ms and an imaging rate that depends on the mobility of the fluorescently labeled molecules (*see* **Note 5**).
5. Add a drop of immersion oil to the objective.
6. While in epifluorescence mode, verify that the laser beam is properly collimated out of the objective by looking at the projection of the laser beam on the ceiling (*see* **Note 6**).
7. Place the chamber containing the transfected cells on the microscope stage.
8. While exposing, tune the angle of the beam to achieve inclined illumination. This is done by first tuning the angle until TIRF illumination is achieved (*see* **Note** 7) and then readjusting back approximately 10° towards vertical illumination. With this inclination it should be possible to visualize several microns into the cells while considerably reducing the background signal.

3.2.2 Acquisition of SMT Data

Two types of single-molecule data need to be acquired: (1) the protein of interest in live cells for measurement of its binding properties and (2) histone H2B or some other protein known to be tightly bound to chromatin to characterize the behavior of chromatin-bound molecules. To permit comparison, both sets of data should be collected under identical acquisition conditions:

1. First, ensure that the laser is shuttered to prevent leakage of laser light through the eyepieces, and then focus on the sample's fluorescence by using the arc lamp.
2. Select the microscope port that deflects the fluorescent light onto the EM-CCD camera, open the laser shutter, and start streaming images in live mode on the screen in order to identify a suitable cell.
3. Choose a nucleus that contains approximately 10–30 fluorescent molecules. This will appear as 10–30 diffraction-limited spots inside of the nucleus (Fig 2a). To achieve the desired frame rate, it might be necessary to collect the signal from a cropped area of the camera sensor instead of acquiring the complete field of view.
4. Focus at approximately 3–4 μm above the coverslip. The coverslip can be recognized as the focal plane where fluorescent speckles (Fig 2a) are visible in the whole field of view.
5. Start the acquisition of the time series. It is advisable to collect at least 500 images per cell.
6. After collecting the movie, acquire a reference image to identify the cell nucleus (Fig 2a). If another fluorescent label is used as a nuclear marker, change the filter cube and the laser excitation to collect the reference image. Otherwise collect a transmitted light image using the lamp.
7. Repeat **steps 1–6** on at least 20 cells.

3.3 Tracking of SMT Data

We provide below an outline of the major steps involved in the analysis of the data as performed by our software TrackRecord. For a detailed description of how to use this software, the user manual for this software package can be consulted (http://code.google.com/p/single-molecule-tracking/):

1. Open Matlab, and from the command window run the `TrackRecord` application. This will launch the graphical interface allowing the analysis of the data. Load the acquired movies into the software (the routines accept ".tif" files) using the Load Stack button (Fig 2b), and set the acquisition parameters by selecting "Set Acquisition Parameters" from the menu "Set Parameters."
2. Noise reduction from each of the acquired images is performed by using the Band-pass Filtering module available in the main GUI.

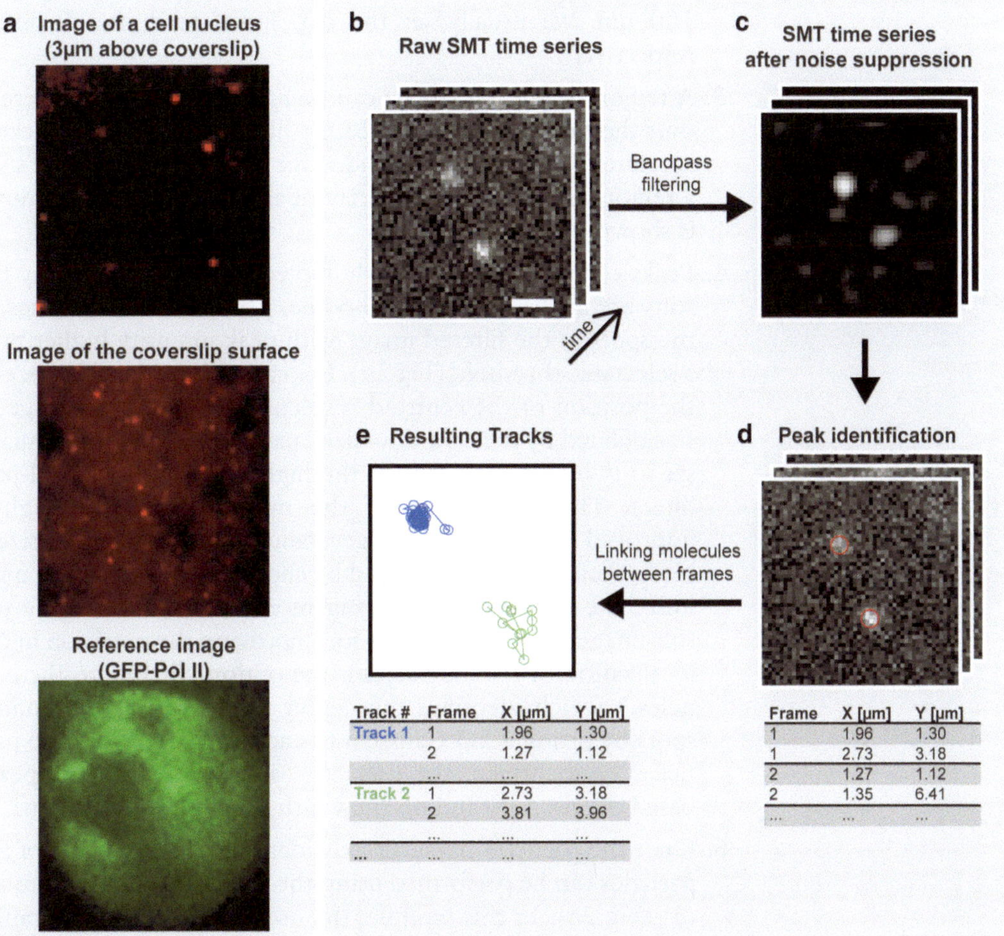

Fig. 2 (**a**) Examples of images for single-molecule tracking of nuclear proteins. We usually select cells that display few (<30) diffraction-limited spots in the nucleus. In this example a HaloTag version of the tumor suppressor p53 was labeled with TMR and collected with an exposure time of 10 ms. The image was taken approximately 3 μm above the coverslip, which can be recognized as the plane with a layer of fluorescent speckles caused by the unliganded TMR deposited on the glass surface. To better identify the nucleus in the image, it is possible to collect a second image of a fluorescently tagged nuclear marker (in this example a GFP version of RNA polymerase II). Scale bar 1 μm. (**b**–**e**) Workflow for the identification and the tracking of single molecules. Starting from the raw movie (**b**), each of the collected images is processed with a band-pass filter (**c**) to suppress the noise and to facilitate the localization of the single molecules. The local maxima of the processed images are then found and used as starting guesses to identify the sub-pixel positions of the single molecules, which are found either by finding the centroids of the diffraction-limited spots or by fitting the raw images with a two-dimensional Gaussian profile (**d**). This localization step produces a list containing the coordinates of the localized particles. These coordinates are then linked into tracks by the tracking algorithm (**e**). Scale bar 1 μm

The thresholds of the band-pass filter determine the smallest and biggest feature of the image that could be tracked (Fig 2c). For tracking diffraction-limited spots with a pixel size of about 100 nm, we usually set the two limits to 1 and 5 pixels, respectively.

3. A region of interest (ROI) can be *optionally* defined, whereby only the particles within the ROI will be identified and tracked. A reference image can be loaded for accurate selection of such a region. Otherwise the overlay of all the frames of the movie is shown.
4. Peaks corresponding to single molecules are identified by the software in the Find Peaks module. The software identifies all the spots in the filtered image with peak intensity higher than a selectable threshold (Fig 2d). For each of the identified peaks, the position of the centroid is calculated over a sub-image of size defined by the "window size" parameter (which we usually set to two pixels larger than the higher limit of the band-pass filter). The localization of the molecules can be slightly improved by selecting a larger value for the "window size." Nevertheless this value should be chosen small enough to minimize the probability of having more than one molecule per sub-image. A better solution for improving the precision in the localization of the molecules is the option "fit PSF to the particles," which identifies the center of the diffraction-limited spots by fitting a 2D Gaussian to each of them. Once the particles are identified, the user can (*optionally*) filter the peaks based on intensity and on the width of the fitted Gaussian.
5. Once the particles are accurately identified, the tracking of the particles can be performed using the "Particle Tracking" module (Fig 2e). In this module, the user can define two parameters to specify how the tracking is performed. The first parameter is the maximum displacement allowed between consecutive frames (in pixels). When tracking the protein of interest, we set this parameter slightly larger than the maximum displacement measured for H2B which defines the motion of the chromatin-bound state. When first measuring this maximum displacement for H2B, we set the maximum displacement parameter at 0.4 μm. The second parameter defines the minimum duration of a valid trajectory. We usually set this parameter to three frames. The user can also activate the "close gaps" option, which considers the possibility that single fluorescent emitters can blink on and off. This is sometimes valuable when it is clear that a molecule remains bound for an extended period, but disappears for one time point simply due to blinking.
6. After tracking has been completed, the tracks can be manually verified using "Check Tracks" under the "Manually Check Tracks" module. This procedure is highly recommended as

even a small error rate in the automatic tracking procedure could produce significant deviations in the final analysis (*see* **Note 8**). The "Check Tracks" button opens up a new window, where the user can edit frame-by-frame the particles' position as well as manage the identified tracks by adding or removing track points and merging, splitting, or deleting tracks altogether.

7. The tracks are preprocessed using the "Preprocess Tracks" under the "Analysis of Tracks" module for subsequent analysis. Essentially, this converts the units of particle positions from pixels to μm, calculates the fluorescence intensity over background of each particle, and allows display of all the tracks on the same image in a new window.
8. At this point the user should save the tracking information derived from the movie by clicking "Save Mat-Track file" in the File menu. Save the .mat files for all of the movies acquired for any given condition.

3.4 Analysis of SMT Binding

The final step is to take the single-molecule trajectories and use them to determine how many molecules are bound to chromatin and for how long. The tracked trajectories can be composed of various segments during which the molecule undergoes either diffusion or binding. The key is to identify the segments of the trajectory that reflect binding. Our procedure accomplishes this by characterizing the behavior of a molecule that is known to be stably bound to chromatin, such as H2B. Thus the first step below (Subheading 3.4.1) is to analyze single-molecule trajectories of H2B. The second step below (Subheading 3.4.2) is to use this information from H2B to select out the segments of each trajectory from the protein of interest that reflect binding. The resultant data are used to determine the fraction of molecules bound and the distribution of residence times on chromatin. The third step below (Subheading 3.4.3) allows for the possibility to fit the residence time distribution to determine whether it can be well described by a single binding state or two distinct binding states and to obtain estimates for the residence time(s).

3.4.1 Quantification of the Mobility of Chromatin-Bound Molecules by the Analysis of Histone SMT Movies

1. Collate the tracks obtained on H2B by choosing "Merge and Analyze Jump Histograms" from the "Further Analysis" menu to select multiple .mat data files containing the H2B tracks. In general, we analyze tracks from approximately 20 cells yielding a total of several hundred tracks. Enough tracks must be analyzed in order to generate sufficiently smooth histograms of displacements and residence times.
2. The software calculates the distribution of displacements for H2B (Fig 3a) and plots a heat map of this distribution—representing the frequency of observing a certain displacement for a certain time (*see* **Note 9**) and an estimate for a maximum

frame-to-frame displacement r_{max} that will be used to identify the chromatin-associated molecules in the following experiments (*see* **Note 10**).

3. The distribution of displacements, together with the results of the analysis, can be copied to the clipboard to be stored in an Excel file.

3.4.2 Determination of Bound Molecules for the Protein of Interest by Objective Thresholding

Binding events for the protein of interest are extracted from the trajectory data by identifying segments of the trajectory for which the distance moved by the molecule between consecutive frames is less than or equal to the maximum displacement (r_{max}) measured for H2B. A second criterion is then applied to these segments, namely, that the number of time points in the bound segment is larger than some minimum value N_{min}. This criterion eliminates short segments where the molecule is freely diffusing but by chance does not move very far. N_{min} can be determined in two ways. First, if the diffusion coefficient of the unbound molecule is known, N_{min} can be calculated as the minimum length of the track that gives a very small (<1 %) probability of mistakenly counting unbound molecules as bound (*see* **Note 10**). Otherwise the N_{min} threshold can be determined more empirically by performing the steps for 3.4.2 and 3.4.3 for a range of N_{min} values from small to large and then plotting the residence time for each as a function of N_{min}. The value of N_{min} at which residence times saturate can be used as the final value of N_{min} for calculating the residence time distribution and bound fraction:

1. Collate the tracks obtained on the protein of interest (from approximately 20 cells) by choosing "Merge and Analyze Jump Histograms" from the "Further Analysis" menu to select multiple .mat data files containing the tracks.
2. The software identifies bound segments of the tracks based on the input values of r_{max} and N_{min} (Fig 3b). The duration of each bound segment is calculated generating a distribution of residence times on chromatin. This distribution can be plotted in two ways: as a conventional histogram or as a cumulative histogram. The cumulative histogram is also known as the survival time distribution $S(t)$ because it corresponds to the probability of having molecules bound longer than a time t. We usually plot the survival time distribution because it allows more direct comparison of different experiments (*see* **Note 11**).
3. The measured residence times could be artifactually reduced due to photobleaching. Thus the software provides an option to correct the measured residence time distribution for photobleaching (*see* **Note 12**). The software displays the photobleaching data together with the best biexponential fit to it, thereby allowing the user to visually inspect the quality of the correction. Also, the survival probability with and without the photobleaching correction is shown.

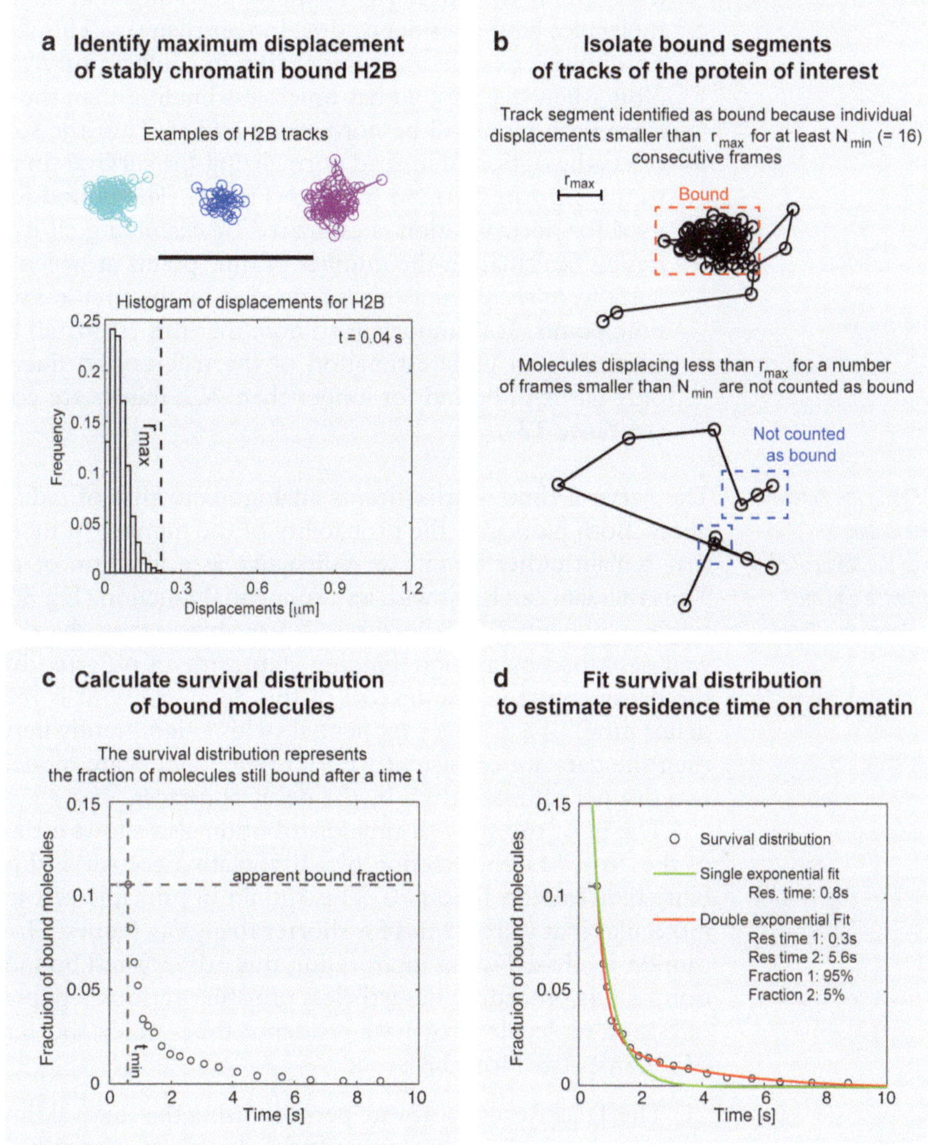

Fig. 3 Identification of bound molecules. (**a**) To identify and quantify the chromatin-associated molecules, it is first necessary to define what is the maximum displacement, r_{max}, expected for bound molecules. It is possible to obtain such a threshold by performing SMT measurements on a protein known to be stably associated with DNA, such as the histone H2B (scale bar 1 μm). (**b**) r_{max} is then used to isolate the bound segments from the tracks obtained on the protein of interest: The molecules that exhibit displacements less than r_{max} for more than a minimum number of consecutive frames N_{min} will be counted as bound. This second threshold N_{min} can be deduced by knowing the diffusion coefficient of free molecules (*see* **Note 7**). (**c**) The duration of each of the binding events is stored and used to populate a cumulative histogram (known as the survival distribution of bound molecules because it represents the probability of having a molecule still bound after a certain time *t*). We usually normalize this histogram for the fraction of bound molecules detected at a time $t_{min} = N_{min}\Delta t$, where Δt is the time between two consecutive images. (**d**) The survival probability of bound molecules can then be fit by a single or a double exponential decay, to obtain estimates of the average residence time(s) and the fraction of molecules in the bound state(s). The user can apply an *F*-test to determine whether the improvement obtained with the biexponential decay is statistically significant

4. By default, the software normalizes the survival time distribution to 1.0 at $t=0$, simply indicating that all bound molecules have a residence time longer than 0 s. This convention however can make it misleading to compare survival time plots when one protein has much less binding than the other, since both plots will be normalized to 1.0. Thus the software offers the option of instead normalizing the survival time plots to the estimated bound fraction (Fig 3c). The bound fraction used for normalization is estimated by examining all trajectories and calculating the number of time points at which molecules were classified as bound divided by the total number of time points. It is important to note that this estimated bound fraction is an underestimation of the true bound fraction as only particles bound for longer than N_{min} frames are counted (*see* **Note 13**).

3.4.3 Fitting the Survival Time Distribution to Determine the Number of Binding States and the Average Residence Time(s)

The survival time distribution is analogous to that of radioactive decay. Both plots give the probability of the number of molecules that remain either bound or radioactive as a function of time t. Thus the data can be fit with an exponential function (Fig 3d). If a single exponential fits the survival probability, then the data are well described by a single binding state with an off rate given by the decay constant (the inverse of this decay constant is the residence time). If a double exponential yields a significantly better fit, then the data are consistent with a two-binding-state model with two distinct off rates given by the decay constants.

The fit of the survival time distribution also allows estimation of the "true" bound fraction by extrapolating the survival probability plot back to time $t=0$. This should in principle account for molecules that were bound for shorter than N_{min} frames. However caution is advised when interpreting this extrapolated bound fraction, as this procedure nevertheless can still overlook a population of bound molecules that have residence times much shorter than N_{min} frames (*see* **Note 13**):

1. The TrackRecord software permits fitting the survival time distribution with both a single and a double exponential. These fits are shown superimposed on the plot of the measured survival probability.
2. The survival time distribution, together with the photobleaching correction and results of the fitting, can be copied to the clipboard to be stored in an Excel file.
3. The user can perform a statistical test (F-test, *see* **Note 14**) to determine if the double exponential provides a significantly better fit than the single exponential, thereby supporting a model for two distinct binding states instead of one.

4 Notes

1. SMT inside of thicker specimens can be done simply by conventional wide-field illumination, but the single-molecule signal is poor due to considerable out-of-focus fluorescence (Fig 1a). This background can be reduced by single plane illumination microscopy (SPIM), but this requires a nontrivial modification of the microscope, namely, the addition of a second objective orthogonal to the first [18, 19]. A simpler approach to reduce out-of-focus fluorescence is HILO illumination, which is obtained by tilting the excitation laser beam of a conventional wide-field microscope. Although HILO produces a thicker illuminated slice compared to SPIM (~2–3 μm vs. 1 μm), HILO can be set up on any microscope which is capable of TIRF illumination. This includes not only TIRF systems, such as the Olympus Cell TIRF, but also the new generation of commercial microscopes for localization-based super-resolution microscopy, such as the Zeiss Elyra P-1, Nikon N-Storm, and Leica GSD.
2. Posttranslational labeling systems are now available from several commercial vendors, including Promega (HaloTag) and New England Biolab (SNAP-tag and CLIP-tag). The principle behind these different labeling systems is very similar: Basically a fusion protein is constructed between a protein of interest and a nonfluorescent receptor. The fusion protein is then expressed in cells, and then the cells are incubated with a fluorescent molecule that has been modified to bind covalently to the expressed receptor. The user can choose between different ligands depending on the dye's spectral characteristics, photostability, and membrane permeability. For single-molecule studies inside the nucleus, we suggest the use of tetramethylrhodamine, as it photobleaches relatively slowly and is cell permeable. Further, the unbound TMR ligand can be easily washed out of the cells (Subheading 3.1 of the protocol). The advantages of HaloTag labeling are that tetramethylrhodamine is much more photostable than GFP or its derivatives, and so is better suited for long-term imaging of single molecules. In addition, the amount of labeled molecules can be easily regulated by adjusting the concentration of the tetramethylrhodamine ligand. This is important because if too many molecules are labeled, it becomes impossible to visualize single molecules (Fig 1b).

   Alternatively, it is also possible to load cells with a small number of molecules labeled with an organic dye if the protein of interest is first isolated, then tagged with the dye, and then finally microinjected into cells (Fig 1b), but this is labor intensive and requires numerous controls to ensure that the microinjected protein is still functional. The HaloTag approach achieves the same endpoint with considerably less effort. Note that the SNAP-tag (New England Biolabs) is a comparable approach to the HaloTag and could also be considered, although we do not have direct experience with it.

3. Multiple HaloTag expression vectors are commercially available differing in both the position of the tag (either N- or C-terminal) and the strength of the promoter driving the expression of the chimera. When expressing a protein in the presence of endogenous copies, we recommend selecting a weak promoter that will result in expression levels significantly lower than endogenous levels. For ectopic expression or expression in a null background (e.g., a knockdown of the endogenous protein), we recommend selecting a promoter that leads to physiological expression levels of the protein of interest.

   Controls are essential to demonstrate that the transfected fusion protein retains its functionality. ChIP can be performed to confirm that the fusion protein binds to DNA targets at comparable levels to the endogenous protein. If a functional GFP fusion is available, FRAP recoveries of the GFP and HaloTag proteins can be compared. After renormalization for bleach depth and observational photobleaching, the two FRAP curves should overlap. Transcriptional regulation can also be tested by measuring mRNA production from target genes after transfection of the fusion protein. Here it may be necessary to use a stronger promoter to drive fusion protein expression such that there is enough present to increase transcription over endogenous levels.

4. To estimate the extent of nonspecific labeling, it is advisable to repeat the TMR labeling protocol on cells that are not transfected with the HaloTag fusions. If the untransfected cells display fluorescent signal in the nucleus, it is advisable to modify the labeling protocol by further reducing the concentration of the fluorescent ligand or by increasing the number and the duration of the rinsing steps.

5. To achieve the best imaging conditions for detecting individual molecules, it is important to optimize the acquisition parameters. First of all, the exposure time should be chosen short enough to minimize the movement of molecules during the acquisition of each frame; otherwise the image of the molecules will appear smeared. In particular, we recommend choosing the exposure time so that the free molecules will move, on average, less than the size of a diffraction-limited spot (~ 250 nm). As nuclear proteins typically have diffusion coefficients of a few $\mu m^2/s$, exposure times of 5–10 ms are suggested.

A second important parameter is the acquisition rate which is the inverse of the time between two acquired images. This rate should be tuned depending on the expected residence time of the bound molecules on chromatin, while also minimizing the effects of photobleaching. For example, with our typical settings for single-molecule imaging (laser illumination synchronized to the camera exposure, power ~0.5 kW/cm^2, exposure time 10 ms), we are capable of observing individual TMR molecules for ~100 frames on average. If we expect to observe an average residence time on the timescale of seconds, we will therefore set the acquisition rate significantly slower than 100 Hz to ensure that the photobleaching timescale will be slower than the rate at which molecules dissociate from DNA. Note, the time interval between two frames (inverse of acquisition rate) should always be longer than the exposure time.

6. Achieving optimal collimation of the laser beam out of the objective is important to ensure that the field of view is uniformly illuminated with an identical incidence angle. Many commercial systems allow tuning the collimation of the illumination beam by changing the position of a relay lens in the excitation optical path. It is possible to adjust the collimation of the beam by observing the projection of the beam on the ceiling: First, make sure that the objective is focused on the focal plane that is normally used to image cells, and then adjust the laser collimation until the size of the beam projection on the ceiling is minimized. In this situation the laser is focused at the back focal plane of the objective lens and has minimal divergence while crossing the sample.
7. Higher inclinations of the excitation beam result in a better signal-to-background ratio. However, total internal reflection of the exciting beam at the interface between the coverslip and the cell medium poses a physical limit of ~60° to the maximum inclination of the incident beam. It is important to note that due to refraction the beam will be further inclined when crossing the glass–water interface, resulting in maximal effective inclinations of more than 75°.
8. Recent algorithms for tracking single molecules can achieve a relatively low error rate (typically smaller than 5 %). However, even this small error rate can introduce severe artifacts in the measurement of the duration of binding, since losing a particle for a few frames in the middle of a track will result in the erroneous assignment of two separate bound tracks, both shorter than the real binding event. For this reason, our software gives the opportunity to hand-check the automatically generated tracks, in order to further reduce the tracking error rate. However, as human intervention could introduce biases in the analysis, we recommend using some simple precaution such as

randomizing the data before proceeding with hand-checking of the tracks.

9. We expect that the distribution of displacements for H2B should be centered at short submicron displacements. If this is not the case, we recommend double checking the parameters used for the tracking followed by hand-checking the automatically generated tracks to identify further errors. Also, it is important that the user select proper tracking thresholds that allow obtaining the full distribution of H2B displacements. If the distribution of displacements is truncated (if it does not go to zero at long displacements), retrack the data allowing for longer frame-to-frame displacements.

10. r_{max} is calculated so that 99 % of the H2B displacements between consecutive frames fall below this value. This sets an upper value for the displacements of chromatin-bound molecules and will be used to identify the bound segments of the single-molecule tracks of the protein of interest. As we explained in the main text, we reject molecules that show small displacements for less than a certain number of frames N_{min}, because they might represent freely diffusing molecules that move slowly for a short time. If the diffusion coefficient of the free molecules D is known, N_{min} can be calculated as the minimum length of the track that gives a <1 % probability of having a free molecule displacing less than r_{max} for more than N_{min} consecutive frames. This probability can be calculated as [12]

$$P(r_{max}, N_{min}) = \left(1 - e^{-\frac{r_{max}^2}{4D\Delta t}}\right)^{N_{min}},$$

where Δt is the time between two consecutive images.

11. The survival distribution of bound molecules $S(t)$ allows comparison of different experiments more easily than a traditional histogram of residence times for two reasons: First, in contrast to a traditional histogram, $S(t)$ does not depend on arbitrary choices about the histogram binning. Second, when multiple populations of molecules with different residence times coexist, it is easier to directly visualize the fraction of molecules in each population because the cumulative histogram is fit with a multiexponential decay, with the amplitude of each exponential corresponding to the fraction size.

12. A single molecule can disappear suddenly because of photobleaching. For this reason photobleaching will shorten the survival distribution of bound molecules and if not corrected will result in shorter estimates of the residence time. The photobleaching of the fluorescent tag is also reflected by the decay in

the number of molecules detected at each frame: When the photobleaching is negligible, the number of molecules that leaves the detection volume is counterbalanced by those entering in the same volume, and the number of detected molecules over time fluctuates around a mean value. In contrast, significant photobleaching will result in a decay of the number of detected molecules over time. Our software calculates the photobleaching curve by plotting the number of detected molecules as a function of time. This curve is then fitted by a double exponential, to obtain the photobleaching trend, and finally the survival distribution of bound molecules is divided by the estimated photobleaching trend. To test the validity of our photobleaching correction, we recommend collecting data at different photobleaching rates (e.g., varying the power of the excitation laser) and then comparing the respective survival distributions of bound molecules, which should be found to overlap after photobleaching correction.

13. As we explained in the protocol, our software estimates an approximate bound fraction $C(t_{min})$ by counting the number of time points at which molecules are bound and dividing it by the total number of time points measured. However this estimate is necessarily an underestimation of the real bound fraction, as our approach ignores all the molecules that are bound for less than a time $t_{min} = \Delta t N_{min}$, where Δt is the time between two consecutive images. To correct for this problem, we suggest extrapolating from the survival probability using a (multi) exponential decay to obtain an estimate of the real bound fraction: First, the survival probability is normalized for the approximate bound fraction as determined above, so that $S(t_{min}) = C(t_{min})$. Then the (multi)exponential fit is extrapolated to $t=0$ yielding $S(0)$ which corresponds to the total bound fraction C_{eq}. It is worth noting that this approach is valid only if the sampled population of bound molecules (those with residence times $t > t_{min}$) is representative of the whole distribution of residence times or in other words if there is no separate population of bound molecules with residence time shorter than t_{min}. Unfortunately, this assumption is difficult to test without using other approaches or without performing additional experiments. In this context, we recently described another approach for the analysis of binding by SMT, which is based on the modeling of the whole distribution of single-molecule displacements with a reaction–diffusion model [12]. This alternative approach does not depend on the selection of thresholds to identify the bound molecules and allows measuring residence times as short as the time between images Δt, but it requires an a priori assignment of the number of distinct binding states and the number of different diffusing states.

14. In order to determine whether the survival probability of bound molecules is better described by a single or by a double exponential, it is possible to compare the resulting models with an *F*-test. In this test, the ratio between the reduced chi-square obtained by the two models is calculated and compared to the value of the *F*-distribution at 95 % confidence. If the ratio is higher, then the improvement obtained when fitting with a biexponential is statistically significant.

Acknowledgements

We are grateful to Drs. Tatiana Karpova and Tatsuya Morisaki for constructive feedback on the manuscript. DM is funded by a Marie Curie International Incoming Fellowship [Grant agreement: 27432].

References

1. Wu B, Piatkevich KD, Lionnet T, Singer RH, Verkhusha VV (2011) Modern fluorescent proteins and imaging technologies to study gene expression, nuclear localization, and dynamics. Curr Opin Cell Biol 23:310–317
2. Phair RD, Misteli T (2001) Kinetic modelling approaches to in vivo imaging. Nat Rev Mol Cell Biol 2:898–907
3. Li G-W, Elf J (2009) Single molecule approaches to transcription factor kinetics in living cells. FEBS Lett 583:3979–3983
4. Carrero G, McDonald D, Crawford E, de Vries G, Hendzel MJ (2003) Using FRAP and mathematical modeling to determine the in vivo kinetics of nuclear proteins. Methods 29: 14–28
5. van Royen ME, Farla P, Mattern KA, Geverts B, Trapman J, Houtsmuller AB (2009) Fluorescence recovery after photobleaching (FRAP) to study nuclear protein dynamics in living cells. Methods Mol Biol 464:363–385
6. Michelman-Ribeiro A, Mazza D, Rosales T, Stasevich TJ, Boukari H, Rishi V, Vinson C, Knutson JR, McNally JG (2009) Direct measurement of association and dissociation rates of DNA binding in live cells by fluorescence correlation spectroscopy. Biophys J 97: 337–346
7. Mueller F, Wach P, McNally JG (2008) Evidence for a common mode of transcription factor interaction with chromatin as revealed by improved quantitative fluorescence recovery after photobleaching. Biophys J 94: 3323–3339
8. Digman MA, Brown CM, Horwitz AR, Mantulin WW, Gratton E (2008) Paxillin dynamics measured during adhesion assembly and disassembly by correlation spectroscopy. Biophys J 94:2819–2831
9. Weidtkamp-Peters S, Weisshart K, Schmiedeberg L, Hemmerich P (2009) Fluorescence correlation spectroscopy to assess the mobility of nuclear proteins. Methods Mol Biol 464:321–341
10. Mueller F, Mazza D, Stasevich TJ, McNally JG (2010) FRAP and kinetic modeling in the analysis of nuclear protein dynamics: what do we really know? Curr Opin Cell Biol 22: 403–411
11. Speil J, Baumgart E, Siebrasse J-P, Veith R, Vinkemeier U, Kubitscheck U (2011) Activated STAT1 transcription factors conduct distinct saltatory movements in the cell nucleus. Biophys J 101:2592–2600
12. Mazza A, Abernathy N, Golob TM, McNally JG (2012) A benchmark for chromatin binding measurements in live cells. Nucleic Acids Res 40(15):e119
13. Selvin PR, Ha T (2007) Single-molecule techniques: a laboratory manual. Cold Spring Harbor Laboratory Press, Cold Spring Harbor
14. Thompson RE, Larson DR, Webb WW (2002) Precise nanometer localization analysis for individual fluorescent probes. Biophys J 82: 2775–2783
15. Grünwald D, Martin RM, Buschmann V, Bazett-Jones DP, Leonhardt H, Kubitscheck U, Cardoso MC (2008) Probing intranuclear environments at the single-molecule level. Biophys J 94:2847–2858
16. Grünwald D, Spottke B, Buschmann V, Kubitscheck U (2006) Intranuclear binding

kinetics and mobility of single native U1 snRNP particles in living cells. Mol Biol Cell 17:5017–5027

17. Elf J, Li G-W, Xie XS (2007) Probing transcription factor dynamics at the single-molecule level in a living cell. Science 316: 1191–1194
18. Ritter JG, Veith R, Siebrasse J-P, Kubitscheck U (2008) High-contrast single-particle tracking by selective focal plane illumination microscopy. Opt Express 16:7142–7152
19. Ritter JG, Veith R, Veenendaal A, Siebrasse JP, Kubitscheck U (2010) Light sheet microscopy for single molecule tracking in living tissue. PLoS One 5:e11639
20. Tokunaga M, Imamoto N, Sakata-Sogawa K (2008) Highly inclined thin illumination enables clear single-molecule imaging in cells. Nat Methods 5:159–161
21. Los GV, Wood K (2007) The HaloTag: a novel technology for cell imaging and protein analysis. Methods Mol Biol 356:195–208
22. Axelrod D (2001) Selective imaging of surface fluorescence with very high aperture microscope objectives. J Biomed Opt 6:6–13
23. Matsuoka S, Iijima M, Watanabe TM, Kuwayama H, Yanagida T, Devreotes PN, Ueda M (2006) Single-molecule analysis of chemoattractant-stimulated membrane recruitment of a PH-domain-containing protein. J Cell Sci 119:1071–1079
24. Crocker JC, Grier DG (1996) Methods of digital video microscopy for colloidal studies. J Colloid Interface Sci 179:298–310

Chapter 10

Single-Particle Tracking for Studying the Dynamic Properties of Genomic Regions in Live Cells

Irena Bronshtein Berger, Eldad Kepten, and Yuval Garini

Abstract

The appropriate functioning of living cells depends on a variety of dynamic processes that necessitate delicate motion, transportation, association, and disassociation in time and space. Different dynamic patterns such as directed motion, normal diffusion, and restricted diffusion take part at different length scales, and their identification serves as a tool for exploring biochemical processes. Here we describe single-particle tracking which is a powerful method that allows the characterization of dynamic processes on the single-molecule or single-particle level with nanometer spatial and sub-second temporal precision. In particular, we describe the cell preparation procedures, microscopy imaging, and image analysis processes for following telomere dynamics in living mammalian cells.

Key words Fluorescence microscopy, Single-particle tracing, Diffusion, Chromatin, Nuclear organization

1 Introduction

Fluorescence microscopy has become an essential tool in biology and the biomedical sciences [1], and the development of fluorescence proteins (FPs) for marking almost any protein of interest expanded the use of fluorescence microscopy, which is now a powerful tool for studying dynamic processes in live cells [2]. Based on the ability to track fluorescent molecules in live cells, a series of dynamic fluorescence microscopy techniques were developed including single-particle tracking (SPT) [3], fluorescence recovery after photobleaching (FRAP) [4], fluorescence correlation spectroscopy (FCS) [5], and continuous photobleaching (CP) [6].

The first step in performing SPT in live cells is to ensure that the entities of interest are fluorescently labeled. This is performed using transfected cells expressing a fluorescent protein (FP) fused to a protein that recognizes the target of interest.

SPT allows finding the position of a single molecule or particle as a function of time with high precision and constructing its time

Yaron Shav-Tal (ed.), *Imaging Gene Expression: Methods and Protocols*, Methods in Molecular Biology, vol. 1042, DOI 10.1007/978-1-62703-526-2_10, © Springer Science+Business Media, LLC 2013

trajectory. The trajectories allow characterizing the type of motion (such as Brownian, constrained, or anomalous) and conclude what is the mechanism that drives the particle's motion.

SPT methodology is based on measuring a series of images at given time intervals [7]. By using appropriate image analysis algorithms, the coordinates of each particle at each time t can be calculated. The x- and y-coordinates can be obtained, as an example, by fitting a two-dimensional (2D) Gaussian function to the fluorescence intensity profile of each particle in a given image. If the fluorescent spot is bright enough and has a high signal to noise ratio, the center position can be determined to be within ~10 nm precision [8]. Precision should not be confused with the spatial resolution limit, which is in the order of 200 nm due to optical diffraction limitations. The three-dimensional (3D) position at each time point can also be found if an appropriate 3D imaging method is used (such as confocal microscopy) and the trajectory $r(t) = [x(t), y(t), z(t)]$ of a moving particle can be found.

Once $r(t)$, the single-particle coordinates, are known as a function of time, the dynamics and more specifically the diffusion processes can be studied [9].

1.1 Random Walk (Brownian Motion)

This model describes motion that has the following features: The location after each step depends only on the location in the previous step, there is no preferred direction, and there is a constant rate at which the particle moves. As an outcome the random walker crosses its previous path many times, and the distance from the origin grows much slower than a directed motion. This type of random walk is the most famous one and leads to normal diffusion with the typical diffusion coefficient D. Since the average distance from the origin does not change with time, it is useful to study random walks with the mean square displacement (MSD) $\langle r^2(t)\rangle$, i.e., the quadratic length of the excursions from the origin (*see* Subheading 3).

The MSD can be studied in two ways—averaging on an ensemble of many particles at a certain time or averaging along a path of a single particle. For the ensemble-averaged MSD, one looks at the average square of the distance that a large group of N particles have traveled since the experiment started:

$$\langle r^2(t)\rangle = \frac{1}{N}\sum_{i=1}^{N}\left(r_i(t) - r_i(0)\right)^2 \quad (1)$$

For the time-averaged MSD, a single track of total time length T is analyzed by taking the average square of all steps with time difference t (we write a bar to designate time averaging):

$$\overline{r^2(t)} = \frac{1}{T-t}\sum_{\tau=0}^{T-t}\left(r(\tau+t) - r(\tau)\right)^2 \quad (2)$$

For a simple Brownian random walk, both MSDs follow a linear dependence with t,

$$\langle r^2(t) \rangle = \overline{r^2(\Delta)} = 2n_d Dt \tag{3}$$

where n_d is the spatial dimension, for example, for a 2D measurement $n_d = 2$. A diffusion process described by Eq. 3 is also called normal diffusion. Each averaging method has strengths and weaknesses, but the comparison is beyond the scope of this assay. As a rule of the thumb, when time averaging is used, only time lags corresponding to $\Delta \le \frac{1}{4}T$ should be calculated. It is possible to increase the accuracy of the time-averaged MSD by averaging over many individual particles, but one must ensure that the particles indeed have the same characteristics.

1.2 Other Random Walk Models: Anomalous Diffusion

Normal Brownian diffusion, where the MSD is linear in t, is only one possible form of diffusion. In cases where the MSD is not linear with t, the diffusion is called *anomalous diffusion*. Contrary to what the name implies, there are many mechanisms and forms to anomalous diffusion, and in recent years it has been shown to be quite abundant [10].

Perhaps the simplest form of anomalous diffusion is that of constrained diffusion in a closed volume. In this case, the particle diffuses normally in space until it reaches the confining volume. From this time onwards the MSD cannot grow and will stay constant. Hence the MSD will be linear at short times, changing to a constant at longer times.

Another large class of MSDs is diffusion for which the MSD grows as a power law in time [11]:

$$\langle r^2(t) \rangle = At^{\alpha} \tag{4}$$

When $0 < \alpha < 1$, the process is called subdiffusion, and when $1 < \alpha$, the process is called super-diffusion. These two classes have very important biological meaning. Subdiffusing particles explore space slower than normal diffusing particles but have a higher chance to interact with nearby targets. Super-diffusing particles on the other hand explore large distances and can reach faraway targets quickly.

Subdiffusion occurs when a particle does not diffuse in a simple homogeneous space. This is most simply interpreted in terms of interactions between the diffusing particles and other mobile or immobile structures. The details of the anomalous diffusion can be used, as an example, to quantify the crowdedness of the cytoplasm at the molecular scale [12] or to identify molecular interactions [13]. In addition, it has been shown that the dynamics of Cajal bodies and telomeres in the nucleus is described by anomalous diffusion [13, 14].

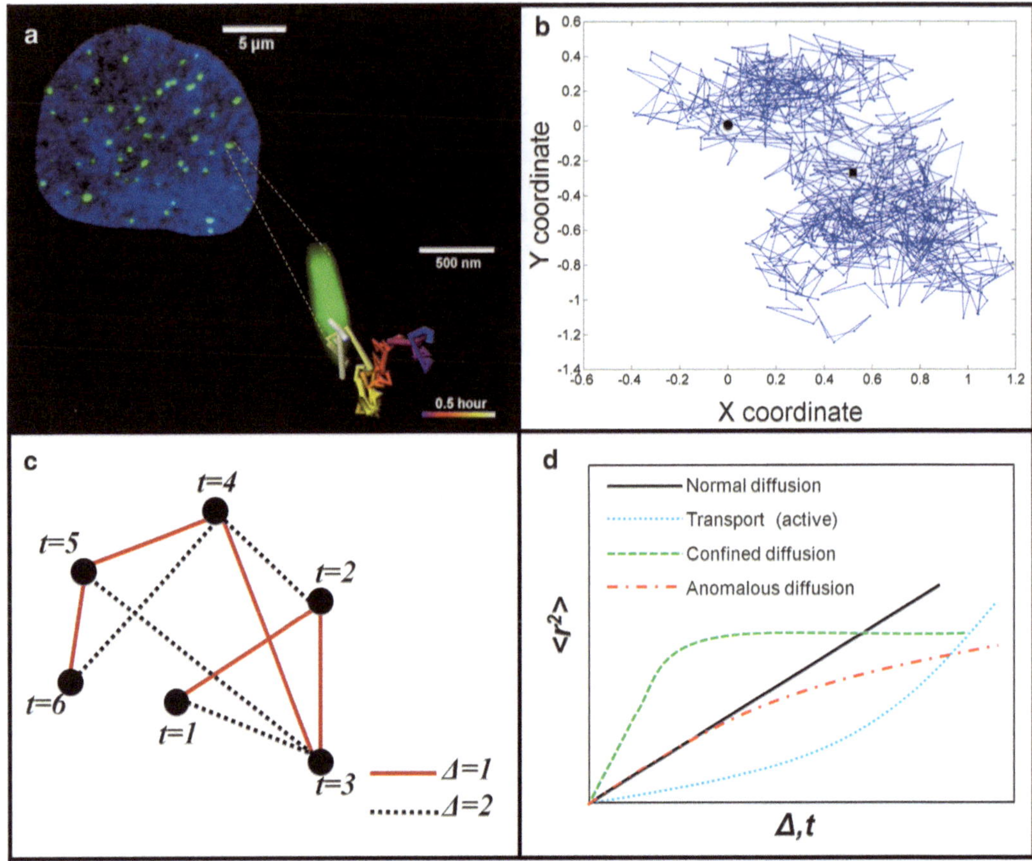

Fig. 1 (**a**) Typical image of the nucleus with chromatin marked in *blue* (Hoechst) and telomeres marked in *green* (TRF2-GFP). The *inset* shows the track of one telomere measured over 30 min. The coordinates were found in 3D and plotted using Imaris software. (**b**) Extracted time track for a single telomere along 1,000 time points shows the random character of motion. Each line segment is a single jump between two consecutive time points. Starting point is designated by a *black square* and end point by a *circle*. (**c**) To calculate the time-averaged MSD, one averages the distance traveled in a certain time lag. Examples for time lag of 1 time step (*red lines*) and 2 time steps (*dotted black lines*) are shown. (**d**) Different possible forms of the MSD. Four common shapes are shown, each telling a different story about the random motion and underlying process

Simple diffusion, constrained diffusion, or anomalous diffusion can be distinguished with the use of plots of the MSD versus Δt. For normal diffusion, the diffusion constant is given by the initial slope of the plot (Fig. 1).

The properties of the measured diffusion are important and have major significance for understanding the underlying biological mechanisms [15, 16]. Understanding these different classes falls beyond the scope of this manuscript, and it is covered by different manuscripts and reviews [17–19].

2 Materials

The following list describes the materials and equipment that are required to perform the SPT on telomeres in eukaryotic cells.

2.1 Cell Culture

1. Dulbecco's low- or high-glucose modified Eagle's medium (depends on type of the cells) (*see* **Notes 1** and **2**).
2. Fetal bovine serum (FBS).
3. 1 % of penicillin and streptomycin antibiotics.

2.2 Cell Culture for Imaging

1. Glass-bottomed tissue culture dishes—3.5 cm.
2. Dulbecco's low-glucose modified Eagle's medium with antibiotics and FBS.
3. Trypsin for detaching cells from tissue culture plate.

2.3 Labeling of Genomic Regions

1. For telomere labeling use the plasmid TRF1-GFP [20] and plasmid GFP-hCNPA for the labeling of centromeres [21]. The lacO–LacI system can be used for the labeling of a unique integrated gene locus using the RFP-LacI protein [22] (*see* **Note 3**).
2. Transfection reagent.

2.4 Time-Lapse Live-Cell Imaging

1. Incubator chamber (e.g., Tokai, Japan) with control for 5 % CO_2 level and temperature both in chamber and at the objective.
2. For 2D fast measurements: an inverted fluorescence microscope (such as an Olympus IX-81) with a CCD camera (such as the Andor DU-885 EMCCD).
3. For 2D and 3D measurements: a confocal microscope (such as an Olympus FV-1000 confocal setup attached to an IX-81 inverted microscope).
4. Objective UPLSAPO 60×, NA = 1.35.
5. The objective is heated to 37 °C.

2.5 Analysis

The analysis can be done with a variety of software packages. We used the following:

1. Imaris software (BitPlane Company).
2. ImageJ software.
3. MatLab software.

3 Methods

3.1 Cell Preparation for the Live-Cell Imaging

1. Split cells (*see* **Notes 1** and **2**) on the glass bottom dishes on the day before transfection. Confluence should be no more than 50–70 %.

2. Transfect cells with plasmids for labeling of entity of interest. The procedure should be done according to the transfection method and instructions of reagent supplier (*see* **Notes 3–5**).
3. Perform imaging experiments day after the transfection.
4. Before performing the imaging, check under the light microscope that the cells have regular shape, that they are attached to the culture dish, and that their density is not too high (*see* **Note 4**). Using a fluorescence microscope, check that the background is low enough (*see* **Note 5**). Sometimes, overexpression of the labeled protein results in a high background level. In this case, the transfection protocol should be modified.

3.2 Imaging

The imaging procedure depends on the total time range that is required for calculating the MSD. If only the fast time range is needed (from ~10 ms up to ~2 s), then the fast measurement mode explained below with the CCD is enough. If only the long time range (~20–500 s) is required, then it is enough to measure using the confocal microscope in 3D measurement mode (*see* **Notes 6–9**).

If, however, one wants to calculate the MSD in the largest possible time range, then it is necessary to combine both the fast CCD imaging and the slow confocal imaging. In this case, actually three different measurements are necessary: (1) fast 2D CCD imaging, (2) slow 3D confocal imaging, and (3) an intermediate time-range 2D confocal measurement that covers the time range in between 2D fast and 3D confocal. 2D confocal imaging can be performed in the time range from 0.4 s up to 100 s. Combination of 2D CCD, 2D confocal, and 3D confocal allows to calculate the MSD in a broad time range of approximately 0.01–500 s (*see* Fig. 2). The measurements' details are described below:

1. The CCD can be used for 2D fast acquisitions in a time range of 10^{-2} to 10^{1} s. The fluorescence signal of the fluorescent particle must be intense enough with respect to the background level, so that the exposure time can be fast enough to achieve the time resolution of ~10 ms. For achieving that, a fluorescent lamp with high intensity is required, and we found that a typical 100 W mercury lamp is enough. If the signal is not strong enough, a longer exposure time must be used, but this will limit the time range for the calculated MSD. The total time of the measurement depends on the bleaching rate and the cell motion. We found that a total time of 10 s is easily achievable without significant bleaching or cell motion. *See also* **Note 8**.
2. The confocal setup can be used to measure 2D data at approximately 2.5 Hz frame rate for 100 s. Longer data acquisition is difficult due to possible out-of-focus motion of the cell and rotation that may not be possible to correct (*see* **Notes 6** and **10**).
3. Long-time-range measurements have to be performed in 3D mode in order to correct for linear and rotational drift of the

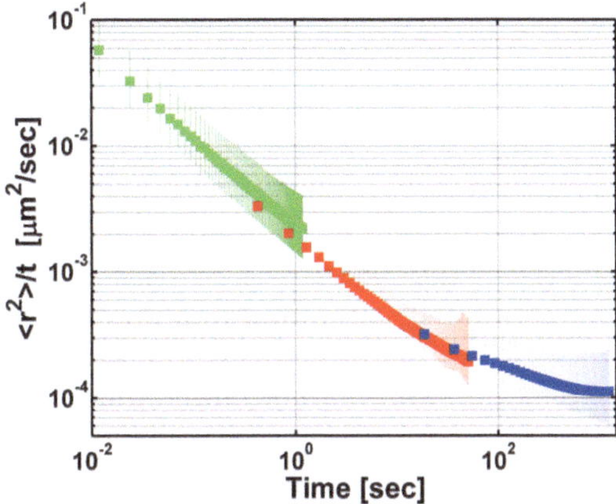

Fig. 2 Quantitative analysis of telomere motion in U2OS cells. Log–log plot of $\langle r^2 \rangle / t$ versus t showing the average MSD and standard deviation. The data were measured and analyzed in three different time ranges (2D CCD (*green square*), 2D confocal (*red square*), and 3D confocal (*blue square*)). The presented data represents an average of 120–1,000 telomeres that are measured in 25–50 cells

nucleus itself, which tends to move along all axes in such a long time range. For the 3D measurements use 35–50 Z-slices of 0.35 μm thicknesses, each with a frame size of 256×256 and pixel size of 0.1×0.1 μm. For these conditions with the FV-1000, the pixel size is achieved with a 63× objective and a zoom mode of 8 (*see* **Notes 11** and **12**). For a 30 min experiment, approximately 60 3D images can be measured every 30 s. The measurement time of a single 3D image takes about 20 s. The laser intensity should not be too strong in order prevent photobleaching and signal saturation (*see* **Notes 13** and **14**). The set of color filters in the confocal has to be adjusted according to the laser excitation wavelength and optimal emission band of the fluorescent molecule.

3.3 Data Analysis

The analysis consists of three major parts: (1) The images are analyzed in order to extract the position of each particle (e.g., telomere). (2) The time path coordinates for each particle are used for calculating the MSD. (3) The MSDs that are calculated for all the particles are used for calculating the average MSD. The MSD can now be used for drawing different possible conclusions. The steps below describe the detailed procedure:

1. The first step, as explained, requires to extract the position coordinates of each particle in each image of the measured sequence of images $r_i(t)$ where the index i represents the particle number. This can be performed by using the Imaris image analysis software package (BitPlane Company) or ImageJ, which both allow the identification of the telomeres as fluorescent spots and find the coordinates as a function of time for each telomere.

(a) Imaris software (BitPlane Company): First of all, upload images. Then, define the studied entity as a fluorescent spot and find the coordinates of the spot. Imaris allows to save the data in Microsoft Excel spreadsheet, where the columns are the coordinates of the spot as a function of time. This file can be used for further analysis in Matlab or other software packages.

(b) ImageJ software: Import the time serial images to ImageJ and use the MTrack plug-in to track fluorescent entity. The data is presented in a table that contains coordinates and intensity values.

(c) In principle, at the end of this step, we do not need the images anymore, and the important data now consists of a table showing the coordinates of each telomere as a function of time (*see* **Note 15**). It may be necessary to correct these coordinates to compensate for the cells motion and rotation; *see* **Note 16**.

2. Using the extracted coordinates $r_i(t)$, the MSD can be calculated using Matlab software, or even an Excel spreadsheet based on Eq. 1. For practical purposes, the MSD can be calculated for each coordinate separately (*x*, *y*, and *z*) by finding the average displacement between each two time points with a time interval $\Delta = n\tau$, where τ is the measurement time interval and *n* is an integer. It can also be calculated on the actual *r* vector (which is easier if one works with Matlab for example). Averaging is performed over the whole measured time,

$$\left\langle r^2(\Delta)\right\rangle = \frac{1}{N-n}\sum_{m=1}^{N-n}\left[r\left((m-1)\tau+\Delta\right)-r\left((m-1)\tau\right)\right]^2 \quad (5)$$

where *r* is the position vector of the particle at each time point and *N* is the total number of measured points.

3. In order to make the qualitative analysis of MSD graphs easier, it is common to present them on a log–log scale (*see* Fig. 2). When this is done, the slope of the presented graph is the anomalous coefficient α. If the diffusion is supposed to be anomalous, or if the MSD curve is clearly not a straight line, it is customary to plot the MSD divided by Δt as a function of Δt in a log–log scale (Fig. 2). In these plots, the slope is $\alpha - 1$. Normal diffusion in this plot has a zero slope, anomalous super-diffusion has a positive slope, and anomalous subdiffusion has a negative slope. *See* also **Note 17**.

4. When using the plot of the MSD divided by Δt as a function of Δt in a log–log scale and the diffusion is normal, then the crossing point of the graph with the *y*-axis in the log–log graph gives the diffusion constant multiplied by the dimension (i.e., 2 in the case of a 2D measurement, 3 in the case of 3D).

5. Another parameter that can be extracted from the path of each telomere is the area (or volume) that is scanned by the particle during the total experiment time. This is a good measure of the dynamics of the particle, and it can be compared between different cells or treatment types. If the coordinates of the particle are given, one can calculate the convex hull function by using Matlab software.
6. It is recommended to check the precision of the system; *see* **Notes 18–20**.

4 Notes

1. According to our experience U2OS, HeLa, MF (Mouse Fibroblast), and MEF (Mouse Embryonic Fibroblast) cell lines are excellent for live-cell imaging studies.
2. Cell lines that loosely attach to the surface and form clumps are sensitive to the small variations in temperature, CO_2 level, and photo damage. Altogether this raises difficulties during live-cell imaging.
3. The fluorescence signal of the target should be high enough with respect to the background in order to achieve high precision in finding the center position of the fluorescence spot. Strong fluorescence signal and low photobleaching are always preferred.
4. Cell confluence on the day of imaging should be no more the 90 %. We find that cell density affects the diffusion ability of the fluorescently labeled targets.
5. For the measurement, choose cells with high fluorescence spot signal and low background. This will improve the precision of finding the coordinates of the fluorescent spot and will allow performing long-term imaging with low light intensity.
6. Before starting a measurement, one should decide if a 2D or 3D SPT measurement is required. There is no simple rule for making the determination, but there are few aspects that help to make a decision:
   (a) If the required dynamics is rather fast, so that in sub-seconds the particle moves a significant distance, a rapid measurement is needed. In such cases it may not be possible to follow the particle in 3D, because measuring each 3D image takes a rather long time. Therefore, in such a case it may be necessary to use a 2D measurement that can reach a frame rate of ~100 images per second (depending on the CCD camera).
   (b) When the particles move rather slowly, the live cells may move by themselves quite a distance. Therefore, a 3D

measurement is preferred, so that the cell motion can be corrected in order to extract the motion of the particles themselves relative to the cell.

(c) In some cases, the nature of the motion is two dimensional, such as the motion on a membrane, and therefore a 2D measurement is preferred, although the particles should be selected so that they move in a membrane (plane) that is perpendicular to the optical axis of the microscope.

(d) Sometimes, it is important to measure the dynamics of entities in a large time range. For such a case, it is necessary to combine the data that is measured for both fast 2D measurements and slow 3D measurements. The measured data should not be necessarily for the same telomeres, as long as a large number of telomeres are measured. For example, it is possible to measure a fast 2D measurement of 400 telomeres that are collected from 40 different cells (approximately 10 telomeres for each cell) and then to measure another set of 400 telomeres in 3D. Even though these are not the same telomeres, assuming that there is no significant variation between the telomeres measured in 2D and 3D, these data can be combined to a single MSD curve that covers a large time range. In such a way, we achieved a time range of almost 6 orders of magnitude for the MSD [13].

7. In case that the long time range is the aim of the study, 2D and 3D confocal measurements have to be in overlapping time resolution between the different time ranges, so that the final data is continuous across the whole time range.

8. Measurement of a large number of time points in the trajectory improves the measurement precision of parameters describing the motion.

9. Specific algorithms that are written in Matlab software can be used to eliminate the rotational and translational movement of living cells, if the motion is significant.

10. The drift correction can be done at minimum on four fluorescence spots. So, once choosing the 2D slice, one has to be sure that there are at least four fluorescence spots with strong signals present in the image.

11. While doing 3D imaging, add a few spare slices above and below the nucleus, to ascertain that the whole nucleus is measured even if it moves out of focus during the long measurement.

12. Cells usually tend to move. Therefore, while performing long data acquisition in 3D, reduce the zoom level so that the nucleus will cover only ~70 % of the image area. This will enable to observe the whole cell in the measured image for a long time, even if it drifts.

13. Phototoxicity may result in enhanced cell motion. To limit the effect, it is better to lower the illumination laser intensity, while still maintaining a good signal to noise.
14. Check that there are no pixels in the image with signal saturation. The presence of such pixels can drastically affect the precision of finding the coordinates of the fluorescence spots.
15. During the image analysis, the algorithms may not be able to find the position of fluorescent-labeled site at each image. This leaves "holes" in the position along the time path and can lead to errors. To prevent that, use only tracks where the position was extracted for all time points.
16. Quantitative motion analysis of particles within a living cell requires the correction of cell motion, rotation, and morphological shape stability over data acquisition. The correction of cell/nucleus rotation can be done by applying a rotation-correction algorithm which requires at least trajectories from four different particles in the image. In case of a limited number of particles of interest, the double labeling can be performed. For example, there are only few Cajal bodies in each nucleus, while 20–40 telomeres can be easily identified by fluorescence labeling. Therefore, by labeling both the Cajal bodies and telomeres, the rotation correction can be done based on data from the telomeres and applied also to the Cajal bodies' coordinates.
17. In order to compare the data that were partially measured in 3D (for the longer times) and partially in 2D, the MSD curves must be extracted only for the 2D motion. This is easily done by taking the projection 3D images on the 2D plane. This simply means to use only the x–y data from the extracted 3D coordinates.
18. The stability of cell or nucleus morphological shape can be checked by extracting the shape of the cell boundaries over the time of data acquisition and comparing it. If large changes in the shape are not observed, it does not affect the measurement. The size of the changes should be compared to the motion of each fluorescently labeled entities of interest (e.g., telomeres) in between images, and "large" refers to this comparison.
19. It is important to estimate the measurement precision of the experiment. A possible efficient method can be performed by measuring the fluorescent spot motion in formaldehyde-fixed cells under identical conditions and repeating the same calculations of the MSD.
20. Few cases may limit the applicability of SPT:
    (a) When the particles diffuse too quickly (e.g., small proteins and peptides) so that they travel over a relatively large

range during the exposure time of a single image, it can result in a weak particle signal that cannot be distinguished over the background.

(b) When the particle density is too high so that individual particles often cross paths, it may become difficult to follow an individual particle and will lead to significant errors. Therefore, the targets that are selected and the protocol that is used for labeling must ensure a concentration that is low enough. It is also related to the typical dynamics. If the particles move very fast, even low particle concentrations may be problematic because they may cross paths in a short time.

Acknowledgements

This work was partially supported by the Israel Science Foundation grants No. 985/08, 1729/08, 1793/07, and 507/07 and the Wolfson Foundation grant 2008.

References

1. Lichtman JW, Conchello J-A (2005) Fluorescence microscopy. Nat Methods 2(12):910–919
2. Lippincott-Schwartz J, Patterson GH (2003) Development and use of fluorescent protein markers in living cells. Science 300(5616): 87–91. doi:10.1126/science.1082520
3. Saxton MJ (2008) Single-particle tracking: connecting the dots. Nat Methods 5(8):671–672
4. Mueller F, Mazza D, Stasevich TJ, McNally JG (2010) FRAP and kinetic modeling in the analysis of nuclear protein dynamics: what do we really know? Curr Opin Cell Biol 22: 403–411. doi:10.1016/j.ceb.2010.03.002
5. Schwille P, Haupts U, Maiti S, Webb WW (1999) Molecular dynamics in living cells observed by fluorescence correlation spectroscopy with one- and two-photon excitation. Biophys J 77:2251–2265. doi:10.1016/S0006-3495(99)77065-7
6. Weidemann T, Wachsmuth M, Ta K, Müller G, Waldeck W, Langowski J (2003) Counting nucleosomes in living cells with a combination of fluorescence correlation spectroscopy and confocal imaging. J Mol Biol 334:229–240. doi:10.1016/j.jmb.2003.08.063
7. Thompson RE, Larson DR, Webb WW (2002) Precise nanometer localization analysis for individual fluorescent probes. Biophys J 82(5):2775–2783
8. Yildiz A, Forkey JN, McKinney SA, Ha T, Goldman YE, Selvin PR (2003) Myosin V walks hand-over-hand: single fluorophore imaging with 1.5-nm localization. Science 300(5628):2061–2065. doi:10.1126/science.1084398
9. Saxton MJ, Jacobson K (1997) Single-particle tracking: applications to membrane dynamics. Annu Rev Biophys Biomol Struct 26(1):373–399. doi:10.1146/annurev.biophys.26.1.373
10. Barkai E, Garini Y, Metzler R (2012) Strange kinetics of single molecules in living cells. Physics Today 65(8):29–35
11. Metzler R, Klafter J (2004) The restaurant at the end of the random walk: recent developments in the description of anomalous transport by fractional dynamics. J Phys Math Gen 37(31):R161
12. Weiss M, Elsner M, Kartberg F, Nilsson T (2004) Anomalous subdiffusion is a measure for cytoplasmic crowding in living cells. Biophys J 87(5):3518–3524
13. Bronstein I, Israel Y, Kepten E, Mai S, Shav-Tal Y, Barkai E, Garini Y (2009) Transient anomalous diffusion of telomeres in the nucleus of mammalian cells. Phys Rev Lett 103:1–4. doi:10.1103/PhysRevLett.103.018102
14. Platani M, Goldberg I, Lamond AI, Swedlow JR (2002) Cajal body dynamics and association

with chromatin are ATP-dependent. Nat Cell Biol 4(7):502–508

15. Golding I, Cox EC (2004) RNA dynamics in live Escherichia coli cells. Proc Natl Acad Sci USA 101(31):11310–11315. doi:10.1073/pnas.0404443101
16. Banks DS, Fradin C (2005) Anomalous diffusion of proteins due to molecular crowding. Biophys J 89(5):2960–2971
17. Saxton MJ (1994) Anomalous diffusion due to obstacles: a Monte Carlo study. Biophys J 66(2 part 1):394–401
18. Metzler R, Klafter J (2000) The random walk's guide to anomalous diffusion: a fractional dynamics approach. Physics Reports 339(1):1–77
19. Saxton MJ (2007) A biological interpretation of transient anomalous subdiffusion. I. Qualitative model. Biophys J 92(4):1178–1191
20. Mattern KA, Swiggers SJJ, Nigg AL, Löwenberg B, Houtsmuller AB, Zijlmans JMJM (2004) Dynamics of protein binding to telomeres in living cells: implications for telomere structure and function. Mol Cell Biol 24:5587–5594
21. Molenaar C, Wiesmeijer K, Verwoerd NP, Khazen S, Eils R, Tanke HJ, Dirks RW (2003) Visualizing telomere dynamics in living mammalian cells using PNA probes. EMBO J 22:6631–6641
22. Brody Y, Neufeld N, Bieberstein N, Causse SZ, Böhnlein E-M, Neugebauer KM, Darzacq X, Shav-Tal Y (2011) The in vivo kinetics of RNA polymerase II elongation during co-transcriptional splicing. PLoS Biol 9:e1000573. doi:10.1371/journal.pbio.1000573

Chapter 11

Microscopic Analysis of Chromatin Localization and Dynamics in *C. elegans*

Christian Lanctôt and Peter Meister

Abstract

During development, the genome undergoes drastic reorganization within the nuclear space. To determine tridimensional genome folding, genome-wide techniques (damID/Hi-C) can be applied using cell populations, but these have to be calibrated using microscopy and single-cell analysis of gene positioning. Moreover, the dynamic behavior of chromatin has to be assessed on living samples. Combining fast stereotypic development with easy genetics and microscopy, the nematode *C. elegans* has become a model of choice in recent years to study changes in nuclear organization during cell fate acquisition. Here we present two complementary techniques to evaluate nuclear positioning of genes either by fluorescence in situ hybridization in fixed samples or in living worm embryos using the GFP-lacI/*lacO* chromatin-tagging system.

Key words Nuclear organization, *C. elegans*, Fluorescence in situ hybridization, Chromatin tagging, GFP-lacI/*lacO*, Microscopy

1 Introduction

Nuclear architecture has been studied in a wide range of models, including lower eukaryotes such as *S. cerevisiae* [1] and invertebrates such as *D. melanogaster*. In recent years, *Caenorhabditis elegans* has also been used, in particular to investigate the relationships between nuclear architecture and cellular differentiation [2–5]. Genome folding has been analyzed either in fixed worms and embryos using fluorescence in situ hybridization (FISH) or in live samples with the *lacO*/GFP-lacI system [2, 3, 5, 6]. Both approaches are highly complementary as each overcomes the limits of the other. The *lacO*/GFP-lacI system has the advantage of allowing the in vivo observation of chromatin position and dynamics. However, the creation of tagged chromatin loci is somehow tedious, and no more than a few loci can be observed simultaneously. In contrast, the main advantage of 3D FISH resides in the ability to detect multiple genomic segments in the same nucleus and, in combination with immunolabeling, to relate their positioning to various nuclear

Yaron Shav-Tal (ed.), *Imaging Gene Expression: Methods and Protocols*, Methods in Molecular Biology, vol. 1042, DOI 10.1007/978-1-62703-526-2_11, © Springer Science+Business Media, LLC 2013

compartments. Here we present recently developed methods and reagents for both techniques and discuss the caveats, advantages, and problems associated with them.

2 Materials

2.1 3D FISH

10× NT buffer: 0.5 M Tris–HCl pH 7.5, 50 mM $MgCl_2$, 0.5 mg/mL BSA. Make 0.1 mL aliquots and store at −20 °C.

100 mM β-mercaptoethanol: 3 μL of 14.3 M β-mercaptoethanol in 0.4 mL of deionized water. Make 0.1 mL aliquots and store at −20 °C.

10× dNTP mix (*see* **Note 1**): 0.5 mM dATP, 0.5 mM dGTP, 0.5 mM dCTP, 0.2 mM dTTP. Make 20 μL aliquots and store at −20 °C.

1 mM labeled dUTP (*see* **Note 2**).

2 U/μL DNAse I.

10 U/μL *E. coli* DNA polymerase I.

0.5 M EDTA.

16 °C water bath.

PCR primers, forward and reverse, 25 μM each.

2 mM dAGC: 2 mM dATP, 2 mM dGTP, 2 mM dCTP.

1.5 mM dTTP.

5 U/μL Taq DNA polymerase.

10× Taq buffer with $MgCl_2$.

PCR cycler.

1.2 % (w/v) agarose in electrophoresis buffer (e.g., TAE 1×).

95 % EtOH, 70 % EtOH (cold).

3 M sodium acetate pH 5.2.

100 % methanol (−20 °C, in a Coplin jar).

1× PBS.

Rubber cement glue.

Deionized formamide.

Coverslips: 18 × 18 mm #1 thickness; 15 × 15 mm #1 thickness; 22 × 22 mm #0 thickness.

Polylysine-treated microscope slides

Metal block in dry ice, −80 °C.

4 % formaldehyde in 1× PBS.

10 mg/mL yeast tRNA stock.

20× saline sodium citrate (SSC), 2× SSC, 0.2× SSC.

50 μg/mL RNAse A in 2× SSC (freshly diluted from a 10 mg/mL stock).

2×SSC/50 % formamide.

Primary antibodies against hapten and secondary antibodies (*see* **Note 3**).

0.5 % Triton X-100 (v/v) in 1× PBS (freshly made).

0.1 N and 0.01 N HCl.

2× hybridization buffer: 20 % (w/v) dextran sulfate in 4× SSC. Make 0.5 mL aliquots and store at −20 °C.

4× SSCT: 80 μL of Tween-20 (cut pipette tip) in 400 mL of 4× SSC.

4× SSCT-BSA: 2 g of bovine serum albumin (fraction V or purer) in 50 mL of 4× SSCT.

1 μg/mL DAPI in 2×SSC.

Vectashield mounting medium (Vector Labs).

2.2 Live Chromatin Imaging

Microscope slides.

Coverslips 20×20 mm, #1.

Agarose.

Laboratory tape.

M9 buffer (KH_2PO_4 3 g/L, Na_2HPO_4 6 g/L, NaCl 5 g/L, $MgSO_4$ 1 mM).

Fine forceps.

Scalpel blade.

Mouth pipette (*see* **Note 12**).

Hourglass.

Molecular biology reagents.

- Gateway® BP Clonase® II mix.
- Gateway® LR Clonase® II mix.
- Competent DH5α.
- pDONR221 *lacO/Cb unc-119* middle vector.
- pDONRP4P1R.
- pDONRP2RP3.
- MosSCI plasmids (*see* https://sites.google.com/site/jorgensenmossci/mossci-reagents) [7–9].

C. elegans strains:

- GW396/AV696 with fluorescent lacI expression: GW396 [*baf-1p* driven somatic expression], AV696 [*pie-1p*-driven germline and early embryo expression].

HT1593 *unc-119* mutant.

Mos insertion strain at locus of interest.

Imaging device:

Laser scanning confocal microscope/spinning disk confocal microscope.

3 Methods

3.1 3D FISH

The success of 3D FISH experiments depends crucially on the labeling of the DNA probe, i.e., the probe should be labeled uniformly. Probes should be highly specific to the target sequence and small in size to allow for efficient penetration in the sample. In this chapter, we describe the two methods of choice for probe synthesis, namely, nick translation and PCR labeling. Equally important are the pretreatment of the sample and the detection of labeled probes.

3.1.1 Probe Synthesis

The choice and synthesis of a target-specific probe is arguably the most crucial step in any 3D FISH experiment. In the case of single-copy targets, the probe should be of sufficient length to give a clear signal. For probes against the *C. elegans* genome, we routinely use fosmids that harbor 30–40 kb of sequence around the chosen genomic target. Individual fosmid clones are identified in WormBase (http://www.wormbase.org) and ordered from Source BioScience (http://www.lifesciences.sourcebioscience.com/clone-products/genomic-dna-clones/c-elegans-fosmid-library-.aspx). Fosmid DNA is labeled by nick translation. In the case of arrayed sequences (e.g., rDNA), much shorter genomic fragments (0.3–0.5 kb) can serve as template for probe synthesis, which is most conveniently done by PCR on genomic DNA. Both methods rely on the incorporation of a modified deoxynucleotide during DNA synthesis to generate labeled probes. The label is either a hapten that can be immunodetected after hybridization (digoxygenin [DIG] and dinitrophenol [DNP]) or a fluorophore. In our hands, biotinylated probes give unacceptably high background in *C. elegans* embryos, and, for this reason, we do not recommend using them.

Nick Translation

As the name implies, the nick translation technique relies on DNAse I to introduce single-strand breaks in the template DNA and on the combined 5′–3′ exonuclease and polymerase activity of DNA polymerase I to introduce labeled nucleotides starting from these pseudo priming sites. The purity of the starting material, the DNAse activity, and the incubation temperature are the most critical parameters:

1. Prepare the following mix, in this order.

DNA	1 μg
10× NT buffer	5 μL
100 mM β-mercaptoethanol	5 μL
10× dNTP mix	5 μL
1 mM labeled dUTP	1 μL (*see* **Note 4**), final concentration (20 μM)
ddH_2O	to 48.5 μL
Mix by pipetting up and down. Put tube on ice. Add the following	
DNAse I (diluted 1:15 in ddH_2O)	1 μL (*see* **Note 5**)
10 U/μL *E. coli* DNA polymerase I	0.5 μL

2. Mix well by pipetting up and down after the addition of enzymes. Do not make air bubbles. Incubate for 90 min at 16 °C.
3. Transfer the reaction on ice and analyze 1/10 of the reaction (5 μL) on a 1.2 % agarose gel (*see* **Note 6**). Include 1 kb and 100 bp ladders.
4. If the probe size is satisfactory (smear between 300 and 600 bp), proceed to the next step. If not, add 1 μL of diluted DNAse I to the reaction, return the tube at 16 °C, and incubate further for 30 min at 16 °C. Repeat **steps 3** and **4**.
5. Add 2 μL of 0.5 M EDTA to stop the reaction. Mix well.
6. *Optional.* Clean the labeled probe using ion-exchange mini spin columns (e.g., Qiagen or Zymo Research). Elute in 50 μL (*see* **Note 7**).
7. Store the labeled probe at −20 °C.

PCR Labeling

1. Prepare the following PCR reaction:

Template	200 ng of genomic DNA
Forward primer	1 μL
Reverse primer	1 μL
2 mM dAGC	5 μL
1.5 mM dT	4 μL (final concentration 120 μM)
1 mM labeled dUTP	3 μL (final concentration 60 μM)
10× Taq buffer	5 μL
ddH_2O	to 49.5 μL

2. Program the following PCR reaction: [94 °C, 3′]; [94 °C, 30″; melting temperature, 30″; 72 °C 45″] 25×; [72 °C, 10′]

3. Heat the PCR cycler to 94 °C. Incubate the probe synthesis mix for 1 min at this temperature. Add 0.5 μL of Taq DNA polymerase (2.5 U). Mix by pipetting.
4. Run the PCR program.
5. Analyze 1/20 of the PCR product (2.5 μL) on a 1.2 % agarose gel.
6. *Optional.* Clean the labeled probe using ion-exchange mini spin columns (e.g., Qiagen or Zymo Research). Elute in 50 μL (*see* **Note 7**).
7. Measure the probe concentration by spectrophotometry. Store at −20 °C.

3.1.2 3D FISH

A few points should be mentioned before detailing our 3D FISH protocol. First, since the goal of the technique is to determine the spatial localization of genomic segments, it is necessary to use fixation conditions that optimally preserve the native structure of the nucleus, i.e., to limit as much as possible the use of dehydrating agents. In the following protocol, a brief incubation in cold methanol is used to wet embryos after freeze cracking of the eggshell before fixing them in cold formaldehyde. Second, we have found it important to perform the hybridization in homemade glass chambers, so as not to compress unduly the relatively thick *C. elegans* samples (~20–30 μm). Third, as mentioned in Subheading 1, 3D FISH can be combined with the immunolabeling of cellular components. To do so, we perform the incubation with antibodies after the Triton X-100 permeabilization step, use a Cy3-labeled secondary antibody, and fix the immunocomplexes with formaldehyde before proceeding with the FISH protocol. Finally, it should be noted that unlike for hybridization to mammalian DNA, it is not necessary to quench the repetitive sequences when performing FISH on *C. elegans* DNA because these make up a much lower proportion of the genome in this organism (the most abundant repeat element in *C. elegans*, the 439 bp long CE000087, covers only 1.32 % of the genome [10]).

Preparation of the Probe Mix

1. Mix the following in an Eppendorf tube, *in this order*.

NT probe 1	*x* μL (typically corresponding to 400 ng of probe)
NT probe 2	*x* μL (typically corresponding to 400 ng of probe)
…	
10 mg/mL tRNA	1 μL (*see* **Note 8**)
Deionized water	to 100 μL
3 M NaOAC pH 5.2	10 μL
95 % EtOH (cold)	400 μL

2. Vortex vigorously. Incubate at −20 °C for at least 2 h.
3. Centrifuge at 13,000 × *g* for 20 min at 4 °C.
4. Decant supernatant. Wash pellet with 0.5 mL of cold 70 % EtOH. Air-dry on the bench for 2–3 min.
5. Add 20 μL of deionized formamide to the pellet (the final probe concentration is 20 ng/μL each). Incubate at 37 °C for 20 min. Pipette up and down to resuspend pellet.

Fixation and Pretreatment of Embryos

1. Isolate embryos according to standard protocols, e.g., by the "bleaching" method.
2. Deposit 12 μL of concentrated embryos in the center of a polylysine-treated slide. Cover gently with an 18 × 18 mm coverslip.
3. Working at low magnification under the stereomicroscope, adsorb liquid using a filter paper until the embryos are slightly compressed between the slide and the coverslip.
4. Place the slide on a metal block in dry ice. Incubate at −80 °C for at least 60 min.
5. Using a razor blade, pop up the coverslip (i.e., "cracking" the eggshell). Proceed immediately to next step.
6. Immediately fix embryos in cold 100 % methanol, 2 min at −20 °C.
7. Rinse slide(s) 1 min in 1× PBS at 4 °C (*see* **Note 9**).
8. Transfer the slide(s) to precooled (4 °C) 4 % formaldehyde in 1× PBS. Fix for 10 min at room temperature.
9. Wash slides in 1× PBS, two times for 2 min at room temperature.
10. Incubate for 5 min in 0.5 % Triton X-100 in 1× PBS.
11. Wash two times for 2 min in 1× PBS.
12. Rinse once in 0.01 N HCl.
13. Incubate for 2 min in 0.1 N HCl at room temperature.
14. Wash once with 1× PBS, 3 min at room temperature. Wash once with 2× SSC, 3 min at room temperature.
15. Treat with 50 μg/mL RNAse A in 2× SSC, 45 min at 37 °C. Perform this step by overlaying the sample with 0.5 mL of the RNAse solution and incubating in a humidified chamber.
16. Wash once with 2× SSC, 2 min at room temperature.
17. Incubate in 2× SSC/50 % formamide for at least 2 h at room temperature.

Hybridization and Post-hybridization Washes

1. Dilute the probe to a concentration of 2–5 ng/μL in 100 % deionized formamide. Add an equal volume of 2× hybridization buffer. 25 μL of probe solution is needed per sample (*see* **Note 10**).

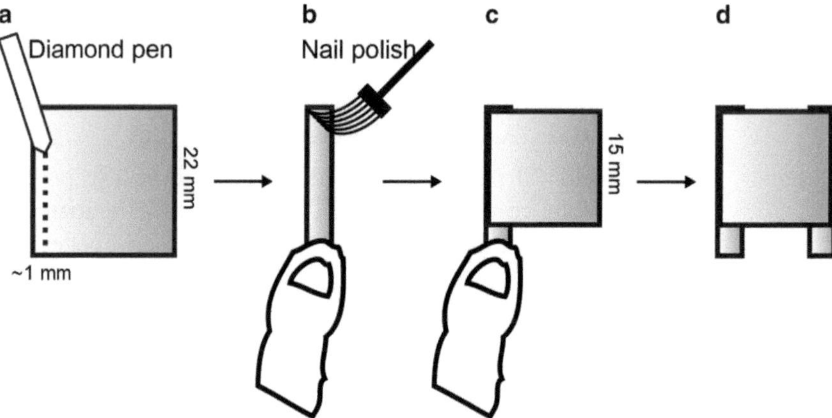

Fig. 1 Making homemade hybridization chambers. (**a**) Using a diamond pen, cut a ~1 mm strip from the side of a 22 mm × 22 mm coverslip (*thickness #0*). (**b**) Apply nail polish on one side of the glass strip. (**c**) Immediately glue the glass strip to the side of a 15 mm × 15 mm coverslip (thickness #1). (**d**) Leave to dry and gently break off the protruding pieces of glass

2. Mix well by pipetting and vortexing. Spin 20 s in a tabletop centrifuge. Keep the probe at room temperature.
3. Take a slide out of the 2× SSC/50 % formamide solution. Gently wipe off excess liquid on either side of the sample using a soft tissue.
4. Prepare a glass chamber: glue with nail polish two small strips of glass of thickness #0 (width of ~1 mm) to the opposite sides of a 15 × 15mm coverslip (*see* Fig. 1).
5. Pipette 25 μL of probe solution on a glass chamber.
6. Using forceps, gently deposit the chamber on the slide, probe facing towards the sample. Let the probe solution spread to the edges of the chamber.
7. Seal the chamber with rubber cement glue. Let dry completely at room temperature.
8. Repeat **steps 3–7** for each slide.
9. Pre-hybridize the slides overnight at 37 °C.
10. Denature the probe and target DNA simultaneously by placing the slide for 5 min on a heating block set at 76 °C.
11. Incubate at 37 °C for 2–3 days.
12. Fill four Coplin jars with 2× SSC. Preheat three of those at 37 °C. Fill two other Coplin jars with 0.2× SSC and preheat at 55 °C.
13. Gently remove the rubber cement glue around the chamber. Do not remove chamber.
14. Place the slide(s) in a Coplin jar filled with 2× SSC at room temperature. The chamber should fall off. If it does not, pull it gently alongside the slide.

15. Wash for 3 × 5 min in 2× SSC at 37 °C.
16. Wash for 2 × 5 min in 0.2× SSC at 55 °C.

Detection

1. If only fluorophore-labeled probes were used, go directly to **step 17**.
2. Set aside a 1 mL aliquot of 4× SSCT-BSA for the dilution of antibodies.
3. Transfer the slide to 4× SSCT-BSA. Incubate for 20 min at room temperature.
4. Dilute the primary antibodies in 4× SSCT-BSA. Use only those antibodies that are needed (*see* **Note 3**).
5. Remove the slide from the blocking solution. Gently wipe off excess liquid on either side of the sample.
6. Pipette 60 μL of primary antibody solution close to the sample. Overlay with an 18 × 18mm coverslip and make sure that the antibody solution covers the sample.
7. Incubate for 2 h at room temperature in a humid chamber, protected from light.
8. Place the slide in a Coplin jar filled with 4× SSCT at room temperature.
9. Wash for 3× 5 min in 4× SSCT at room temperature.
10. Dilute the secondary antibodies in 4× SSCT-BSA. Use only those antibodies that are needed.
11. Centrifuge the secondary antibody solution for 2 min at 13,000 × *g*. Transfer the supernatant to another microtube.
12. Remove a slide from the washing solution. Gently wipe off excess liquid on either side of the sample.
13. Pipette 60 μL of secondary antibody solution close to the sample. Overlay with an 18 × 18 mm and make sure that the antibody solution covers the sample.
14. Incubate for 1 h at room temperature in a humid chamber, protected from light.
15. Place the slide in a Coplin jar filled with 4× SSCT at room temperature.
16. Wash for 2× 5 min in 4× SSCT at room temperature.
17. Wash for 1× 5 min in 4× SSC at room temperature.
18. Gently pipette 100 μL of DAPI solution (final concentration is 1 μg/mL) on the sample. Incubate for 2 min at room temperature, protected from light.
19. Rinse in 2× SSC and mount with Vectashield. Seal with nail polish.

The protocols presented here have been applied successfully to perform 3D DNA FISH on embryos up to the 150- to 200-cell stage, after which time the signal decreases from the outside to the inside of the embryo due to poorer probe and antibody penetration. Samples are usually imaged on a laser scanning confocal microscope using a 63× plan apochromat oil objective (numerical aperture of 1.4), with optical sections taken at intervals of 300–700 nm. We have observed that some fosmid probes (about 1 in 5) give a signal that consists of several dots of variable intensities throughout the nucleus, especially when direct-labeled with fluorophores. The reason for this high background remains unknown. In any case, the simplest solution to this problem is to replace the bad fosmid with a neighboring or even overlapping one. Typical 3D DNA FISH results are shown in Fig. 2. In this experiment, we used the centrosomes as extranuclear reference points in order to be able to compare gene positioning in the same blastomere from

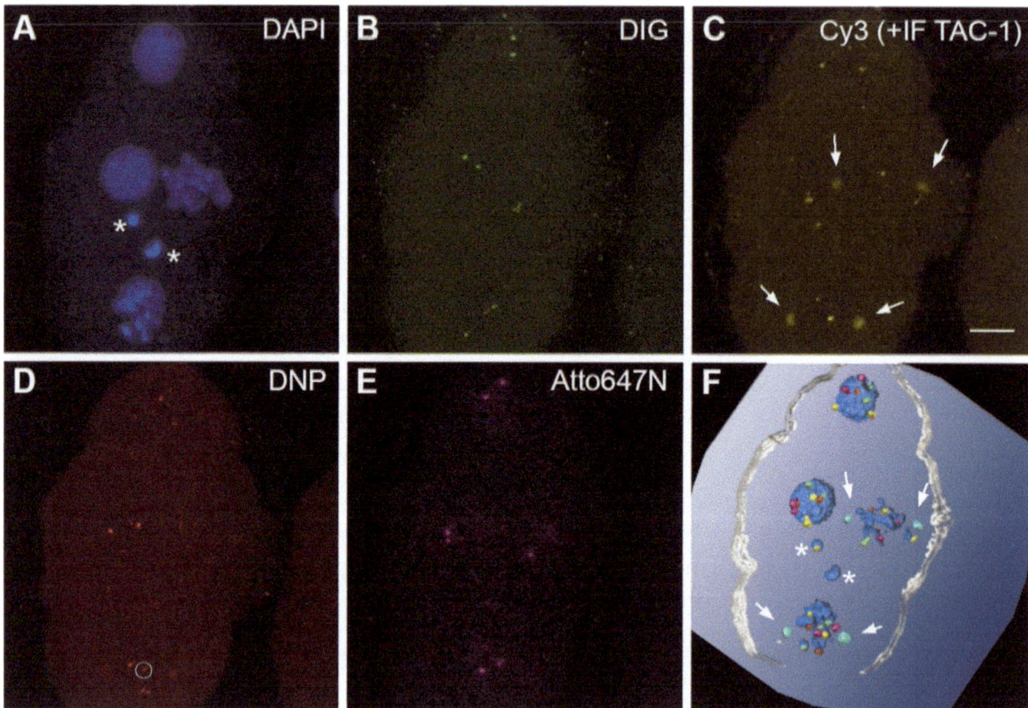

Fig. 2 Analysis of gene positioning in early *C. elegans* embryo using immuno-DNA FISH. Raw data (maximal projections, **a**–**e**) and 3D reconstruction (**f**) of a 4-cell stage embryo that was hybridized to the following probes. (**b**) Fosmid WRM0634bG05 (chr. X) labeled with DIG and detected with mouse anti-DIG and anti-mouse FITC. (**c**) Fosmid WRM0619aE04 (chr. V) labeled with Cy3, TAC-1 protein labeled with mouse anti-TAC-1 and anti-mouse Cy3 prior to FISH. (**d**) Fosmid WRM0628cE09 (chr. II) labeled with DNP and detected with rabbit anti-DNP and anti-rabbit Texas Red. (**e**) Fosmid WRM0623bE05 (chr. III) labeled with Atto647N. Remnants of polar bodies (*asterisks*) hybridize poorly due to high level of DNA condensation. Centrosomes are labeled weakly but clearly (*white arrows*). Bleeding of strong Cy3 signals in the Texas Red channel is sometimes observed (*circled* in (**d**))

different embryos. As expected, the background is somewhat higher when using hapten-labeled probe (DIG and DNP), due to the use of primary/secondary antibodies in the detection protocol, but signals are nonetheless clear.

3.2 Live Chromatin Imaging

In *C. elegans*, a number of laboratories have successfully validated the *lacO*/lacI system for tagging loci in vivo [2, 3, 5, 11–13]. The technique is a two-step process. The first step involves the expression of a fusion between a DNA-binding protein, a fluorescent protein, and a nuclear localization signal. Second, *lacO* repeats are inserted into the genome either as high-copy injected plasmid arrays, as low-copy bombarded transgenes, or as a single-copy insertion. Strains expressing different lacI fusions can be obtained from the laboratories which created them (*see* below). We therefore focus on the creation of *lacO* insertions, in particular at a given locus using a modified MosSCI transposon-mediated homologous recombination procedure [7]. In the last part, we describe how to mount live embryos for microscopy and the imaging setups adapted for *C. elegans*.

3.2.1 Strains for Expression of GFP-lacI/lacI-GFP

The DNA-binding Lac repressor is expressed as a fusion with green fluorescent protein (GFP). Expression levels of the protein have to be kept low, as overexpression elevates the background fluorescence, enhances nonspecific binding, and can cause slow-growth/sick animals. Two strains expressing GFP-lacI from integrated arrays have been published to date. The first one is based on injected integrated arrays, where GFP-lacI is expressed from the housekeeping promoter *baf-1* (GW396 [2]). The second is based on a bombarded construct in where GFP-lacI is transcriptionally regulated by a *pie-1* promoter (AV696 [5]). In the strain GW396, GFP-lacI is visible from about the 20-cell stage to adulthood due to germline silencing (note that this array has a *vit-5::GFP* intestinal marker expressed from the late L4/adulthood, which hinders GFP-lacI observation). In the AV696 strain, fluorescent signal decreases rapidly at the beginning of embryogenesis, but is useful to locate chromatin in the germline or in early embryos. The two types of expression constructs cannot be used in conjunction due to germline silencing in *trans* of the *pie-1* promoter construct [14]. Both constructs do not contain *lacO* sites in the plasmids used to create them and therefore do not create a GFP-lacI spot.

3.2.2 Creation of lacO-Tagged Transgenes and Insertion of lacO into the Genome

To allow visualization of a given transgene or locus, a number of binding sites for the bacterial repressor have to be integrated into the genome. Most of the *C. elegans* published arrays obtained by gonadal injection have been created using plasmids that contain a single *lacO* site (the 17 bp consensus sequence recognized by lacI; e.g., all the Fire library plasmids contain this sequence as a single copy). Due to the high-plasmid-copy number in the injected arrays (several hundreds of copies), they are readily detected by GFP-lacI and create a visible spot, usually at the nuclear periphery linked to their silencing [2, 4]. Smaller transgenes (in the range of 10–50 copies) bound by lacI can be created by co-bombarding plasmids of interest with *lacO* repeats (256 repeat, pSR1 [2, 15]). Although the *lacO* repeats do not get integrated each time, this method leads to about 50 % co-integration rate, with transgenes visible when GFP-lacI is expressed in *trans*. These low-copy transgenes are not subject to silencing and display usually a random localization in early embryos [2, 4]. Finally, single targeted genomic insertions of *lacO* arrays can be achieved using MosSCI and derivatives [6, 7]. As this method is new, a detailed description is given below.

As few as 24 lacI binding sites (*lacO*) are sufficient to allow the formation of a visible spot, although the signal-to-noise ratio depends on the expression level of the fluorescently tagged binding protein lacI. We use 256 *lacO* repeats (about 12 kb in size); however, the actual number of sites which get integrated is likely to be lower. Plasmids with binding site repeats are intrinsically recombinogenic in bacteria. It is therefore highly recommended to use recA strains (DH5α/XL1 blue) or recB recJ (SURE) for amplification/cloning. When defreezing these strains, it is good practice to isolate single colonies and test the length of *lacO* repeats as the repeat stretch has a tendency to shrink in size.

To insert *lacO* repeats at a site of interest, one needs a *Mos* insertion located next to the site of interest. Due to the resolution of optical microscopes, "near" means in the next 20–40 kb [16]. Many *Mos* insertions (about 13,000) have been created by the NemaGENETAG project (http://elegans.imbb.forth.gr/nemagenetag/). These are available as invalidated or validated insertions. All available insertions can be browsed at http://pbil.univ-lyon1.fr/segalat/data/index.php and ordered at http://ums3421.univ-lyon1.fr/spip.php?article14. This is the most cost-effective manner to create *lacO*-tagged loci, as compared to ZFN/TALEN methods (Transcription Activator-Like Effector Nuclease [17]), in which the nuclease is engineered to create a double-strand break at a given sequence, providing a template for homologous recombination. MosSCI however requires a *Mos* insertion to be available in the region of interest. However, given that 20 kb genomic distance cannot be resolved using light microscopy, several kilobases can separate the region of interest and the *Mos* insertion.

Insertion of the *lacO* repeats is achieved using Mos Single-Copy Insertion (MosSCI [7]) and requires homology between the insertion site and the plasmid that serves as a template for homologous recombination. To efficiently achieve the creation of templates with homology surrounding *lacO* repeats, we designed a triple-plasmid Gateway system with a middle *lacO/Cb unc-119* plasmid. This is used to create a plasmid where the *lacO* repeats are flanked with homology on both sides.

1. Prepare the *C. elegans* recipient strain by crossing it to an *unc-119(ed3)* mutant and selecting for unc worms homozygous for the *Mos* insertion of interest. At this point, it is good to cross in the transgene for the expression of GFP-lacI to have a visual confirmation of *lacO* integration later (*see* above for strain names).
2. Design primers to amplify left and right sequences located 5′ and 3′ of the *Mos* insertion site (Fig. 3a). 1.5 kb is sufficient for efficient recombination [18]. Primers are designed for Gateway BP cloning.

   Use following 5′ extensions for the following primers from a to d (in 5′ → 3′ direction):

   (a) GGGG ACA ACT TTG TAT AGA AAA GTT G—locus-specific sequence

   (b) GGGG AC TGC TTT TTT GTA CAA ACT TG—locus-specific sequence (reverse complement)

   (c) GGGG ACA GCT TTC TTG TAC AAA GTG G—locus-specific sequence

   (d) GGGG AC AAC TTT GTA TAA TAA AGT TG—locus-specific sequence (reverse complement)

   To minimize amplification errors, we use Phusion polymerase (NEB) according to the manufacturer's instruction with the following cycling conditions: [98 °C, 30″]; [98 °C, 10″; 45 °C, 30″; 72 °C 30″/kb] 5×; [98 °C, 10″; 50 °C, 30″; 72 °C, 30″/kb] 20×; 72 °C, 10′ 12 °C ∞.

   Test amplicon length and perform BP cloning according to the manufacturer's manual (Invitrogen), except that it is preferable to transform and plate at least half of the cloning reaction to get a sufficient number of colonies. In our hands, BP cloning has been highly efficient, with at least seven out of eight colonies tested positive for the desired insertion.
3. Perform the Gateway LR reaction with the 5′, middle [see note above about *lacO* repeats plasmid], and 3′ clones (Fig. 3b). Transform at least half of the LR reaction to get a sufficient number of colonies. Test the plasmids by PCR/miniprep.
4. Test the length of *lacO* sites on the destination vector. These should be about 10 kb in size to make a visible spot in vivo. Normally the *lacO* repeats length does not change drastically during the LR recombination, as recombination uses the Clonase mix and transformation is carried in DH5α strains.
5. Carry on MosSCI according to the original protocol [7, 8] (Fig. 3c). In our hands, integration at certain loci was more difficult to achieve than at others. The indirect, heat-shock procedure was preferred in such cases.

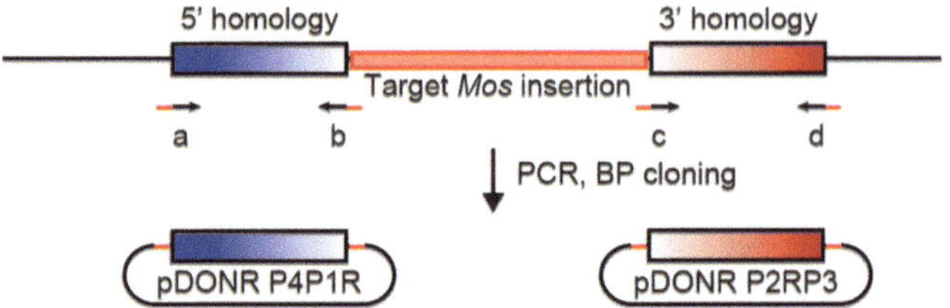

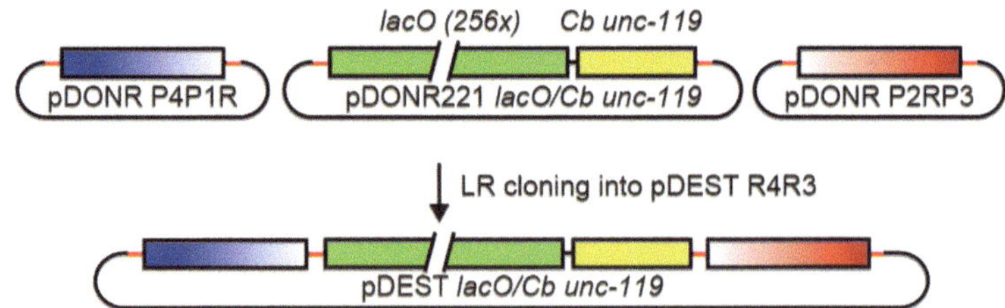

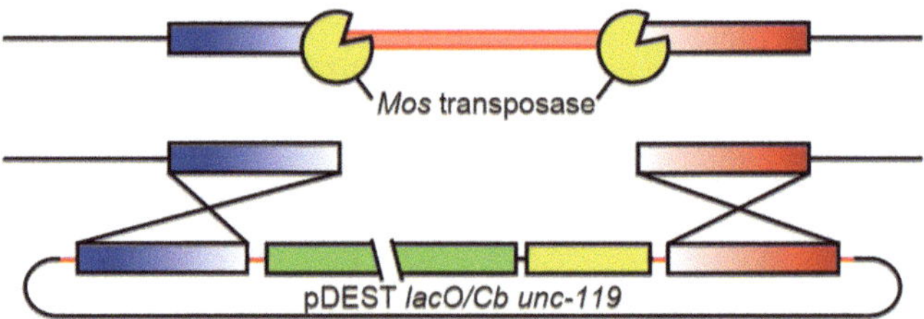

Fig. 3 Method for insertion of *lacO* repeats at a given *Mos* insertion site. (**a**) Regions flanking the *Mos* insertions are amplified by PCR using primers a, b, c, and d with Gateway-specific overhangs (*red*, see sequence in the text). Amplified homology stretch is recombined in the Gateway 5′ and 3′ vectors using BP. (**b**) The template for recombination is created by LR recombination of the vectors created in (**a**) with the middle *lacO/Cb unc-119*. (**c**) *Mos*-mediated single-copy integration is carried out in the strain carrying the *Mos* insertion of interest in an *unc-119(ed3)* background, which leads to the integration of the *lacO/Cb unc-119* template at the *Mos* locus. (**d**) Insertion is tested by PCR using primers located in the *lacO/Cb unc-119* insertion and outside of the homology stretch

6. Check insertion by PCR using primers outside of the homology region and in the inserted transgene (Fig. 3d). Also, check for the presence of a visible spot under the microscope. During the homologous recombination procedure, some *lacO* repeats might get lost, which therefore makes the GFP-lacI spot difficult or impossible to see.

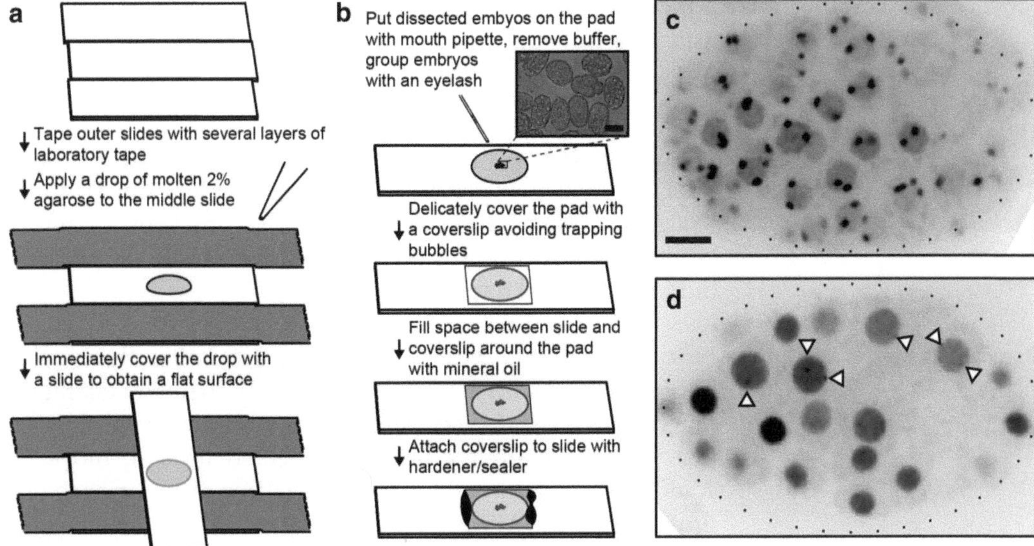

Fig. 4 Imaging live embryos carrying *lacO* inserts. (**a**) Preparation of the agarose pad, using three microscope slides. The outer two slides are taped to the bench using thick laboratory tape. Additional layers of tape will determine the thickness of the agarose pad. A drop of molten 2 % agarose is placed in the center of the middle slide and immediately covered with a perpendicular slide, creating a flat pad. (**b**) Embryos dissected in an hourglass are transferred to the agarose pad using a mouth pipette. Excess buffer is removed with the pipette, and embryos are grouped with an eyelash (bar = 20 μm). After covering the embryos with a coverslip, the space around the pad is filled with injection oil, and the coverslip is sealed to the slide using Vaseline/lanoline/paraffin or Vaseline alone. (**c**) Axial z projection of a stack of optical slices of an embryo expressing GFP-lacI from an integrated large array with *lacO* (GW76, *gwIs4[baf-1::GFP-lacI; myo-3::RFP]X* [2]). Two large spots corresponding to the homolog chromosomes are visible in each nucleus (bar = 5 μm). (**d**) Axial z projection of a stack of optical slices of an embryo expressing GFP-lacI from an integrated large array without *lacO* and an insertion of *lacO* repeats created using the technique described in Fig. 1 (GW392, *gwIs39[baf-1::GFP-lacI; vit-5::GFP]III gwSi13[lacO/Cb unc-119@ttTi9115]V* [4]). The GFP-lacI-expressing array is not visible as it carries no *lacO* sites. The single-copy insertion forms very small spots (*arrows*)

3.2.3 Preparing Worms and Embryos for Imaging

Preparing Agarose Pads for Microscopy

1. Prepare the molding bench (Fig. 4a).

   Put three microscope slides next to each other on a flat surface. Stick both outer slides with tape to the surface, putting thick laboratory tape on top of the slides. The thickness of the pad will depend on the thickness of the tape glued on the outer slides. Put two to three additional layers of tape on the slides to increase the thickness of the pad.

2. Prepare 2 % agarose in water and melt at 95 °C. Agarose can be aliquoted in 0.5–1 mL tubes for future use and stored at room temperature. Remelting of aliquoted agarose is achieved by putting the solution at 95 °C for 5 min followed by brief vortexing. The molten agarose can be kept longer at 70 °C instead of 95 °C. At high temperatures, agarose hydrolyzes and does not harden anymore. To image (moving) worms instead of embryos, add 0.5 μL 10 % NaN_3 to the agarose to

inhibit movement. Alternatively, anesthetizing worms with 400 mM ethanol has been recently shown to deliver similar results [19].

3. Prepare the agarose pad (Fig. 4a). Put a drop (50–100 μL) of molten agarose on a slide placed in the middle of the two taped slides. Immediately place a second slide, perpendicularly to the first one, creating a flat surface. This can be kept for several hours if the upper slide is not removed (otherwise the pad dries out and becomes unusable).

Preparing Embryos for Live Imaging

1. Pick gravid adults with fine forceps, and move them to an hourglass filled with M9 (*see* **Note 11**). Try to avoid taking too much bacteria as *E. coli* will stick to embryos once dissected and usually has high autofluorescence. If needed, move the adults with a platinum loop to bacteria-free plates prior to putting them in the M9.
2. Cut adults in two in the middle of the worm (approximately where the vulva is). Squeeze delicately the head and tail parts to release the embryos from the adult carcasses.
3. Using a mouth pipette, aspirate the embryos (*see* **Note 12**).
4. Transfer the embryos to the agarose pad (Fig. 4b). Make sure to hold the tip of the mouth pipette vertically during the transfer as failure to do so will lead to dripping of embryos out of the capillary. Once over the agarose pad, blow in the mouth pipette to expel the embryos, allow them to settle on the agarose (20 s), and aspirate delicately excess M9 with the mouth pipette. The M9 layer should be at a minimum height, so that embryos do not move anymore when an eyelash is put on the pad.
5. Group embryos together using an eyelash tip (a flexible eyelash fixed on a tip/toothpick with nail polish). This is very helpful for finding the embryos under the microscope and minimizes the search time at high magnification.
6. Put a coverslip on top of the pad/embryos, making sure to avoid trapping too many bubbles, in particular close to the embryos as they diffract light. The coverslip should be laid at an angle on one side of the pad and then slowly lowered on the embryos, pushing away bubbles.
7. For longer acquisition times, it is recommended to minimize evaporation (which would flatten the pad, damage embryos, and shift focus). We use injection oil (mineral oil, Sigma M5904) to hinder water exchange between the pad and the exterior. Injection oil has the advantage to allow oxygenation, and embryos can develop and hatch if kept overnight in such a chamber. Put enough oil to fill the gap between the slide and the coverslip around the agarose pad.

8. In order to avoid movement of embryos and sliding of the coverslip/pad, glue the coverslip to the slide using a 1:1:1 mixture of vaseline/lanoline/paraffin (on two sides only to avoid impairing gas exchanges). Alternatively, Vaseline can be used alone but is less solid. Both types of glue have to be melted.

3.2.4 Imaging Living Embryos

As for any live imaging, it is good practice to evaluate whether the development of the worms was not disturbed by the laser exposure. This is especially important when studying chromatin dynamics, as animals will experience repeated illumination over an extended period of time. Evaluating whether the movements observed correspond to normal behavior or to DNA repair-linked activities is important. An easy way to check this is to compare illuminated and control embryos at approximately the same stage at time of fluorescent imaging and follow their development over time using DIC imaging.

Imaging *lacO*-tagged chromatin from embryos is challenging: embryos are relatively thick (about 20 μm) and diffract light; the *lacO*/GFP-lacI spot can be very small and highly dynamic; moreover, the fluorescent signal is low as overexpression of lacI is deleterious for the animals and *C. elegans* embryos are highly sensitive to focused light. Several imaging systems were assayed and the best system for this application was a spinning disk confocal microscope. Confocality is essential as for single-copy *lacO* repeats, the observed structure is small (Fig. 4d); out-of-focus haze would make it invisible. Two systems can achieve confocal images: point scanning and spinning disk confocal microscopes. In the first system a focused laser beam is moving over the sample at high speed; emitted light is filtered for out-of-focus photons by passing through a pinhole and acquired by a photomultiplier. In these systems, maximum laser power as well as laser damage is concentrated on a single point. An alternative solution which minimizes excitation/damage is spinning disk confocal devices. In spinning microscopy, the laser beam is split in hundreds of focused sub-beams, rotating on the sample at high speed. The laser power on each point as well as the time spent by the laser on each point of the sample is minimized but repeated several hundred times per second. Emitted light is filtered through the pinholes, and photons are acquired using a highly sensitive CCD camera. The whole field of view can be acquired in as little as 30 ms; acquisition speed is limited by the frame reading rate of the camera. Using a correct combination of lenses/objective/camera, sampling will meet the Nyquist rate (about 90 nm/pixel). The thickness of the embryos (about 20 μm) and the small size of *lacO* transgenes make it very difficult to obtain images of entire embryos while keeping them alive (in the *z*-axis). Emitted light from objects far from the objective is diffracted by the structures above. Hence, when acquiring image stacks of embryos, we limit our acquisition to the 10 μm closer to the objective if embryos have to be imaged repeatedly for a longer period of time. Alternatively, if only one time point is to be acquired, we apply high laser power to be able to record light emitted from nuclei located far away from the objective.

Typical images acquired with a spinning disk confocal (TILL Photonics) of embryos carrying two different types of *lacO*-tagged chromatin are shown in Fig. 4. Large arrays obtained by gonadal injection and chromosomal integration (about 300–500 copies of the injected plasmids, 3–5 Mb in size) create large heterochromatic structures, which are located at the nuclear periphery in relation with their epigenetic silencing (Fig. 4c [4]). In Fig. 4d, an embryo with 256 *lacO* repeats integrated as a single copy using the technique described above is shown (Subheading 3.2.2). In each nucleus, the GFP-lacI spot is smaller and emerges from background nuclear fluorescence created by unbound GFP-lacI. Finally, it has to be noted that chromatin inside the nucleus is dynamic; hence precise positioning of the GFP-lacI/*lacO* spots necessitates a fast acquisition device. Typically, imaging should not be more than 100 ms per optical slice.

4 Notes

1. Due to the AT richness of the *C. elegans* genome (66 %), we increase the final concentration of unlabeled dTTP in the nick translation reaction to 20 μM. The molar ratio of labeled dUTP to unlabeled dTTP is 1:1 (compared to 2:1 or higher in most protocols).
2. Our preferred labeled nucleotides for a 4-color 3D DNA FISH experiment are DIG-dUTP, Cy3-dUTP, DNP-dUTP, and Atto647N-dUTP. We have also used successfully FITC-dUTP and Cy5-dUTP. However, labeling with Atto594-dUTP and Atto488-dUTP proved to be inefficient.
3. We use the following primary antibodies at a 1:200 dilution to detect hapten-labeled probes: mouse anti-digoxygenin (Jackson ImmunoResearch) and rabbit anti-DNP (Sigma). The secondary antibodies (1:400 dilution) are highly adsorbed FITC-conjugated goat anti-mouse and Texas Red-conjugated goat anti-rabbit, both from Jackson ImmunoResearch.
4. While the quantity of labeled dUTP given here works well in most cases, we found it had to be increased to 3 μL for efficient labeling with FITC-dUTP.
5. The optimal dilution of DNAse I has to be determined empirically and will result in the generation of fragments of 300–600 bp in length after a 90 min incubation. The activity is mainly influenced by the purity of the template DNA. Initial experiments should include a range of dilutions, from 1:500 to 1:10.

6. DIG-labeled fragments migrate slower than unlabeled ones of similar length. DNP- and Cy3-labeled probes are poorly stained by DNA-intercalating agents (e.g., ethidium bromide, SYBR Green). Note that the reaction can be stored at −20 °C at this point. If the reaction needs to be resumed after storage, a fresh aliquot of *E. coli* DNA polymerase I should be added to the reaction.
7. The cleaning of labeled probes (e.g., on ion-exchange spin columns) is optional. However, we have found that performing this step leads to an improvement in signal-to-noise ratio in experiments involving *C. elegans*, and we recommend doing it.
8. We use tRNA as a carrier during probe precipitation instead of ssDNA, which sticks to the eggshell and thus gives undesirable background upon DAPI staining.
9. We have found that slides have to be transferred from cold methanol (−20 °C) to cold PBS and then cold formaldehyde (4 °C) in order to minimize the loss of embryos. By doing so, we also found that subsequent incubations and washes could be performed in Coplin jars without significant loss of embryos from the slide.
10. The final probe concentration is 1–2.5 ng/μL. In our hands, this range of concentration gives excellent results. The *C. elegans* probes are repeat poor and can therefore be used at a lower concentration than the ones that are used in FISH experiments on mammalian samples.
11. In our hands, GFP fluorescence is always higher when worms have been grown at temperature above 20 °C. This is likely due to GFP variants optimized for mammalian expression at 37 °C, as we also observed that fluorescence is higher after heat shock at 34 °C.
12. Mouth pipettes are made using capillaries. (a) A 10 μL pipette is heated in the flame of an ethanol burner. (b) Once the glass is soft, remove it from the flame and quickly pull apart the ends. (c) Break the ends apart to create a pipette with an end with a diameter of ~40 μm. (d) Place the pipette in a mouth pipette aspirator (Sigma A5177 or building plans are available upon request).

Acknowledgements

We thank Darina Korčeková for expert help in developing 3D DNA FISH protocols, the Meister laboratory, Susan Gasser, and the Gasser laboratory for continuous support and helpful discussions. This work was funded in part by programs of the Charles University in Prague (UNCE 204022 and Prvouk/1LF/1) as well as by the Czech Science Foundation (grants P302/11/1262 and

P302/12/G157), the Swiss National Foundation (SNF assistant professor grant PP00P3_133744), and the Fondation Suisse pour le Recherche sur les Maladies Musculaires.

References

1. Taddei A, Schober H, Gasser SM (2010) The budding yeast nucleus. Cold Spring Harb Perspect Biol 2(8):000612, doi:cshperspect. a000612 [pii] 10.1101/cshperspect.a000612
2. Meister P, Towbin BD, Pike BL, Ponti A, Gasser SM (2010) The spatial dynamics of tissue-specific promoters during C. elegans development. Genes Dev 24(8):766–782, doi:24/8/766 [pii] 10.1101/gad.559610
3. Yuzyuk T, Fakhouri TH, Kiefer J, Mango SE (2009) The polycomb complex protein mes--2/E(z) promotes the transition from developmental plasticity to differentiation in C. elegans embryos. Dev Cell 16(5):699–710, doi:S1534-5807(09)00127-0 [pii] 10.1016/j.devcel. 2009.03.008
4. Towbin BD, Meister P, Pike BL, Gasser SM (2010) Repetitive transgenes in C. elegans accumulate heterochromatic marks and are sequestered at the nuclear envelope in a copy-number- and lamin-dependent manner. Cold Spring Harb Symp Quant Biol 75:555–565. doi:10.1101/sqb.2010.75.041
5. Yuen KW, Nabeshima K, Oegema K, Desai A (2011) Rapid de novo centromere formation occurs independently of heterochromatin protein 1 in C. elegans embryos. Curr Biol 21(21):1800–1807. doi:10.1016/j.cub. 2011.09.016
6. Towbin BD, Gonzalez-Aguilera C, Sack R, Gaidatzis D, Kalck V, Meister P, Askjaer P, Gasser SM (2012) Step-wise methylation of histone H3K9 positions heterochromatin at the nuclear periphery. Cell 150(5):934–947. doi:10.1016/j.cell.2012.06.051
7. Frokjaer-Jensen C, Davis MW, Hopkins CE, Newman BJ, Thummel JM, Olesen SP, Grunnet M, Jorgensen EM (2008) Single-copy insertion of transgenes in Caenorhabditis elegans. Nat Genet 40(11):1375–1383, doi:ng.248 [pii] 10.1038/ng.248
8. Frokjaer-Jensen C, Davis MW, Ailion M, Jorgensen EM (2012) Improved Mos1-mediated transgenesis in C. elegans. Nat Methods 9(2):117–118. doi:10.1038/nmeth. 1865
9. Zeiser E, Frokjaer-Jensen C, Jorgensen E, Ahringer J (2011) MosSCI and gateway compatible plasmid toolkit for constitutive and inducible expression of transgenes in the C. elegans germline. PLoS One 6(5):e20082
10. Stein LD, Bao Z, Blasiar D, Blumenthal T, Brent MR, Chen N, Chinwalla A, Clarke L, Clee C, Coghlan A, Coulson A, D'Eustachio P, Fitch DH, Fulton LA, Fulton RE, Griffiths-Jones S, Harris TW, Hillier LW, Kamath R, Kuwabara PE, Mardis ER, Marra MA, Miner TL, Minx P, Mullikin JC, Plumb RW, Rogers J, Schein JE, Sohrmann M, Spieth J, Stajich JE, Wei C, Willey D, Wilson RK, Durbin R, Waterston RH (2003) The genome sequence of Caenorhabditis briggsae: a platform for comparative genomics. PLoS Biol 1(2):E45. doi:10.1371/journal.pbio.0000045
11. Carmi I, Kopczynski JB, Meyer BJ (1998) The nuclear hormone receptor SEX-1 is an X-chromosome signal that determines nematode sex. Nature 396(6707):168–173
12. Kaltenbach LS, Updike DL, Mango SE (2005) Contribution of the amino and carboxyl termini for PHA-4/FoxA function in Caenorhabditis elegans. Dev Dyn 234(2):346–354. doi:10.1002/dvdy.20550
13. Gonzalez-Serricchio AS, Sternberg PW (2006) Visualization of C. elegans transgenic arrays by GFP. BMC Genet 7:36
14. Robert VJ, Sijen T, van Wolfswinkel J, Plasterk RH (2005) Chromatin and RNAi factors protect the C. elegans germline against repetitive sequences. Genes Dev 19(7):782–787
15. Rohner S, Gasser SM, Meister P (2008) Modules for cloning-free chromatin tagging in Saccharomyces cerevisiae. Yeast 25(3):235–239
16. Meister P, Gehlen L, Varela E, Kalck V, Gasser SM (2010) Visualizing yeast chromosomes and nuclear architecture. Methods Enzymol 470:537–569. doi:10.1016/ S0076-6879(10)70021-5
17. Wood AJ, Lo TW, Zeitler B, Pickle CS, Ralston EJ, Lee AH, Amora R, Miller JC, Leung E, Meng X, Zhang L, Rebar EJ, Gregory PD, Urnov FD, Meyer BJ (2011) Targeted genome editing across species using ZFNs and TALENs. Science 333(6040):307. doi:10.1126/science.1207773
18. Robert V, Bessereau JL (2007) Targeted engineering of the Caenorhabditis elegans genome following Mos1-triggered chromosomal breaks. EMBO J 26(1):170–183
19. Woock AE, Cecile JP (2011) Inhibiting C. elegans movement with ethanol for live microscopy imaging. Worm Breeder's Gazette 19(1):5

Chapter 12

Measuring the Dynamics of Chromatin Proteins During Differentiation

Arigela Harikumar and Eran Meshorer

Abstract

Chromatin-protein interactions are important in determining chromosome structure and function, thereby regulating gene expression patterns. Most chromatin associated proteins bind chromatin in a transient manner, with residence times on the order of a few seconds to minutes. This is especially pertinent in mouse embryonic stem cells (ESCs), where hyperdynamic binding of chromatin associated proteins to chromatin is thought to regulate genome plasticity. In order to quantitatively measure binding dynamics of such chromatin proteins in living cells, a combination of GFP-fusion proteins and photobleaching-based assays such as fluorescence recovery after photobleaching (FRAP) and fluorescence loss in photobleaching (FLIP) are advantageous over other existing biochemical assays, because they are applied in living cells at a single cell level. In this chapter we describe a detailed protocol for performing FRAP and FLIP assays for measuring structural chromatin protein dynamics such as Heterochromatin Protein 1 (HP1) and linker histone H1 in mouse ESCs and during ESC differentiation.

Key words GFP-fusion proteins, FRAP, FLIP, Chromatin binding proteins, Embryonic stem cells, Live imaging

1 Introduction

Gene expression, the fundamental cellular process, is chiefly regulated by chromatin structure and its interaction with chromatin binding proteins [1]. Apart from the core histones themselves, which are strongly associated with DNA, the interactions of proteins with chromatin are highly dynamic in nature [2]. This dynamicity is particularly high in embryonic stem cells (ESCs), where hyper-dynamic binding of structural chromatin proteins is considered to be partly responsible for genome and functional plasticity [3]. Therefore, studying the dynamic association between chromatin proteins and chromatin is very informative, especially since the methods that are used to decipher the binding kinetics are applied in single, living cells. This is even more pertinent when dealing with ESCs, since, as noted above, chromatin proteins,

Yaron Shav-Tal (ed.), *Imaging Gene Expression: Methods and Protocols*, Methods in Molecular Biology, vol. 1042, DOI 10.1007/978-1-62703-526-2_12, © Springer Science+Business Media, LLC 2013

including heterochromatin protein 1 (HP1), linker histone H1 and core histones, all have a higher turnover rate (dynamic exchange rates) on chromatin, compared to normal differentiated cells [1, 3]. To study the dynamic association of chromatin binding proteins with chromatin, photobleaching-based assays such as Fluorescence Recovery After Photobleaching (FRAP) and Fluorescence Loss In Photobleaching (FLIP) are normally used. These assays provide powerful spatial and temporal information regarding the protein's turnover rate (on/off rates), its mobile fraction, immobile fraction and whether or not it shuttles between compartments within cells or nuclei.

To carry out photobleaching-related experiments, the desired gene is fused to fluorescent proteins such as GFP, YFP, mCherry etc., and expressed in living cells [4]. Additionally, GFP gene tagging in BACs using recombineering [5] or CD-tagging of endogenous genes directly with GFP/YFP exons can also be used for this purpose [6, 7]. These methods are preferable, as the fusion protein is driven by an endogenous promoter, but usually are unavailable. It should be noted that a fluorescent tag could alter the normal function of the protein. Thus, to ensure precision, the protein should be monitored and compared to a normal protein.

In FRAP assays, after successful expression of the fusion protein of interest inside the cells, an intense focused laser beam is used to bleach a relatively small region of interest (ROI), where the laser wavelength is selected according to the fluorescent protein used for fusion. Time lapse imaging is then used to monitor the recovery of the bleached molecules, which are replaced with unbleached molecules following the irreversible photobleaching event. FLIP is a complementary technique that measures signal decay rather than recovery. It not only determines the protein's mobility, but also by bleaching one compartment and measuring the fluorescence decay in another determines whether the protein shuttles between different compartments [8–12].

For a successful FRAP experiment, certain conditions must be fulfilled. First, the fluorescent signal to be bleached must be clearly visible over any background. Second, the photo-bleaching should be fast enough so that recovery during the bleaching itself is negligible, allowing good temporal resolution and enabling to calculate the recovery half-time. To achieve this, the laser used for photobleaching must be powerful enough (usually in the range of 50 mW). A laser with sufficient power ensures high spatial resolution and allows the bleaching of small, submicron, cellular areas. Finally, it is essential to maintain the homeostasis of the cells. This can be achieved by mounting an environmental chamber on the confocal microscope. To minimize photodamage due to prolonged imaging, a spinning disk confocal microscope equipped with a photobleaching module is recommended.

In the present chapter we describe the basic requirements and methodology to perform FRAP and FLIP assays, particularly in mouse ESCs, for the study of chromatin protein dynamics in undifferentiated cells and during ESC differentiation.

2 Materials

2.1 Cloning Your Gene of Interest (GOI) into a Vector Expressing a Fluorescent Protein

1. Plasmid: For N-terminal fusions (the GFP will be at the C-terminal end), use pEGFP-N1 plasmid (*see* **Note 1**) and for C-terminal fusions (the GFP will be at the N-terminal end) use pEGFP-C1 plasmid (*see* **Note 2**) (all available from Clontech).
2. Restriction enzymes for which a recognition sequence is present at the multiple cloning site (i.e. BglII, XhoI, HindIII, BamHI).
3. 5 units/μl T4 DNA ligase.
4. TOP 10 chemically competent *E. coli* or equivalent for transformation.

2.2 Cell Culture and Plasmid Transfection for Transient Expression of the GFP Fusion Protein

1. Dulbecco's Modified Eagle's Medium (DMEM).
2. ESC-grade fetal bovine serum (FBS).
3. 1 mM sodium pyruvate.
4. 0.1 mM nonessential amino acids.
5. 0.1 mM β-mercaptoethanol.
6. 1,000 U/ml leukemia inhibitory factor (LIF).
7. 0.25 % trypsin for detaching cells from tissue culture plate.
8. Phosphate-buffered solution (PBS) for washing cells.
9. GFP fusion plasmid DNA.
10. Opti-MEM reduced serum free medium.
11. TransIT-LT1 transfection reagent (Mirus).
12. R1 mouse embryonic stem cells (or equivalent) and mouse embryonic fibroblasts.
13. 0.5 μM All-trans Retinoic Acid (ATRA).

2.3 Time-Lapse Live-Cell Imaging of Fluorescent Protein Dynamics in Living Cells

1. Confocal microscope of choice (laser scanning or spinning disk confocal) equipped with a 60× NA = 1.4 (or higher) oil objective. We recommend a Revolution spinning disk confocal system (Andor) with the Yokogawa CSU-X spinning disk head. This system is equipped with a dual capacity to photobleach with a point scanning system and to switch back the laser light to collect images using the spinning disk. GFP, YFP and mCherry are the fluorescent proteins generally used in FRAP/FLIP experiments. A ~488 nm laser is required for GFP and

YFP, and a ~560 nm laser is required for mCherry. Solid state lasers are preferable but gas-medium lasers are also commonly used. Since experiments are conducted on living cells, an environmental microscope chamber is required to maintain humidity, CO_2 and temperature (*see* **Note 3**). Photobleaching is performed under maximal laser intensity (100 %) and imaging under minimal intensity that allows proper signal to noise ratio (around 10 %) to prevent phototoxicity and photobleaching during imaging.

2. 8-well μ-Slides (ibidi; Munich, Germany) or chambered cover glasses (Lab-Tek; Rochester, NY) or glass-bottom culture dishes (MatTek; Ashland, MA).
3. Image analysis software such as: Imaris (http://www.bitplane.com/), Metamorph (Molecular Devices, Downingtown, PA), or ImageJ (NIH, Bethesda, MD; http://rsb.info.nih.gov/ij/).

3 Methods

3.1 Construction of the Expression Vector Containing Your Gene of Interest (GOI) Fused with a Fluorescent Protein

Clone your gene of interest (GOI) into the expression vector selected (pEGFP-C1 or pEGFP-N1, see above for details). To express a fluorescently tagged protein, the coding region of the gene should be inserted in frame with a vector containing a fluorescent protein of choice such as GFP, YFP or mCherry etc. (*see* **Notes 2** and **3**). The presence of an antibiotic resistance gene in the vector will help in generating stable cell lines.

3.2 Transfecting ES Cells

1. Coat the 8-well μ-Slides with gelatin and let stand for 15–30 min.
2. Aspirate gelatin and seed 22,000 MEFs/well in 250 μl total volume of DMEM (supplemented with 10 % FBS). Grow the cells in a tissue culture incubator (37 °C, 5 % CO_2). MEFs are required to maintain ES cell pluripotency at the undifferentiated state.
3. After 6 h, aspirate the media, and seed 15,000 R1 ES cells/well in each MEF-coated well in 250 μl ESC media (supplemented with 10 % ESC-grade FBS, 1 mM sodium pyruvate, 0.1 mM nonessential amino acids, 0.1 mM β-mercaptoethanol, and 1,000 U/ml LIF), to reach 50 % confluence the following day.
4. After 18 h, replace the ESC media with 250 μl/well of fresh ESC media.
5. In a 1.5-ml Eppendorf tube, dilute 10 μl TransIT-LT1 transfection reagent (Mirus) in 100 μl reduced serum free media (Opti-MEM). Mix gently with pipette and incubate at room temperature for 5–20 min.

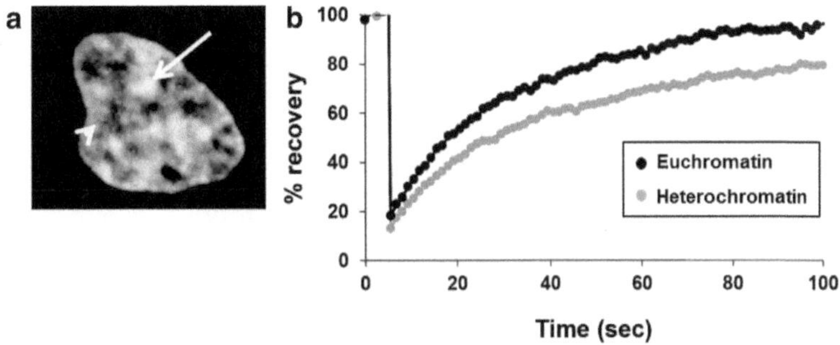

Fig. 1 (**a**) A representative embryonic stem cell expressing H1e-GFP fusion protein, where heterochromatin (*arrow*) and euchromatin (*arrowhead*) are readily distinguishable. (**b**) A representative FRAP curve of H1e-GFP fusion protein in embryonic stem cells, where heterochromatin (*black*) and euchromatin (*grey*) were analyzed separately. Shown is an average of 20 cells

6. Add 1.5 μg GFP fusion plasmid DNA to the diluted transIT-LT1 reagent and mix by gentle pipetting and incubate at room temperature for 15–30 min.
7. Add 13.5 μl of transfection complexes to each well containing cells and media. Swirl the 8-well μ-Slides to ensure even dispersal.
8. After 24 h, replace the old ESC media with 250 μl of fresh ESC media.
9. For differentiation, treat ESCs with 1.0 μM Retinoic Acid for 2–5 days in ESC media without LIF, and repeat experiments in differentiating cells (*see* **Note 4**).

3.3 Performing FRAP and FLIP

1. Mount the 8 well μ-slide inside the microscope. Find the cells expressing the fluorescently labeled protein and adjust the imaging conditions such that clear fluorescence signal can be detected with minimal background (Fig. 1a). Ensure proper localization of the fluorescent proteins. When expression levels are too high, “spilling” of the protein into additional cellular compartments, such as the nucleolus, can occur. Long-term exposure to fluorescent lamps and to high intensity lasers should be avoided to reduce photobleaching of the fluorescent proteins and photodamage to the cells in general.
2. Experimental parameters required for time-lapse imaging including exposure time, number of fields, and the number of repeats should be adjusted as needed. For highly mobile proteins, i.e. HP1, use 3–4 frames/s for 20–40 s. For proteins with intermediate mobility, i.e. H1, use 1–2 frames/s for around a minute. For immobile proteins, i.e. core histones, use 6–12 frames/min for 5–10 min. ESCs are notoriously difficult

to image for extended periods due to motion. Light intensity and exposure time used in the experimental procedure should be as low as possible to prevent photobleaching, but sufficient to distinguish signal over background. Do not use frame averaging. The quality of the image itself is secondary.

3. For FRAP, set the following protocol: collect 3–5 frames before photobleaching, then photobleach, then collect an additional 60–120 images at the desired intervals according to the protein investigated. Bleach spot should be minimal. Appropriate photobleaching reveals a black hole in the GFP fluorescence signal that slowly "recovers" or fills up with unbleached fluorescent proteins from the surroundings. Repeat for 20–30 cells in three different experiments.
4. For FLIP, set the following protocol: Collect 3–5 frames before bleaching, then repeatedly bleach the same spot while simultaneously collecting images. In FLIP experiments, the decay in fluorescence is slower than the recovery in fluorescence in FRAP experiments and the interval between images should be longer. For proteins with intermediate mobility, bleach and collect image every 5 s, and for highly mobile proteins bleach and collect image every 1–2 s. Repeat for 20–30 cells in three different experiments (*see* **Note 5**).

3.4 FRAP and FLIP Data Analysis

1. FRAP. In each one of the FRAP frames collected (before and after bleaching), measure the fluorescence intensity in the bleached region (ROI_b), in a non-bleached region within the same cell (ROI_{nb}) and in a background region outside the cell (ROI_{bg}). For each frame, use the following formula:

$$F = \left(ROI_b - ROI_{bg}\right) / \left(ROI_{nb} - ROI_{bg}\right) / \left({}_{pb}ROI_b - pbROI_{bg}\right)$$

$/\left({}_{pb}ROI_{nb} - {}_{pb}ROI_{bg}\right)$,where pb denotes "pre-bleached". The approximate value must be 1 for the first few pre-bleached images. The first image after the bleach gives the value of the bleach depth which is subtracted from 1. The experiment is repeated for every cell for 20–30 cells (*see* **Note 6**). A representative FRAP curve (an average of ten cells) is shown in Fig. 1b.

2. FLIP. Here, the fluorescence intensity is a measure of the non-bleached nuclear area (ROI_{nb}) and the background area (ROI_{bg}). The calculation of the FLIP data is similar to the FRAP curve, where only the analyzed area (ROI_{nb}) must be different compared to the actual bleached area and is not used in the calculations $(ROI_{nb} - ROI_{bg})/({}_{pb}ROI_{nb} - {}_{pb}ROI_{bg})$. Here, a neighboring cell (ROI_n) can be used for normalization according to the formula:

$$F = \left(ROI_{nb} - ROI_{bg}\right) / \left(ROI_n - ROI_{bg}\right) / \left({}_{pb}ROI_{nb} - {}_{pb}ROI_{bg}\right)$$

$$/\left({}_{pb}ROI_n - {}_{pb}ROI_{bg}\right).$$

Using curve fitting, one can estimate with good proximity the mobile fraction, the immobile fraction and the half-maximum. The distance between the bleach depth and the recovered signal on the *Y*-axis refers to the mobile fraction, especially when the kinetics reaches a plateau. The immobile fraction is the distance between the recovered signal and the pre-bleach (100 %) signal on the *Y*-axis. The immobile fraction represents the molecules that are tightly bound and are not exchanged within the timescale of the experiment.

Mathematical models are available to fit the data. If only a single population of molecules is present, all sharing the same mobility parameters, the data can fit a single exponential equation, where t is time, A is the mobile fraction, $1 - A$ is the immobile fraction and k_{off} is the dissociation constant. In most cases, we find that the data fits a two component exponential equation $I(t) = A_1(1 - e^{-K_{1\text{off}} \cdot t}) + A_2(1 - e^{-K_{2\text{off}} \cdot t})$, indicating two mobile populations with different kinetics.

4 Notes

1. Use pEGFP-N1, pEGFP-N2 or pEGFP-N3 according to the available reading frame.
2. Use pEGFP-C1, pEGFP-C2 or pEGFP-C3 according to the available reading frame.
3. If only short-term experiments are planned, an air blower which maintains a constant temperature is also optional. In this case, it is preferable to use a pH-maintaining buffer.
4. Many different differentiation protocols are available. The advantage of using retinoic acid, although the resulting cells are not completely homogenous, is that the cells are grown as monolayers and do not require transfer.
5. Since the shape and size of the bleached region influences the recovery dynamics, identical protocols must be followed for each protein and experiments comparing between different conditions should be carried out on the same day since laser power and other conditions may vary.
6. An alternative to FRAP is inverse-FRAP (or iFRAP), where the entire nucleus is bleached except for the ROI and the fluorescence decay is measured at the ROI. This can be used to measure k_{on}.

Acknowledgements

We thank the Abisch Frenkel Foundation, the Israel Cancer Research Foundation (ICRF), the Human Frontiers Science Foundation (HFSP), the Israel Science Foundation (ISF 657/12;

1252/12), the Israel Ministry of Science, and the European Research Council (ERC-281781) for financial support. AH is Marie Curie Nucleosome4D fellow.

References

1. Meshorer E, Yellajoshula D, George E et al (2006) Hyperdynamic plasticity of chromatin proteins in pluripotent embryonic stem cells. Dev Cell 10:105–116
2. Phair RD, Misteli T (2000) High mobility of proteins in the mammalian cell nucleus. Nature 404:604–609
3. Melcer S, Hezroni H, Rand E et al (2012) Histone modifications and lamin A regulate chromatin protein dynamics in early embryonic stem cell differentiation. Nat Commun 3:910
4. Nissim-Rafinia M, Meshorer E (2011) Photobleaching assays (FRAP & FLIP) to measure chromatin protein dynamics in living embryonic stem cells. J Vis Exp 52:e2696
5. Poser I, Sarov M, Hutchins JR et al (2008) BAC TransgeneOmics: a high-throughput method for exploration of protein function in mammals. Nat Methods 5:409–415
6. Sigal A, Danon T, Cohen A et al (2007) Generation of a fluorescently labeled endogenous protein library in living human cells. Nat Protoc 2:1515–1527
7. Cohen AA, Geva-Zatorsky N, Eden E et al (2008) Dynamic proteomics of individual cancer cells in response to a drug. Science 322: 1511–1516
8. Ellenberg J, Siggia ED, Moreira JE et al (1997) Nuclear membrane dynamics and reassembly in living cells: targeting of an inner nuclear membrane protein in interphase and mitosis. J Cell Biol 138:1193–1206
9. Dundr M, Misteli T (2003) Measuring dynamics of nuclear proteins by photobleaching. Curr Protoc Cell Biol Chapter 13, Unit 13 5
10. Phair RD, Misteli T (2001) Kinetic modelling approaches to in vivo imaging. Nat Rev Mol Cell Biol 2:898–907
11. Mueller F, Mazza D, Stasevich TJ et al (2010) FRAP and kinetic modeling in the analysis of nuclear protein dynamics: what do we really know? Curr Opin Cell Biol 22:403–411
12. Lenser T, Weisshart K, Ulbricht T et al (2010) Fluorescence fluctuation microscopy to reveal 3D architecture and function in the cell nucleus. Methods Cell Biol 98:2–33

Chapter 13

Electron Spectroscopic Tomography of Specific Chromatin Domains

Liron Even-Faitelson, Eden Fussner, Ren Li, Mike Strauss, and David P. Bazett-Jones

Abstract

The eukaryotic genome is packaged within the nucleus as poly-nucleosome 10 nm chromatin fibres. The nucleosome core particle, the fundamental chromatin subunit, consists of a DNA molecule wrapped around a histone octamer. Biochemical modifications of both the DNA and histone proteins have been characterized that influence chromatin structure and function. These modifications include DNA methylation, histone variants and posttranslational modifications of the core histone protein tails. An outstanding area for investigation in the field of nuclear cell biology is the characterization of the functional relation between these biochemical modifications and the underlying chromatin structure and nuclear subcompartmentalization. Electron spectroscopic tomography is a high-resolution microscopy technique that facilitates visualization of individual 10 nm chromatin fibres in three dimensions. The method, therefore, has a role to play in exploring the relationships of the epigenome and nuclear organization. Correlating immunofluorescence microscopy with electron spectroscopic tomography provides a powerful approach to relate epigenetic marks with high resolution chromatin organization.

Key words Electron spectroscopic imaging, Tomography, Chromatin, Correlative microscopy, Histones, Nucleus

1 Introduction

It is now well established that the nucleus is a highly organized cellular compartment containing the genetic material, proteins and ribonucleic acids. These functional nuclear macromolecular components are arranged in both specific nuclear bodies and chromosome territories within the interphase nucleus [1, 2]. Chromatin is a nucleoprotein complex, which packages the genome and facilitates its regulation. In eukaryotes, nucleosomes, which are comprised of a histone octamer, form the protein core of the chromatin subunit. These histones undergo extensive and diverse posttranslational modifications on their amino-termini, thereby leading to various regulatory effects. Moreover, besides posttranslational

Yaron Shav-Tal (ed.), *Imaging Gene Expression: Methods and Protocols*, Methods in Molecular Biology, vol. 1042, DOI 10.1007/978-1-62703-526-2_13, © Springer Science+Business Media, LLC 2013

modifications of histones, variants of the histone molecules contribute to the properties of the chromatin protein core and its effects. These effects include direct alteration of the physical properties of nucleosomes, direct influence on contacts between nucleosomes and on chromatin structure, and recruitment of non-histone effector complexes [3]. In concert, these effects comprise the "histone code" hypothesis, which postulates that specific combinations of histone modifications are a predictor of genome silencing or activation [4].

Chromatin can be loosely defined as either euchromatin or heterochromatin, where euchromatin refers to transcribed or "open" regions of the chromatin, whereas heterochromatin describes "silent" or compact or "closed" chromatin domains. These terms are often correlated with certain histone modifications, e.g. H3K9me3 is a heterochromatin-associated modification, whereas histone H3 acetylation is typically associated with euchromatin [5]. The increasing amount of data from both biochemical and microscopy approaches has elucidated some correlations between specific histone posttranslational modifications and genomic compaction and transcription potential. However, we still lack a comprehensive understanding of the precise relationship between chromatin modifications, structure and function.

Microscopy has played a pivotal role in our understanding of chromatin organization and transcription potential [6]. Several visible light microscopy techniques have been described that enable detection of specific molecules in situ, either in fixed cells by indirect immunofluorescence, or in living cells by expression of fluorescently tagged proteins [7]. These methods allow visualizing the localization of certain histone variants or specific histone modifications inside the nucleus [8]. They also allow the study of transcription of specific loci [9]. However, in order to relate nucleosome-level resolution of chromatin fibres and organization, a higher-resolution imaging technology is required. To this end, conventional transmission electron microscopy (CTEM) has contributed, together with biochemical data [10], to the discovery of the nucleosome [11], and has enabled the visualization of actively transcribing ribosomal genes [12]. CTEM can be used to detect compact chromatin domains using heavy atom stains. These stains, however, preclude high-resolution visualization of individual chromatin fibres and, therefore, CTEM is not well suited for studying decondensed chromatin domains. In contrast, the technique of electron spectroscopic imaging (ESI), based on energy filtered transmission electron microscopy (EFTEM), permits such high resolution imaging that enables visualization of unstained single 10 nm chromatin fibres. ESI enables such resolution by deriving contrast from the elemental content (nitrogen and phosphorus) of the chromatin itself and avoids the use of heavy atom stains, such as uranyl acetate and/or lead citrate that are typically used in CTEM [13].

These stains not only may obscure fine structural details, but may also generate ambiguities due to the non-uniform and unpredictable staining characteristics of different biochemical components of the biological specimen.

ESI is based on the principle of electron energy-loss spectroscopy and combines an electron energy-loss imaging spectrometer with a conventional transmission electron microscope to enable the direct quantitative imaging of elements within the specimen. Mapping of specific elements is achieved by generating images from inelastically scattered electrons (i.e. incident electrons that have lost energy due to ionization events of atoms while passing through the specimen) that have lost a particular amount of energy in a suitably thin specimen. Inside the cell's nucleus, the nitrogen signal offers information of the total distribution of biomolecules, as it is found in both proteins and nucleic acids, whereas phosphorus serves as a specific marker for the distribution of nucleic acids. The advantage of ESI in visualizing chromatin is that the distribution of nucleic acids can be resolved and delineated from the protein component of macromolecular complexes by comparing the high-resolution phosphorus images with the nitrogen distribution. However, ESI suffers from some of the same limitations of conventional TEM in that structural ambiguities may arise from overlapping biomolecules which are projected onto a two-dimensional plane in TEM micrographs. This limitation must be overcome by generating three-dimensional images in order to visualize the in situ assembly of numerous overlapping and inter-mingling chromatin fibres. To achieve this goal, and to preserve the quantitative information provided by ESI, it is possible to combine ESI with electron tomography.

Electron tomography (ET) is ideally suited to study 3-D structures in biological specimens [14, 15]. ET generates 3-D structural data by acquiring a series of images while tilting the specimen through a range of angles. When combined with phosphorus mapping by ESI, this method facilitates visualization of three-dimensional structures of chromatin fibres in situ at remarkably high spatial resolution. Nucleosomes, linker sequences and even the gyres of DNA wrapping around the nitrogen-rich protein histone core can be visualized in three-dimensions [16–18]. ESI-tomography has advanced our understanding of global chromatin organization in both heterochromatin and euchromatin domains in mouse nuclei. These studies predict that compact domains are characterized by 10 nm fibres that have closer fibre-fibre packing and are more frequently bent. It would be of great utility to combine these high-resolution structural studies to specific histone or DNA modifications associated with gene silencing or activation to thoroughly uncover the relationship between chromatin structure and function.

To achieve the final goal of correlating biochemical markers of chromatin with ultrastructural data, the biological specimen must first be labeled with an antibody for the biochemical marker in question. It is then possible to detect this biochemical marker with a fluorescently labeled secondary antibody and perform correlative ESI of the desired region, as described previously [19, 20]. The addition of tomography to the region of interest subsequently provides 3-D structural data of the chromatin containing the relevant biochemical marker. Alternatively, one can label the thin sections of the specimen, directly, with the desired antibody and use secondary gold antibodies that also can be visualized using ESI. When applying the latter technique, it is possible to place mirror sections on different grids, label with two different biochemical markers and visualize the relation between the two markers, as demonstrated in Fig. 1. Moreover, direct labeling of the sections may help avoid penetration problems by the antibodies in some biological samples. Imaging of the specimen using ESI and detection of the gold-labeled marker facilitates recognition of the chromatin domains that are of interest and can be followed up by ESI-tomography in order to get a 3-D structural view of all of the chromatin fibres within this particular domain, as illustrated in Fig. 2.

2 Materials

2.1 Sample Preparation, Dehydration, Embedding and Serial Sectioning

1. Mouse spleen (*see* **Note 1**).
2. 1× Phosphate-buffered saline (PBS) (*see* **Note 2**).
3. 16 % Paraformaldehyde (PFA) stock solution. Store at 4 °C. Dilute with PBS to the working concentration just before use.
4. Anhydrous ethyl alcohol: make a series of dilutions in distilled water: 30, 50, 70, 90, 100 %.
5. LR white embedding resin (cat. 14381-CA; Electron Microscopy Sciences, Hatfield, PA, USA). Store at 4 °C.
6. Gelatin embedding capsules (cat. 70104; Electron Microscopy Sciences, Hatfield, PA, USA).
7. Ultramicrotome, Ultracut UCT (Leica Microsystems, Vienna, Austria).
8. 400-mesh copper electron microscope grids with parallel bars containing a single perpendicular centre bar (cat. G400PB-Cu; Electron Microscopy Sciences, Hatfield, PA, USA).

2.2 Immunolabeling and Carbon Film

1. 1× PBS.
2. Parafilm M.
3. PBS-T: 0.5 % Triton X-100 in PBS.

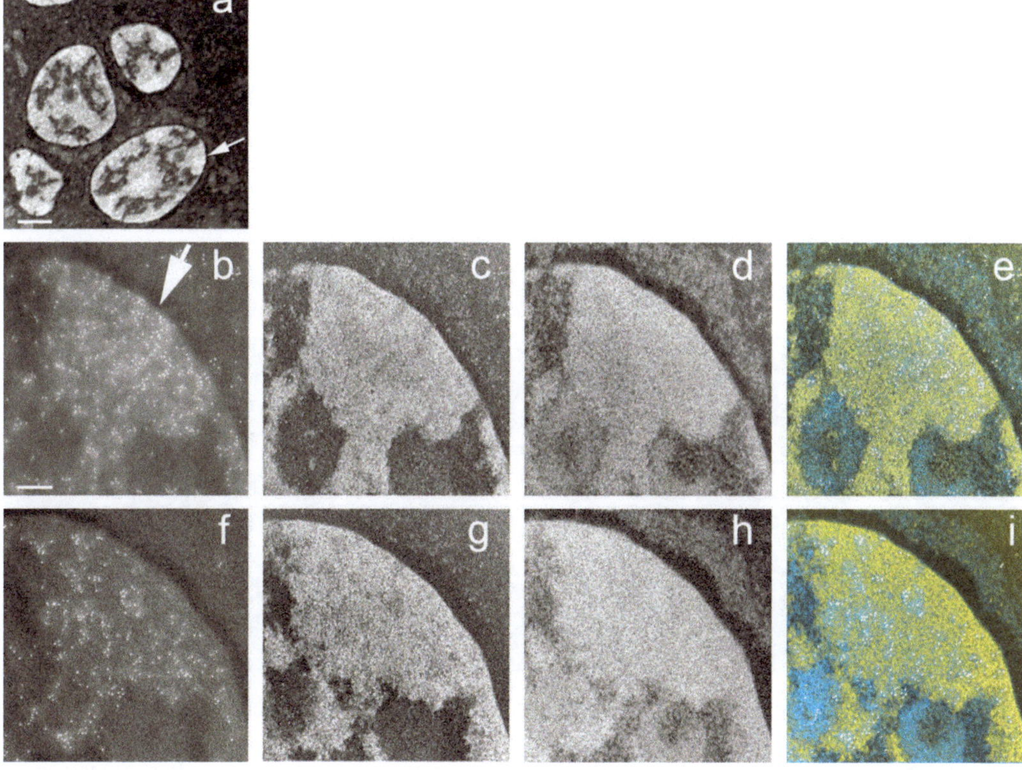

Fig. 1 Electron spectroscopic images correlated with on-section labelling of 70 nm-thick mirror sections of a mouse splenocyte nucleus. (**a**) A low-magnification EM image of mouse splenocytes nuclei within a mouse spleen. The *arrow* points to an area, which is shown in high magnification in micrographs (**b**–**i**). Two mirror sections were placed on separate grids and labelled on-section using anti-H3K9me3 (**b**–**e**) or anti-H3K27me3 antibody (**f**–**i**). (**b**) and (**f**) show the gold map representing the gold-conjugated secondary antibody on the two sections. (**c**) and (**g**) show the phosphorus map of the two sections. (**d**) and (**h**) show the nitrogen map of the two sections. (**e**) and (**i**) are overlays in which, the gold signal is shown in *white*, the phosphorus signal is *yellow*, representing chromatin, and in *blue* is the nitrogen signal from which the phosphorus signal has been subtracted, thus representing protein. Scale bar, 1.4 μm in (**a**), and 200 nm in (**b**–**i**)

4. Sodium Citrate Buffer pH 6.0: 10 mM Sodium Citrate, 0.05 % Tween-20.
5. Water-bath at 60 °C.
6. 0.05 % Trypsin in 0.1 M $CaCl_2$.
7. Humid chamber (*see* **Note 3**).
8. Blocking solution: 2 % bovine serum albumin, 5 % donkey serum (*see* **Note 4**) in PBS-T.
9. Glass coverslips.
10. Flexible plastic adhesive (LePage; Henkel Canada Corporation, Mississauga, ON, Canada).

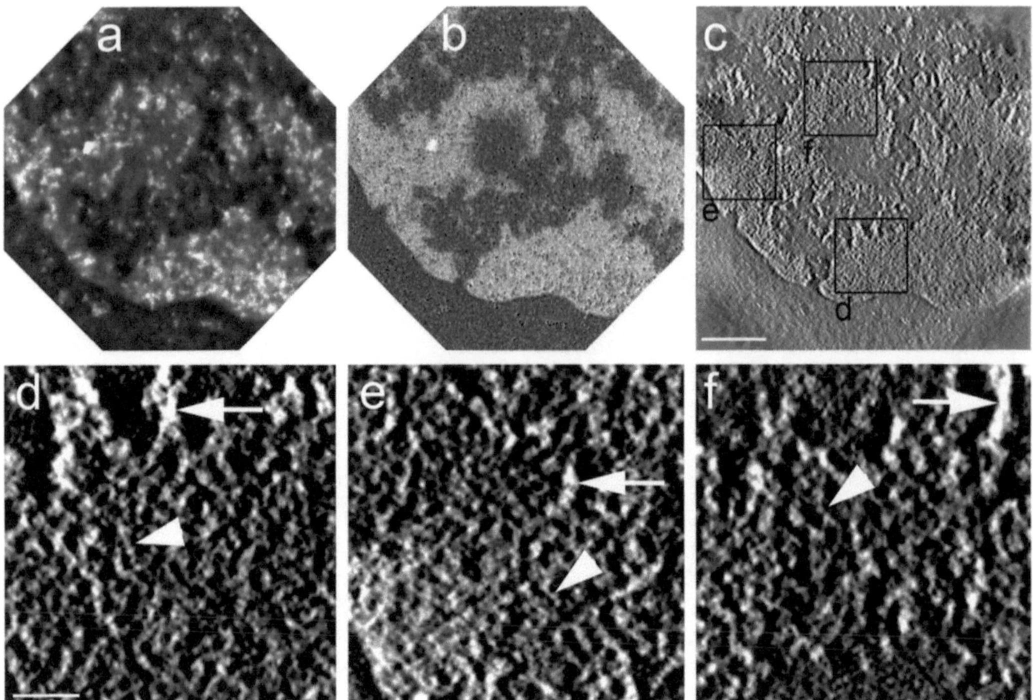

Fig. 2 Electron spectroscopic tomogram of a H3K9me3-labelled 70 nm-thick section of a mouse splenocyte nucleus. (**a**) A pre-edge phosphorus image showing the gold signal from the gold-conjugated secondary antibody bound to the anti-H3K9me3 antibody. (**b**) A phosphorus map of the zero-tilt image from the tomogram. (**c**) A central slice from a tomographic reconstruction. (**d–f**) Digitally zoomed-in images of chromatin domains marked on (**c**). These zoomed-in images clearly show 10 nm chromatin fibres in H3K9me3 labelled domains (marked with *arrowheads*). *Arrows* point to RNP based structures at the periphery of the compact chromatin domain, i.e. transcription that occurs at those sites. Scale bar: 200 nm in (**a–c**), and 40 nm in (**d–f**)

11. Primary antibody: rabbit anti-H3K9me3 (*see* **Note 5**) diluted in blocking solution.
12. Secondary antibody: 10 nm gold-conjugated anti-rabbit antibodies diluted in PBS (Electron Microscopy Sciences, Hatfield, PA, USA).
13. Ultrapure distilled water.
14. Optional: Expired secondary antibodies conjugated with 10 nm gold particles.
15. Carbon evaporator, Auto306 (BOC Edwards, Crawley, West Sussex, UK).

2.3 Imaging and Tomography

1. FEI Tecnai T20—Transmission electron microscope equipped with a post-column Gatan imaging filter (GIF).
2. SerialEM software (Boulder Laboratory for 3-Dimensional Electron Microscopy of Cells, University of Colorado) [21].

3. IMOD software (Boulder Laboratory for 3-Dimensional Electron Microscopy of Cells, University of Colorado) [22].
4. ImageJ software [23].

3 Methods

The functional consequences of chromatin organization and the regulation of this organization remain crucial and outstanding questions in the field of cell biology. Furthermore, many biochemical markers, such as histone variants and different histone post-translational modifications, have been described, yet their relations to gene activity and chromatin structure remain to be resolved. Tri-methylation on lysine 9 of histone H3 (i.e. H3K9me3) is a well-characterized modification that is often used as a marker for heterochromatin and silenced chromatin [24]. Here, this histone modification is being used to demonstrate the relationship between this biochemical marker and the ultrastructure of chromatin labelled with this marker using ESI-tomography correlated with on-section labeling of the specimen for this specific histone modification.

3.1 Sample Preparation, Dehydration, Embedding and Serial Sectioning

1. Sacrifice a mouse using standard procedures and remove the spleen (*see* **Note 1**). Quickly continue to the fixation step to avoid destruction of the molecular ultrastructure.
2. Fixation: place a fresh mouse spleen in ice-cold 4 % PFA. In order to preserve the original structures, it is crucial to perform the fixation steps on ice. As quickly as possible, cut a 1 mm-thick slice of the spleen and place in ice-cold 4 % PFA for 1 h. Following 1 h of fixation, cut the spleen slice into small (~1 mm^3) pieces and place a piece in a small glass vial with 3.5 ml of ice-cold 4 % PFA for 24 h at 4 °C. It is important to note that PFA should be disposed of in a designated hazardous waste container.
3. Wash the fixed tissue with PBS. Replace the 4 % PFA with 3.5 ml PBS and place on a shaker for 1 h at room temperature. Repeat the wash step four times, for a total of 4 h.
4. Dehydrate the tissue sample by replacing the PBS with 3.5 ml of 30 % ethanol and incubate for 30 min at 4 °C. Repeat this step with 50, 70, and 90 % ethanol. For the final dehydration step, replace the 90 % ethanol with 3.5 ml 100 % ethanol and incubate for at least 1 h at 4 °C. Replace the 100 % ethanol four times, incubating for at least 1 h at 4 °C each time.
5. Embed the tissue in LR white resin. Replace the 100 % ethanol with 3.5 ml LR white resin and rotate on a shaker at room temperature for 2 h. Repeat this step four times for a total of

8 h. Following the last incubation, place the piece of tissue in a gelatin capsule and fill up the capsule with LR white resin. Since LR white does not polymerise when in contact with air, take care to fully fill the capsule with resin and close the capsule carefully, including as little air as possible. Incubate the capsule in an oven at 50 °C for 48 h.

6. Remove capsule from oven and remove gelatin capsule from polymerised resin by carefully cutting the capsule with a razor blade and peeling the capsule. The resin around the tissue should then be trimmed with a razor blade, thereby leaving a small area for sectioning in the ultramicrotome.
7. Section the embedded sample with an ultramicrotome to generate a ribbon of 5–15 serial sections on the surface of water. The thickness of the sections should be between 70 and 100 nm (*see* **Note 6**).
8. Use a 400-mesh copper electron microscope grid to pick up the sections.

3.2 Immunolabeling and Carbon Coating

1. Place the grid with the sections in a 150 μl drop of PBS placed on parafilm, and incubate for 15 min in room temperature.
2. Transfer the grid into a 150 μl drop of PBS-T on parafilm, and incubate for 10 min at room temperature.
3. Wash with PBS by transferring the grid to a 150 μl drop of PBS on parafilm, and incubate for 5 min at room temperature.
4. Transfer the grid to sodium citrate buffer, pre-warmed to 60 °C in a water-bath, for 30 min at 60 °C. Then allow the specimen to cool down by transferring the grid in the sodium citrate buffer to 4 °C for 30 min. This step assists in antigen retrieval following the epitope masking that occurs as a result of the fixation step, and increases the binding capacity of the primary antibody to the sample.
5. Wash the grid by transferring it into a 150 μl drop of PBS on parafilm, and incubate for 5 min at room temperature.
6. Transfer the grid to a 150 μl drop of 0.05 % trypsin solution for 10–20 min at 37 °C inside a humid chamber.
7. Perform washes by transferring the grid to a 150 μl drop of PBS placed on parafilm, and incubate for 5 min at room temperature. Repeat this step three times for a total of three washes.
8. Blocking: transfer the grid to a 150 μl drop of blocking solution on parafilm, and incubate for 20 min at 37 °C inside a humid chamber.
9. Primary antibody labeling: transfer the grid with the sections facing down onto a 10 μl drop of anti-H3K9me3 diluted in blocking solution. Place a coverslip over the grid and seal the

edges with flexible plastic adhesive to avoid dehydration of the antibody. The flexible plastic adhesive allows sealing of the grid with the antibody, while allowing easy removal following the incubation. Incubate over night at 37 °C inside a humid chamber.

10. Transfer the grid to a 150 μl drop of PBS on parafilm, and incubate for 5 min at room temperature. Repeat this step three times for a total of three washes.
11. Secondary antibody labeling: transfer the grid with the sections facing down on to a 10 μl drop of anti-rabbit gold antibody diluted in 5 % donkey serum. Place a coverslip over the grid and seal the edges with a flexible plastic adhesive to avoid dehydration of the antibody. Incubate for 1 h at 37 °C.
12. Transfer the grid to a 150 μl drop of PBS on parafilm, and incubate for 5 min at room temperature. Repeat this step three times for a total of three washes.
13. Transfer the grid to a 150 μl drop of ultra-pure distilled water on parafilm, and incubate for 5 min at room temperature. Repeat this step three times for a total of three washes.
14. Air-dry the grid.
15. Optional step (*see* **Note 7**): Gold particles from the on-section labeling procedure can be used as fiducials in the tomographic reconstruction. Alternatively, you may choose to apply expired secondary antibodies conjugated with 10 nm gold particles diluted in water to a single surface of the sectioned samples. To do so place the grid in a 50 μl drop of diluted gold secondary antibodies for 5 min at room temperature, followed by three washes in water for 2 min each, at room temperature. Ideally, 20–30 gold particles per field at the imaging magnification are required for optimal reconstructions (*see* **Note 8**).
16. In order to provide stability to the sample and eliminate deformation due to the electron beam, a 5 nm carbon film should be evaporated directly onto the grid (*see* **Note 9**).

3.3 Imaging and Tomography

3.3.1 Microscope Alignments

1. Adjust the eucentric height—using the alpha wobbler at a magnification between 5.7 and 19k until the specimen drift during tilting is minimized.
2. Focus on the sample in bright field.
3. Align the objective aperture in diffraction mode.
4. Align the beam tilt using the direct alignment mode. Usually this alignment only requires an adjustment of the X and Y rotation centre alignment.
5. Align the beam shift (also using the direct alignment). This alignment is particularly crucial as SerialEM uses beam shift

calculations to centre the specimen and the region of interest throughout the tilt series.

6. Begin working in the EFTEM mode. Centre beam over GIF CCD with the direct beam shift alignment and focus on sample.
7. Tune the GIF at each magnification beginning with 5.7k and up to the working magnification (usually 10 or 19k). Tune the filter in an area on the grid that does not contain a region of interest in the specimen. Some energy filters are more stable than others and may require less frequent tuning.
8. Prepare a gain reference image.

3.3.2 SerialEM Calibrations

1. Open SerialEM in administrator mode.
2. Use a high contrast sample for calibrations. We recommend a uranyl acetate/lead citrate contrast enhanced tissue sample.
3. Focus beam on sample and centre the beam using the direct beam shift alignment. Adjust the specimen position as necessary to centre the region of interest. Align zero loss at 5.7k. Beginning at this magnification run the Image shift calibration. Check the SerialEM log and determine the calculated distance between the direct measurement and the newly calculated one. The program will alert if the difference between the two measurements is too great and will prompt the operator to repeat the calibration. Repeat until a satisfactory calibration is obtained (*see* **Note 10**).
4. Repeat step three at each magnification between 5.7k and the working magnification (usually 10 or 19k).
5. At the working magnification very accurately focus on the sample. Centre the beam around the region of interest using the direct beam shift alignment. Check that the exposure settings provide moderate electron counts during a zero loss image acquisition (3000–6000 counts/pixel). Align the zero loss peak. Several repetitions may be necessary before a satisfactory calibration is obtained. Occasionally it is necessary to select an alternative region to attempt this calibration if the contrast of the sample is inadequate, or if minor instabilities in the sample are interfering with these calculations.
6. *Save Calibrations.* Ensure that the new calibrations are being saved in a .txt file in the user version of the program in a calibration folder (*see* **Note 11**).

3.3.3 Data Acquisition

1. Launch the user version of SerialEM.
2. Find the region of interest and make sure that there is a sufficient number of gold particles on the sample surface (ideally, 20–30 gold particles per field).

3. Coarsely adjust the eucentric height using the wobbler.
4. Align the zero loss peak.
5. Bring the sample into focus, and centre the beam using the direct beam shift alignment.
6. Using the SerialEM functions align the eucentric height (both fine and coarse alignment). After the eucentric height alignment check the image to ensure the sample and region of interest are still centred. If not, adjust the beam shift by using the direct alignment setting. The sample and the beam should now be automatically re-centred. If the sample moved or the beam shifted during this alignment repeat until only modest movements before and after the alignment can be detected.
7. Align the zero loss peak and accurately focus the region of interest at an energy offset corresponding to one of the acquisition settings (e.g. 155 eV post-edge phosphorus). Measure the defocus at zero energy loss. This value should be fairly small (generally it is between ±1 and 2 μm when focus is found at an energy loss offset). Recheck the value and if it is fairly consistent carry on. If a large difference is measured repeat the defocus measurement until the values are within 0.2 μm.
8. Using the "extra output" function in the tomography acquisition tab, specify the energy offsets to be used in the tilt series (e.g. 155 eV and 120 eV for post- and pre-edge phosphorus) and the exposure times that generate pre- and post-edge images with sufficient electron counts at zero and high tilt angles.
9. Open two new .mrc files for the zero loss and energy loss images respectively.
10. Begin your tilt series. There are several parameters that can be manipulated in the start-up menu (and these functionalities differ between SerialEM versions). We acquire our datasets with 2° between tilt angles and ±60°. We keep our exposure constant throughout the data acquisition and do not re-measure the eucentric height during the setup. Once all these parameters have been chosen, begin the tomogram data acquisition.
11. *Terminate* the tomogram and close both .mrc files. Unfortunately, failure to properly terminate the tomogram results in corrupt files that cannot be used for a reconstruction of the hard-earned dataset.

3.3.4 Tomographic Reconstruction

1. Split the energy offset dataset into separate pre- and post-edge images and convert to .st files
2. Launch etomo (part of the IMOD suite) and select one of the two datasets. Use the dual tilt functionality to simultaneously align both the datasets in IMOD.

3. ESI-tomography is particularly susceptible to X-rays due to the long exposure times. It may therefore be necessary to increase the stringency of the X-ray detection routines.
4. Cross-correlate the pre- and post-edge images at each tilt angle.
5. Choose 20 gold fiducial particles, on the zero-tilt image, as evenly distributed across the field as possible on one of the image pair datasets for optimal reconstruction. Utilize the "light fiducial markers" and "local tracking optional settings" for the energy-loss datasets.
6. Fix the fiducial model by opening the model by selecting "fix fiducial model" and fill in any gaps and move inaccurately automatically placed points within specific tilt pairs. Track the fiducials and check the alignment statistics. Ensure that the residual errors are below 1. The IMOD help pages and manuals can assist with these steps. Transfer the fiducial model to the other energy-loss offset image dataset.
7. Conservatively define the tomographic volume for final reconstructions. However, it may be necessary to adjust the default tomogram thickness when defining the tomographic boundaries. Be sure to have the entire section contained within the boundaries. Having defined the reconstruction volume in the pre-edge tilt series transfer these values to the post-edge data set and select "done".
8. Proceed with the tomographic reconstruction using IMOD until the final processing stage and use SIRT reconstruction methods [22].
9. Create phosphorus ratio tomograms in ImageJ with the MRC reader by dividing the post-edge image by the pre-edge image using the image calculator and generate the jump ratio map in a new 32-bit float window. Save the new mrc file in a SIRT folder and change the file extension to .ali
10. Save the 5th, 10th, 15th and 20th iteration for analysis. Although it is widely established that up to 20 iterations provides a faithful reconstruction, we generally use the 10th or 15th iteration for our measurements and display. The iterative methods can lead to excessive noise at higher iterations.

3.3.5 Displaying Tomograms

1. Rotate the tomographic volume about the *x*-axis to view the fibres running parallel to the image plane using the command: 3dmod –Y *.srec*

   Then rotate the best iteration using the command: clip rotx in.srec out.mrc
2. Average 4–5 of the central slabs of the generated tomogram and reduce noise using a median or Gaussian filter. Adjust

levels so the background noise is reduced while preserving the range of intensity present within the phosphorus-rich chromatin structures.

3. Display using either ImageJ [23] with the mrc reader plugin enabled or with Chimera (available as freeware from UCSD) [25]. It is also possible to filter/render the entire reconstruction in these programs.

4 Notes

1. This technique can be performed on any tissue or tissue culture cells. If using another tissue, the procedure remains the same. It is important that the specimen size is no more than 1 mm^3 in order to allow penetration and adequate fixation, dehydration and embedding. For tissue culture cells sample preparation grow the cells on coverslips, as described in Ahmed et al. (2009) and follow the steps described above, using the methodology described in Ahmed et al. (2009) for working with coverslips [20].
2. All solutions should be prepared in deionized and filtered water that has a resistivity of 18.2 MΩ-cm or better. Water purity is most critical in washing steps following secondary antibody labeling (Subheading 3.2).
3. A humid chamber is a plastic container with wet paper towels meant to keep the solutions that the grids are incubated with at 37 °C from dehydrating.
4. We use donkey serum for blocking since our secondary antibody is produced in donkey. The blocking serum used should correspond to the secondary antibody being used.
5. To demonstrate the technique we used H3K9me3 as a biochemical marker for a specific chromatin domain. One could use any biochemical marker, providing the existence of a good antibody that recognises this marker. It is important to first test out the antibody using immunofluorescence labelling to make sure it is specific and adequate for the assay.
6. Thicker sections have the advantage of containing more biological material in the z-dimension. However, at high tilt angles, plural scattering can cause artifacts in the energy loss images. If the accelerating voltage of the microscope being used is 200 kV, individual sections should be restricted to 70–100 nm in thickness.
7. A major caveat in using the on-section labeling for gold fiducial markers is that too much gold can impede the reconstruction and may result in artifacts in the final phosphorus reconstruction.

8. To get the correct number of gold particles per field one would need to experiment with the dilution of the antibody and/or incubation time. The dilution and/or incubation time may also need to be calibrated for tomograms acquired at different magnifications.
9. Sample stability is crucial in ESI-tomography. Uneven sample depression of the specimen section, for example, between widely spaced grid bars, can have pronounced effects on the ability of SerialEM to track the sample correctly.
10. The SerialEM manual and help pages can assist during calibration.
11. The particular microscope being used may have slightly different settings or magnifications from the FEI T20, but the principles remain the same.

Acknowledgements

The research is funded by operating grants from Canadian Institutes of Health Research, and the Natural Sciences and Engineering Research Council of Canada. D.P.B.-J. holds the Canada Research Chair in Molecular and Cellular Imaging.

References

1. Spector DL (2006) SnapShot: cellular bodies. Cell 127:1071
2. Cremer T, Cremer C (2001) Chromosome territories, nuclear architecture and gene regulation in mammalian cells. Nat Rev Genet 2:292–301
3. Campos EI, Reinberg D (2009) Histones: annotating chromatin. Annu Rev Genet 43:559–599
4. Jenuwein T, Allis CD (2001) Translating the histone code. Science 293:1074–1080
5. Bannister AJ, Kouzarides T (2011) Regulation of chromatin by histone modifications. Cell Res 21:381–395
6. Rapkin LM, Anchel DR, Li R, Bazett-Jones DP (2012) A view of the chromatin landscape. Micron 43:150–158
7. Giepmans BN, Adams SR, Ellisman MH, Tsien RY (2006) The fluorescent toolbox for assessing protein location and function. Science 312:217–224
8. Kimura H, Hayashi-Takanaka Y, Yamagata K (2010) Visualization of DNA methylation and histone modifications in living cells. Curr Opin Cell Biol 22:412–418
9. Shav-Tal Y, Singer RH, Darzacq X (2004) Imaging gene expression in single living cells. Nat Rev Mol Cell Biol 5:855–861
10. Kornberg RD (1974) Chromatin structure: a repeating unit of histones and DNA. Science 184:868–871
11. Olins AL, Olins DE (1974) Spheroid chromatin units (nu bodies). Science 183:330–332
12. Miller OL Jr, Beatty BR (1969) Visualization of nucleolar genes. Science 164:955–957
13. Bazett-Jones DP, Hendzel MJ (1999) Electron spectroscopic imaging of chromatin. Methods 17:188–200
14. Frank J (2006) Electron tomography: methods for three-dimensional visualization of structures in the cell, 2nd edn. Springer, New York
15. McIntosh R, Nicastro D, Mastronarde D (2005) New views of cells in 3D: an introduction to electron tomography. Trends Cell Biol 15:43–51
16. Fussner E, Djuric U, Strauss M et al (2011) Constitutive heterochromatin reorganization during somatic cell reprogramming. EMBO J 30:1778–1789
17. Fussner E, Strauss M, Djuric U et al (2012) Open and closed domains in the mouse genome are configured as 10-nm chromatin fibres. EMBO Rep. 13:992–996
18. Aronova MA, Kim YC, Harmon R et al (2007) Three-dimensional elemental mapping of

phosphorus by quantitative electron spectroscopic tomography (QuEST). J Struct Biol 160:35–48

19. Dellaire G, Nisman R, Bazett-Jones DP (2004) Correlative light and electron spectroscopic imaging of chromatin in situ. Methods Enzymol 375:456–478
20. Ahmed K, Li R, Bazett-Jones DP (2009) Electron spectroscopic imaging of the nuclear landscape. Methods Mol Biol 464:415–423
21. Mastronarde DN (2005) Automated electron microscope tomography using robust prediction of specimen movements. J Struct Biol 152:36–51
22. Kremer JR, Mastronarde DN, McIntosh JR (1996) Computer visualization of three-dimensional image data using IMOD. J Struct Biol 116:71–76
23. Rasband WS (1997–2012) ImageJ U.S. National Institutes of Health, Bethesda, Maryland, USA. http://imagej.nih.gov/ij/
24. Jenuwein T (2001) Re-SET-ting heterochromatin by histone methyltransferases. Trends Cell Biol 11:266–273
25. Pettersen EF, Goddard TD, Huang CC et al (2004) UCSF Chimera—a visualization system for exploratory research and analysis. J Comput Chem 25:1605–1612

Chapter 14

BAC Manipulations for Making BAC Transgene Arrays

Nimish Khanna, Qian Bian, Matt Plutz, and Andrew S. Belmont

Abstract

Chromosome tagging using lac or tet operator repeats for in vivo visualization of chromosome dynamics has now become a standard methodology used in a range of organisms. One variation of this approach has been to build transgene arrays creating artificial chromosome blocks to study various aspects of chromatin structure, transcription, replication, or DNA repair. Previously, plasmid transgenes with or without subsequent gene amplification have been used to build these arrays. However, plasmid arrays typically show heterochromatic properties, while gene amplification typically results in chromosome instability of the amplified regions. To avoid these problems, we are now building transgene arrays from large genomic DNA inserts cloned in bacterial artificial chromosomes (BAC). These BAC transgenes show transcriptional levels within several fold of endogenous genes while also exhibiting targeting to specific nuclear compartments similar to the targeting of the endogenous genes. Here we describe Tn5 transposition and BAC recombineering methods used to retrofit BACs for their use in building BAC transgene arrays. This includes insertion of operator repeats and selectable markers into these BACs as well as targeted insertion or deletion of BAC sequences.

Key words Transgene arrays, BAC retrofitting, BAC recombineering, *galK* selection, Large-scale chromatin structure, Lac operator

1 Introduction

Our laboratory has used three generations of approaches to build large transgene arrays or engineer large blocks of chromosomes [1].

At first we used plasmid transgenes, containing a 256mer lac operator repeat plus a cDNA DHFR (dihydrofolate reductase) minigene, followed by gene amplification using methotrexate selection. This first generation approach yielded gene amplified regions of varying sizes and chromatin compaction levels. However, whereas the amplified regions were stable in their characteristic compaction levels, they were unstable in size and chromosome location. In our second generation approach, we used plasmid constructs containing anywhere from 32 to 256 copies of the lac operator, a selectable marker, and also endogenous promoters and/or cis-regulatory elements of specific genes driving reporter

Yaron Shav-Tal (ed.), *Imaging Gene Expression: Methods and Protocols*, Methods in Molecular Biology, vol. 1042, DOI 10.1007/978-1-62703-526-2_14, © Springer Science+Business Media, LLC 2013

gene expression or intact transgenes. Multi-copy plasmid transgene arrays were stable in both size and chromosome location; however, they typically showed condensed chromatin and variegated expression of transgenes contained in these arrays.

In our third generation approach, we used BACs rather than plasmids to generate transgene arrays. By using BACs containing large ~100–250 kb insertions of mammalian DNA, we generated more normal chromatin compaction [2], with BAC transgenes forming large-scale chromatin fibers indistinguishable to the surrounding euchromatin regions as visualized by electron microscopy [3]. These BAC transgene arrays show stable size and chromosome location, with transgene expression levels within several fold of endogenous genes [2, 4]. Moreover, both Hsp70 and beta globin BAC transgenes recapitulated the observed targeting of their endogenous gene counterparts to nuclear speckles and the nuclear periphery, respectively. Together our results suggest that BAC transgene arrays recapitulate the behavior of endogenous chromatin much better than the previously used gene amplification and plasmid-based systems. Current work in our laboratory is focused on using BACs to create transgene arrays in order to study large-scale chromatin structure and chromatin dynamics and their relationship to transcriptional regulation. Technically, this requires inserting operator or MS2 repeats into the BAC to tag DNA or RNA sequences, inserting reporter genes into the BAC to assay transcriptional competence, and deletion analysis to dissect cis-regulatory elements regulating the large-scale chromatin structure of BAC transgenes.

The large size of these BACs, however, precludes traditional cloning methods. Instead, we have been using Tn5 transposition to retrofit BACs with operator arrays and/or selectable markers and BAC recombineering to add or delete specific DNA sequences at defined locations.

Tn5 transposition is a relatively simple and fast method for introducing DNA sequences into a BAC. In vitro incubation of a DNA transposon, constructed to contain the DNA element to be inserted, with the BAC DNA and a commercially available Tn5 transposase enzyme is followed by an *E. coli* transformation to select colonies containing BACs with transposon insertions. Although insertion is random, a large number of bacterial clones, each containing a BAC with a different insertion site, are obtained in a single reaction. DNA sequencing can then be used to screen for a suitable BAC insertion site.

Methods for homologous recombination allow for insertion or deletion of DNA sequences at specific sites within the BAC. However, homologous recombination methods involving transient expression of RecA lead to recombination among direct repeats such as lac operator or MS2 arrays. We therefore instead have been using BAC recombineering with the λ-red system [5].

With this system recombination is limited typically to just the several hundred bps at the ends of linear DNA fragments, sparing internal fragment sequences and, in particular, direct repeats inserted into the BACs.

The λ-red system is a powerful bacterial recombination system as it requires only short homology arms to perform recombination at high fidelity. The three λ-red phage genes (Exo, Bet, and Gam) required for recombineering are integrated into the *E. coli* genome and their expression is under tight control of the temperature-sensitive λ-*cI857* repressor. At 32 °C the λ-cI857 repressor suppresses expression of the λ-red recombination machinery. When the temperature is shifted to 42 °C for 15 min, the repressor is inactivated and the recombination machinery is expressed at a level allowing high frequency of recombination.

For many applications, the removal of the selectable marker after the first round of recombineering is necessary. Both positive and negative selection is possible using the *E. coli galK* as a selectable marker. The *galK* gene encodes galactokinase, which is essential for utilizing galactose, allowing positive selection using galactose as the carbon source. Negative selection is provided by growth in the presence of the galactose isoform 2-deoxy-galactose (DOG), as *galK*-catalyzed phosphorylation of DOG generates the toxic 2-deoxy-galactose-1-phosphate. To employ *galK* gene as the selectable marker, the endogenous *galK* gene was removed from the bacterial genome to generate a specialized strain. The BAC to be modified needs to be introduced into this strain. A linear DNA fragment containing *galK* gene can then be integrated into the BAC by homologous recombination. The recombinants are selected on minimal medium in which galactose is the only carbon source. *GalK* can then be removed by another round of recombination using negative selection.

We initially adopted the *galK* recombineering system [5] for insertion of DNA sequences, which uses a two-step procedure in which the *galK* selectable marker is first inserted and then replaced with the desired DNA fragment to be inserted. However, the primary application of this system apparently was for introduction of point mutations or very small DNA insertions. We observed a pronounced size dependence of recombination efficiency for larger fragments. When trying to insert a 4 kb DNA fragment, all colonies selected based on loss of *galK* expression contained large BAC deletions spanning the region with the *galK* marker or mutations in the *galK* gene. We reasoned that the large size of our DNA fragment led to a frequency of homologous recombination in the small percentage of cells transformed with the DNA fragment significantly lower than the spontaneous background frequency of BAC mutation and deletion occurring in the vast excess of plated *E. coli* cells.

To solve this problem, we revised the scheme of the two-step recombination to achieve the "seamless" insertion of large DNA

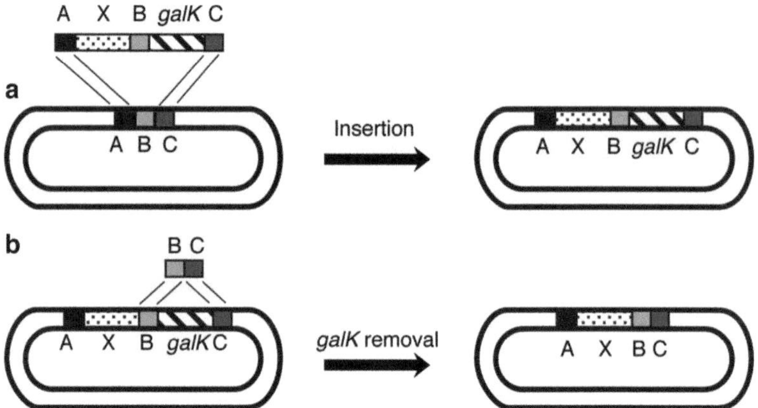

Fig. 1 Inserting large DNA sequences into BACs using *galK* selection. (**a**) A linear DNA-targeting construct contains three homology regions, A, B, and C: X, the DNA sequence to be inserted, is placed between A and B, and *galK*, the selectable marker, is placed between B and C. In bacteria, recombination between homology boxes A and C results in insertion of X and *galK*, separated by homology region B, into the BAC. (**b**) A DNA fragment corresponding to homology boxes B and C is used to delete *galK* by homologous recombination using negative selection against *galK*

sequences into BACs (Fig. 1). A targeting fragment containing three homology arms A, B, and C, each with a typical size of 45–50 bp, is constructed. The *galK* selectable marker is then inserted between homology boxes B and C and the DNA sequence, X, to be inserted is inserted between homology boxes A and B. This fragment is then introduced to bacteria carrying the target BAC leading to recombination between homology boxes A and C and insertion of both *galK* and X. These recombinants are efficiently selected for using *galK*-positive selection. To remove *galK*, a DNA fragment containing homology boxes B and C is generated by PCR using the BAC DNA as template and used for *galK* deletion by homologous recombination. Because of the small fragment size, recombination is sufficiently efficient to compete with the background rate of spontaneous loss of *galK* by mutation or large, internal BAC deletions. As a result, 20–50 % of total colonies after DOG selection are correct recombinants, which can be screened relatively easily.

Here we describe three protocols used in our laboratory to insert DNA sequences into BACs or to make deletions at specific sites. Protocol 1 is for random insertion of DNA sequences by Tn5 transposition. Protocol 2 is for addition of DNA sequences using BAC recombineering. In this second protocol we have modified the original λ-red recombineering approach to facilitate the insertion of large DNA fragments into BACs. Protocol 3 is for making deletions in BACs using BAC recombineering.

2 Materials

2.1 Transposition Reaction

1. EZ-Tn5™ kit (Epicentre Technologies).
2. pMOD-2 vector (Epicentre Technologies).
3. BAC maxiprep kit (we use Qiagen's large construction kit).
4. Restriction enzymes.
5. Electrocompetent cells (we use Epicentre's TransforMax EC100 Electrocompetent *E. coli*).
6. LB and SOC media.
7. Electroporator (we use Bio-Rad Gene Pulser II).
8. Cuvette for electroporation (we use the 0.2 cm electrode gap cuvette from Bio-Rad).

2.2 BAC Recombineering

1. *E. coli* SW102 strain.
2. p*galK* construct.
3. 1× M9 Medium (1 l): 6 g Na_2HPO_4, 3 g KH_2PO_4, 1 g NH_4Cl, 0.5 g NaCl. Autoclave the medium.
4. 5× M63 medium (1 l): 10 g $(NH_4)_2SO_4$, 68 g KH_2PO_4, 2.5 mg $FeSO_4{\cdot}7H_2O$. Adjust to pH 7 with KOH and autoclave.
   - M63v-Gal minimal plates: M63, 45 mg/ml L-Leucine, 1 mg/ml D-Biotin, 0.2 % D-(+)-Galactose, 15 g/l Agar, 12.5 µg/ml Chloramphenicol.
5. M63-DOG minimal plates: M63, 45 mg/ml L-Leucine, 1 mg/ml D-Biotin, 0.2 % Glycerol, 0.2 % 2-Deoxy-D-galactose, 15 g/l Agar, 12.5 µg/ml Chloramphenicol.
6. MacConkey plate: 1 % D-galactose, 15 g/l MacConkey agar, 12.5 µg/ml Chloramphenicol.
7. Low salt LB medium (1 l): 10 g Bacto-tryptone, 5 g yeast extract, 5 g NaCl.

Adjust to pH 7 with NaOH and autoclave.

8. Autoclaved conical flasks, centrifuge tubes.
9. Autoclaved ddH_2O.
10. Shaking water baths (32 and 42 °C).
11. 32 °C incubator.
12. Antibiotics (Chloramphenicol, Kanamycin, etc.).
13. SOC medium.
14. Electroporator (we use Bio-Rad Gene Pulser II).
15. Cuvette for electroporation (we use the 0.2 cm electrode gap cuvette from Bio-Rad).

3 Methods

3.1 Protocol 1. Transposon Insertion Reaction (Adapted from Epicentre)

1. Prepare the custom transposon by cloning the DNA of interest into the Multiple Cloning Site (MCS) of the pMOD-2 vector.
2. Isolate the functional EZ-TN5 transposon by either restriction enzyme digestion with *Psh*A1 or *Pvu*II (*see* **Note 1**) or PCR amplification.
3. Isolate the target BAC DNA by maxi preparation (*see* **Note 2**).
4. Prepare the transposon insertion reaction mixture by adding in the following order:

   0.5 μl EZ-Tn5 10× reaction buffer.

   0.1 μg target BAC DNA.

   *x*μl molar equivalent [Kan/Neo-8.32] transposon.

   *x*μl sterile water to bring reaction volume to 4.5 μl.

   0.5 μl EZ-Tn5 Transposase.

   5 μl total reaction volume.

   Calculation of μmole target DNA.

   μmole target DNA = μg target DNA/[(# base pairs in target DNA) × 660].
5. Incubate the reaction mixture for 2 h at 37 °C.
6. Stop the reaction by adding 0.5 μl of 10× Stop Solution. Mix and heat for 10 min at 70 °C.
7. Proceed to Transformation and Recovery (**steps 8** and **9**) or store the reaction mixture at −20 °C.
8. Transform 1 μl of the transposon insertion reaction mixture into 50 μl of a *recA*⁻ and *endA*⁻ electrocompetent bacterial strain (*see* **Note 3**).For electroporation we use the 0.2 cm electroporation cuvette and the Bio-Rad electroporator with the following settings: Resistance = 200 Ω, Capacitance = 25 μF and Voltage = 2.5 kV.
9. Recover the electroporated cells by adding 950 μl SOC medium to the electroporation cuvette. Tap gently to mix. Transfer to a tube and incubate on a 37 °C shaker for 1 h to facilitate cell growth.
10. Plate 1/10 and 9/10 volume of cells onto selective medium (*see* **Note 4**) as dictated by the transposon insert and grow plates overnight at 37 °C.
11. Inoculate several individual colonies into 5 ml LB containing antibiotics. Use 4 ml of the overnight culture for mini-prep. Check the restriction digestion patterns of the DNA to make sure there are no unwanted rearrangements within the BAC. Freeze down the clones with correct digestion patterns.

12. Sequence the BACs with insertions from both ends of the transposon to identify the transposon insertion site and to verify that no BAC sequence was lost during insertion.

3.2 Protocol 2: BAC Recombineering for Addition of DNA Sequences

3.2.1 Transform BAC into E. coli SW102 Strain

1. Purify the BAC DNA by mini-prep protocol. The integrity of the BAC DNA should be checked by restriction digestion pattern generated by appropriate restriction enzymes.
2. The day before transformation, inoculate a 2 ml LB culture with SW102 cells. Grow the culture overnight in a 32 °C shaking water bath.
3. Inoculate a 25 ml low salt LB culture in a flask with 0.5 ml of the overnight culture and grow the culture at 32 °C to an OD600 value of 0.55–0.6 (which takes about 3–4 h).
4. Prepare an ice-water bath before the culture is ready. Cool down ddH_2O and the centrifuge tubes on ice.
5. When the culture is ready, incubate it on ice for 5–10 min. Then aliquot 12.5 ml culture to each centrifuge tube.
6. Centrifuge the culture at 3,000 × *g* for 5 min at 4 °C. Pour off the supernatant. Put the tubes back into the ice-water bath. Then add 1 ml of ice-cold ddH_2O to each tube. Swirl the tube in the ice-water slurry until the pellet is completely resuspended. Add 9 ml more ice-cold ddH_2O to the tube to bring the total volume to 10 ml, then swirl briefly.
7. Repeat **step 6**. The pellet will become much looser after this centrifugation. The supernatant should be removed with a 10 ml serological pipette with caution. Remove as much supernatant as possible without disturbing the pellet. Add 9 ml of ice-cold ddH_2O to the tube and swirl briefly.
8. Centrifuge again. After this centrifugation, remove as much supernatant as possible; this usually leaves 0.5–1 ml of supernatant in the tube. Resuspend the pellet with a pipetman and transfer the cell suspension from each tube to the autoclaved 1.5 ml microcentrifuge tube on ice.
9. Centrifuge the suspension using a benchtop centrifuge machine in a 4 °C room at 1,000 × *g* for 5 min. Remove supernatant to reduce the volume to about 100 μl per tube. These electrocompetent cells are ready for the transformation.
10. Add 1–2 μl (50–100 ng) of the mini-prepped BAC DNA to 50 μl of the SW102 competent cells. Mix DNA with competent cells by gently tapping the tube.
11. Perform electroporation following the procedure used in Protocol 1.
12. Recover the cells by adding 950 μl of SOC medium. Shake in a 32 °C water bath for 1 h.

13. Plate 100 μl (1/10 of the total cells) on a LB plate containing 12.5 μg/ml chloramphenicol. Incubate the plate in a 32 °C incubator. The colonies should be ready to pick after ~24 h incubation. We typically obtain 10^4–10^5 colonies per μg DNA starting from mini-prep BAC DNA (which is not as pure as BAC DNA from a maxi prep using endonuclease treatment).
14. Inoculate several isolated colonies into 5 ml LB cultures containing chloramphenicol (12.5 μg/ml). Use 4 ml of the overnight culture for mini-prep DNA preparation. Check the restriction digestion patterns of the DNA to make sure there are no unwanted rearrangements within the BAC. Freeze down clones with correct digestion patterns.

*3.2.2 Generation of a DNA Fragment Containing the Homology Regions, the **galK** Selection Cassette, and the Sequence to be Inserted*

To simplify cloning, an oligonucleotide containing homology regions A, B, and C and appropriate restriction sites can be synthesized. Then *galK* and the DNA sequence of interest (X) can be inserted at these restriction sites (Fig. 1a).

1. Determine the desired site for insertion of the DNA sequence of interest in the BAC. The 50 bp regions flanking this targeting site can be picked as homology regions A and B. The 50 bp region downstream of B can be picked as homology region C. Check the restriction sites within the regions A, B, and C and make sure they do not contain sites that will be used later for cloning. Then add the sequences of the restriction sites between the homology regions (we typically use *Eco*RI restriction site to insert *galK* cassette between B and C). In addition, flank both ends of the oligonucleotide sequence with a rare restriction site that can be used to excise this DNA fragment.
2. Transform the commercially synthesized oligonucleotide, cloned within a vector, to an appropriate *E. coli* strain (e.g., DH5α, Stbl2). Purify the plasmid DNA. Insert the *galK* selection cassette, which can be digested from the p*galK* plasmid (we typically use *Eco*RI restriction), into the site between homology regions B and C. Screen for correct clones.
3. Insert the sequence X into the site(s) between homology regions A and B. Screen for correct clones.
4. Purify the plasmid DNA. Digest the plasmid with restriction enzymes that cut the ends of the oligonucleotide to release the five-element fragment.
5. Gel purify the five-element fragment. The concentration of the purified fragment should be at least 20 ng/μl.

3.2.3 Insertion of the Sequence of Interest into the BAC Followed by Gal Selection

1. The day before recombination, inoculate a 2 ml LB culture containing antibiotics with the SW102 bacteria clone carrying the BAC to be modified. Grow the culture overnight in a 32 °C shaking water bath.

2. Inoculate a 25 ml low salt LB culture with 0.5 ml of the overnight culture and shake at 32 °C. Grow the culture to an OD600 of ~0.55–0.6.
3. After the culture reaches the optimal density, split the culture in half. Transfer 12.5 ml of the culture into a 30 ml centrifuge tube and leave it on ice. This will be used as the uninduced control for recombination. Heat shock the other half of the culture at 42 °C for 15 min in a shaking water bath. After heat shock, cool down the culture on ice for 5–10 min, and then transfer it to a 30 ml centrifuge tube.
4. Follow **steps 6–9** in Subheading 3.2, **step 1**.
5. Transform 5 μl of the DNA fragment prepared in Subheading 3.2 to 50 μl of both induced and uninduced SW102 competent cells. Recover the cells with 950 μl of SOC medium and shake in 32 °C water bath for 1 h. The recombination reaction will occur in the bacteria during this period.
6. After recovery, transfer the 1 ml cultures to autoclaved Eppendorf tubes. Spin the cells down with a benchtop centrifuge at top speed for 1 min.
7. Carefully remove the supernatant. Resuspend the pellet thoroughly with 1 ml of M9 minimal medium. Then centrifuge again.
8. Repeat **step 6**.
9. Remove the supernatant and resuspend the pellet with 1 ml M9 minimal medium. Plate 1/10 and 9/10 of the induced and uninduced samples on M63-Gal minimal plates.
10. Incubate the plates in a 32 °C incubator. The plates should be placed into humid containers as they may dry out. The colonies should become obvious by day 3 of selection and reach a desirable size at day 3 or 4 (*see* **Notes 5** and **6**). We typically get tens to hundreds of colonies per plate for the 9/10 dilution.

3.2.4 Screening for Clones Carrying Correctly Recombined BACs

After drug selection, colony PCR can be used to confirm whether the DNA sequence of interest is inserted into the correct site within the BAC. The PCR primers that bind to the regions flanking homology boxes can be used to amplify the regions between them in the engineered BAC. The size of the product will indicate whether there is an insertion in the desired location. Alternatively, when the insertion is too large for efficient PCR, we use two primer pairs to check both ends of the insertion. Each primer pair consists of one primer binding to a region outside of the homology box within the BAC and one primer binding to a region flanking the insertion site.

After PCR verification of homologous recombination, restriction digests should be performed to rule out BAC DNA rearrangements or deletions and, when operator repeats are present, shortening of these repeats.

3.2.5 Generation of DNA Fragment for Removing galK

A second round of recombination followed by negative selection can be used to remove *galK*. Since *galK* is inserted between homology regions B and C, a DNA fragment consisting of homology regions B and C can be generated by PCR and used for recombination (Fig. 1b):

1. Design a pair of primers that match the ends of homology region B and C and perform standard PCR using the unmodified BAC DNA as the template.
2. Purify the PCR product using a Qiagen PCR purification kit. The concentration of purified fragment should be at least 20 ng/μl.

3.2.6 Removal of galK Using Negative Selection with DOG

1. Streak the glycerol stock *galK*-positive clone onto a MacConkey plate to obtain single colonies (*see* **Note** 7). Incubate the plate at 32 °C. Bright-red colonies should become obvious after 24 h.
2. Pick one bright-red colony from the MacConkey plate and inoculate a 2 ml culture containing antibiotics. Shake overnight at 32 °C.
3. Inoculate a 25 ml culture with 0.5 ml of overnight culture. Grow the culture to an OD600 of 0.55–0.6.
4. Heat shock the culture at 42 °C for 15 min.
5. Follow **steps 6–9** in Subheading 3.2, **step 1**.
6. Electroporate 50 μl of the electrocompetent cells with 5 μl of the DNA fragment. As a control, electroporate another 50 μl of cells with 5 μl of water (*see* **Note 8**). Recover the cells in 5 ml LB for 4.5 h at 32 °C.
7. As in Subheading 3.2, **step 3**, pellet 0.5 ml of the recovered culture and wash twice with 1xM9 salts. Then make two serial dilutions with 1xM9 salts. Plate 100 μl of the resuspended pellet (1/100 of the total cells), 100 μl of a 1:10 dilution (1/1,000 of the total cells), and 100 μl of a 1:100 dilutions (1/10,000 of the total cells) on DOG plates. Incubate at 32 °C for 3 days. We typically see tens to hundreds of colonies at the 1:1,000 dilution. The number of colonies on the recombination plates may not be significantly different to the control plates (could vary from 1:1 to 100:1). Screen 20–30 colonies from the recombination plate.

3.2.7 Screening for Clones Carrying Correctly Recombined BACs

Following *galK* removal, recombinants should again be screened using colony PCR. Following PCR screening, restriction digest fingerprinting should be performed to confirm the correct arrangement of the BAC sequences. By choosing appropriate restriction enzymes, the loss of the ~1.5 kb *galK* fragment should be obvious.

3.3 Protocol 3: Deletion of Specific BAC Sequences Using galK as the Dual Selectable Marker

BAC recombineering provides a convenient system for deleting specific sequences from the BAC. Using this method we have deleted regions ranging from 3 to 116 kb. The use of *galK* as the selectable marker allows the subsequent removal of *galK* by negative selection allowing multiple, sequential deletions to be made.

3.3.1 Generation of a DNA Fragment Containing Homology Regions and galK Using PCR

The recombination fragment for making deletions can be conveniently generated using PCR. The sequences flanking the region to be deleted are chosen as the homology boxes (Fig. 2a). The PCR primers should consist of two parts, the 17 bp sequence homology to the template p*galK* and the 43 bp homology box, similar to the primers described in Subheading 3.2, **step 2**.

The 17 bp sequences we typically use to PCR amplify gal*K* are:

Forward primer: 5′-cgacggccagtgaattg-3′ and reverse primer: 5′-tgcttccggctcgtatg-3′

It is critical to completely remove the intact p*galK* template from the PCR product by doing both DpnI digestion and gel extraction.

3.3.2 Deletion of Sequences Followed by Gal Selection

The experimental procedure is exactly the same as Subheading 3.2, **step 3**. The uninduced control should give nearly zero colonies, while the one tenth dilution of the induced sample should give dozens to hundreds of colonies after the selection.

3.3.3 Screening for Clones Carrying Correctly Recombined BACs

Colony PCR can be used to screen the recombinants. The PCR primers flanking the deleted region should be used, with the intact BAC before the deletion serving as a negative control.

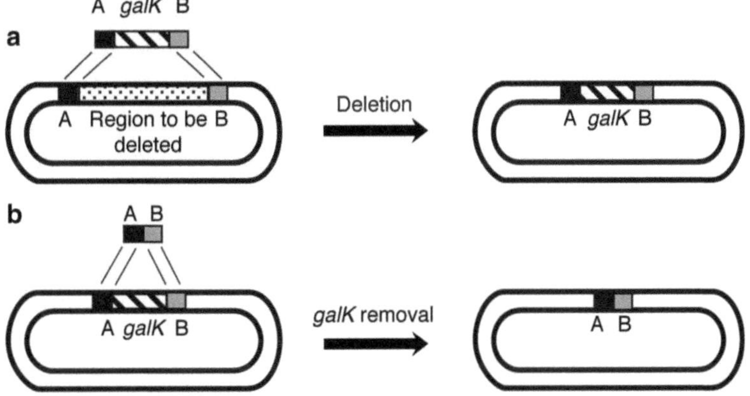

Fig. 2 Deleting DNA sequences within a BAC using *galK* selection. (**a**) Linear DNA-targeting construct is formed by inserting *galK* between homology regions A and B which flank the region to be deleted. Recombination with the target BAC in bacteria results in the deletion of the DNA sequence between homology regions A and B with insertion of *galK* between regions A and B. (**b**) Using a DNA-targeting construct consisting of homology regions A and B, galK can be removed by homologous recombination, allowing another round of deletions, if necessary

Following this PCR screening, restriction digests should be performed to confirm the deletion and the integrity of the BAC. Freeze down the colonies with correct digestion patterns.

3.3.4 Generation of the DNA Fragment for Removing galK

DNA fragments for removal of *galK* are generated by PCR using 60 bp primers partially complementary to each other. Each of the 60 bp primers consists of a 52 bp homology region flanking the *galK* marker and a 8 bp sequence complementary to the last 8 bp of the other homology region. This will give rise to a 16 bp complementary region between the two primers. For PCR, no extra template is needed (Fig. 2b).

3.3.5 Removal of galK Followed by Negative Selection with DOG

The experimental procedures are exactly the same as Subheading 3.2, **step 6**.

4 Notes

4.1 Transposon Insertion Reaction

No colonies on the selective plate after transposon insertion reaction:

1. Disruption of the mosaic ends (ME) of the isolated custom transposon:
   (a) *Psh*AI restriction enzyme, which is one of the enzymes used for isolating the transposon, might show star activity if incubated for too long at 37 °C, disrupting the efficiency of transposition. To avoid this possibility, we typically digest with *Psh*AI restriction enzyme at 25 °C for the least possible time.
   (b) To test the custom transposon preparation, perform the control transposon insertion reaction of the isolated transposon with the control target provided with the kit.
   (c) Sequence the transposon construct to verify the sequence of the ME and the flanking regions.
2. Sheared or contaminated target BAC DNA: to rule out the possibility of sheared BAC or the presence of inhibitors in the BAC DNA, perform the control transposon insertion reaction of the maxi prep BAC DNA with the control transposon provided with the kit.
3. Low competence of the electrocompetent cells used for transformation: if competent cells have a transformation efficiency $<10^8$ cfu/μg DNA, one may not obtain sufficient colonies after transposon insertion on the selective plate.
4. High concentration of the selective drug in plates: survival of the successfully transposed colonies might require lower than the conventionally used antibiotic concentration. For example, for kanamycin selection we get colonies only after reducing the concentration to 20 μg/ml.

4.2 BAC Recombineering

4.2.1 galK Selection

5. Colonies in the uninduced control:
   (a) Contamination from the uncut plasmid: uncut plasmid harboring *galK* transforms very efficiently and comprises most of the background. Make sure that there is no contamination from the uncut plasmid.
   (b) Incorrect composition of the selective plates: there should not be any other source of carbon in M63-Gal minimal plates apart from galactose.
6. No colonies in the induced sample:
   (a) Low recombination efficiency: for insertion of long or repetitive DNA sequences, increasing the length of the homology arms might help in increasing the recombination frequency. This is also true when there are internal sequences in the DNA fragment to be inserted with homology to other regions in the BAC outside the targeted region.
   (b) Absence of prophage needed for recombineering: the SW102 bacteria need to be grown at 32 °C. They should never be grown at 37 °C. SW102 bacteria die at 37 °C; only those SW102 bacteria that are defective for recombineering survive at this temperature.

4.2.2 galK Counterselection

7. Carry-over of SW102 cells containing the unmodified BAC: for the positive selection using galactose as the only carbon source, cells not expressing *galK* stop growing but are not killed. Therefore, when colonies are picked from the *gal* plate, small numbers of "hitchhiker" cells might also get picked, which can recover later in normal media and survive the DOG selection. To eliminate the contamination of these "hitchhikers," the *galK*-positive clone should be streaked onto a MacConkey indicator plates and only isolated, bright-red colonies should be picked for subsequent removal of *galK*.
8. High background: Apart from the desired removal of *galK*, loss-of-function mutations of the *galK* gene can also give rise to clones surviving the DOG selection. As a control for *galK* removal, ddH_2O is transfected into a heat-shocked sample to evaluate the spontaneous mutation frequency of *galK*. Even with a high background level, screening large numbers of colonies (15–20) by colony PCR should enable identification of correct *galK* deletions.

Acknowledgements

This work was supported by NIH grants GM42516, GM58460, and GM98319 to A.S.B.

References

1. Belmont AS, Hu Y, Sinclair PB, Wu W, Bian Q, Kireev I (2010) Insights into interphase large-scale chromatin structure from analysis of engineered chromosome regions. Cold Spring Harb Symp Quant Biol 75:453–460
2. Hu Y, Kireev I, Plutz MJ, Ashourian N, Belmont AS (2009) Large-scale chromatin structure of inducible genes- transcription on a linear template. J Cell Biol 185:87–100
3. Kireev I, Lakonishok M, Liu W, Joshi VN, Powell R, Belmont AS (2008) In vivo immunogold labeling confirms large-scale chromatin folding motifs. Nat Methods 5:311–313
4. Sinclair P, Bian Q, Plutz M, Heard E, Belmont AS (2010) Dynamic plasticity of large-scale chromatin structure revealed by self-assembly of engineered chromosome regions. J Cell Biol 190:761–776
5. Warming S, Costantino N, Court DL, Jenkins NA, Copeland NG (2005) Simple and highly efficient BAC recombineering using galK selection. Nucleic Acids Res 33:e36

Part III

Imaging Nuclear Processes and Sub-nuclear Structures

Chapter 15

Spatiotemporal Visualization of DNA Replication Dynamics

Marius Reinhart, Corella S. Casas-Delucchi, and M. Cristina Cardoso

Abstract

The ability of cells to copy their DNA allows them to transmit their genetic information to their progeny. In such, this central biological process preserves the instructions that direct the entire development of a cell. Earlier biochemical analysis in vitro and genetic analysis in yeast laid the basis of our understanding of the highly conserved mechanism of DNA replication. Recent advances on labeling and live-cell microscopy permit now the dissection of this fundamental process in vivo within the context of intact cells. In this chapter, we describe in detail how to perform multiple DNA replication labeling and detection allowing high spatial resolution imaging, as well as how to follow DNA replication in living cells allowing high temporal resolution imaging.

Key words DNA replication, Fluorescent protein, Immunofluorescence staining, Live-cell microscopy, Nucleotide pulse labeling

1 Introduction

The accurate duplication of the genome is the basis for cell proliferation. The process of DNA replication takes place during S-phase and is organized both spatially and temporally [1], so that the activation of single replication origins throughout S-phase results in conserved in situ labeling patterns that change as S-phase progresses [2–4]. Nevertheless, the mechanism by which active replication spreads along a chromosome remains unclear.

First hints about the organization of active replication sites were obtained by pioneering experiments using radioactively labeled nucleotides, which are incorporated during the DNA synthesis process. These studies provided very valuable data on the localization of active replication sites along single stretched DNA fibers and further presented the first in situ data on DNA replication in mammalian cells [5, 6]. However, detailed spatial

Marius Reinhart and Corella S. Casas-Delucchi have contributed equally to this work.

Yaron Shav-Tal (ed.), *Imaging Gene Expression: Methods and Protocols*, Methods in Molecular Biology, vol. 1042, DOI 10.1007/978-1-62703-526-2_15, © Springer Science+Business Media, LLC 2013

information about the in situ organization of replication sites had to wait for the development of the first specific antibodies against halogenated nucleotides in the 1980s [2, 7]. In situ detection of incorporated nucleotides and later of replication factors [8, 9] provided quite detailed spatial information; however, they represent single snapshots of the dynamic replication process. The combination of two pulses with differently modified nucleotides, on the other hand, for the first time ascertained the organization of DNA replication in situ both spatially and temporally: each snapshot gives detailed spatial information and the correlation between both provides valuable information on the way active sites of DNA replication progress in situ [10, 11]. Later, the sequential use of directly labeled nucleotides made it possible to visualize sites of DNA synthesis in living cells at different time points. However, fluorescent dUTPs are not permeable through the cell membrane and need to be delivered by microinjection of cells or scratch loading ([12] and references therein). However, these elegant methods have the drawback that it is technically difficult to get even one nucleotide into the cell and incorporated into DNA making it not widely used and difficult to extend to multiple labelings. Finally, the development of fluorescent proteins for cell biological applications [13] made the next step possible, namely, the visualization of DNA replication progression in vivo over longer periods of time [14, 15]. Following fluorescently tagged PCNA in vivo, Sporbert et al. showed that new replication foci are always activated adjacently to already active ones [16] and proposed a domino model to explain the propagation of active DNA replication, where active replication results in its own propagation by, for instance, destabilizing chromatin/DNA and thereby facilitating firing of nearby origins [17].

Here, we present in detail two very useful methods, both allowing visualization of DNA replication dynamics in situ using fluorescence microscopy. The first, consisting of time-lapse imaging of living cells expressing fluorescently tagged replication proteins, provides both 3D spatial information and especially extensive temporal information. The second, a further development of pulse and chase experiments using modified nucleotides to label replication in situ [18], allows the visualization of up to four combined snapshots of replication sites active at selected time intervals. While the former approach can be used to visualize replication progression continuously, the latter facilitates the study of the spatial progression throughout chromatin domains thanks to the simultaneous visualization of all four replication time points. Like the Heisenberg uncertainty principle [19], achieving the highest possible spatial resolution is often incompatible with acquiring the most detailed temporal information.

2 Materials

All solutions and materials used for cell culture and live-cell microscopy must be sterile.

2.1 Live-Cell Visualization of DNA Replication Dynamics

1. Cell lines: for live-cell microscopy, cells should grow adherently. While the cell line to be used depends on the interest of the scientist, there are certain considerations simplifying the acquisition of data (*see* **Note 1**). In general, transiently expressed fluorescently tagged replication factors are necessary. However, there are also stable cell lines available [15].
2. Growing medium: use the standard medium required for the cell line to be imaged.
3. Pre-warmed PBS containing 0.5 mM EDTA and 0.05 % trypsin.
4. Plasmids: mammalian expression vectors coding for the replication factor of interest tagged to a fluorescent protein. The fluorescent marker should be chosen according to the wavelengths that can be imaged using the microscope available. It is possible to combine different fluorescent markers, also depending on the microscope setup. The most standard marker, which can typically be imaged in most microscopes, is green fluorescent protein (GFP) [20]. Replication factors most commonly used to label sites of ongoing replication in living cells are PCNA and DNA Ligase 1 [14, 15, 21].
5. Transfection reagents: nucleofection system from Amaxa (Lonza), nucleofection solution V, cuvette, and pipette (*see* **Note 2**).
6. Microscopy dishes: the form and size depends on the optical table inset available in the microscope. The bottom has to be thin enough for higher magnification objectives to be able to image through to the sample. Material can be glass or optical plastic. Glass lids are recommended for optimal contrast images (*see* **Note 3**).
7. Microscope: for high-resolution imaging of living cells, we recommend the use of a spinning disk confocal microscope, characterized by high-speed acquisition and low phototoxicity to cells. The stage should be motorized to allow the acquisition of 3D stacks at several points in one experiment.
8. Incubation chamber: the incubation chamber on the microscope must keep a constant temperature, CO_2, and humidity imitating the normal cell growth conditions (*see* **Note 4**).

2.2 Polytemporal DNA Replication Staining

1. Growth medium.
   Human cervical cancer cells (HeLa) and *M. cabrerae* fibroblasts [22] are cultivated in Dulbecco's Modified Eagle's Medium (DMEM) supplemented with 10 % fetal calf serum (FCS), 2 mM L-glutamine, and 25 mg/l gentamicin.

2. Denaturation.
   (a) Enzymatic denaturation: DNaseI (Roche), 2× DNase buffer: mix 60 mM Tris–HCl pH 8.1, 0.66 mM $MgCl_2$ and 1 mM mercaptoethanol in ddH_2O.
   (b) Acid denaturation: mix 336 μl 12 N HCl with 10 μl Triton X-100 and 654 μl ddH_2O (freshly prepared).
3. PBST wash buffer.

   Mix 1× PBS with Tween 20 to a final concentration of 0.01 %.
4. PBSTE DNAse stop buffer.

   Add 1 mM EDTA to 1× PBST.
5. Blocking buffers.

   4 % bovine serum albumin (BSA) in PBS.

   0.2 % fish skin gelatin (Sigma) in PBS.
6. Primary antibodies/chemical detection.
   (a) Click-iT EdU Alexa Fluor 488 Imaging Kit (Invitrogen).
   (b) Rat anti-bromodeoxyuridine (BrdU), reacts weakly with chlorodeoxyuridine (CldU) (Gentaur Molecular Products, catalog no. OBT0030CX).
   (c) Mouse anti-bromodeoxyuridine (BrdU), reacts also with iododeoxyuridine (IdU) (Becton Dickinson, catalog no. 347580).
   (d) Anti-DNA Ligase 1 rabbit polyclonal antibody [14].
7. Secondary antibodies (*see* **Note 5**).
   (a) CF405M-conjugated goat anti-mouse IgG (H+L) highly cross-adsorbed, 2 mg/mL (Biotium).
   (b) Cy5-conjugated AffiniPure donkey anti-rat IgG (H+L) highly cross-adsorbed, 1.5 mg/mL (Jackson ImmunoResearch Europe).
   (c) Cy3-conjugated AffiniPure donkey anti-rabbit IgG (H+L) highly cross-adsorbed, 1.5 mg/mL (Jackson ImmunoResearch Europe).
8. Mounting.

   Add 8 g Moviol 4-88 (Polyscience) to 40 ml 0.2 M Tris–HCl pH 8.5 and dissolve by heating to 50–60 °C with occasional stirring. After cooling down, add 20 ml glycerol and 1–2.5 % DABCO (anti-fading agent) and spin at 3,800 × *g* for 15 min, aliquot supernatant and store at −20 °C.

Alternatively, use Vectashield mounting medium (Vector Laboratories, Inc.).

3 Methods

Below, we provide detailed protocols for (1) replication labeling by live-cell microscopy for mouse embryonic fibroblasts and (2) fixed cell polytemporal replication imaging in human cervical cancer cells and cabrera's vole fibroblasts, a rodent species endemic from the Iberian peninsula [22].

3.1 Live-Cell Visualization of DNA Replication Dynamics

Here, we present a detailed protocol to image active sites of DNA replication and, simultaneously, highlight specific nuclear regions, in this case using MaSat-GFP, a polydactyl zinc finger protein that specifically binds the major satellite repeats enriched at mouse constitutive heterochromatin (*see* Fig. 1 and Movie S1).

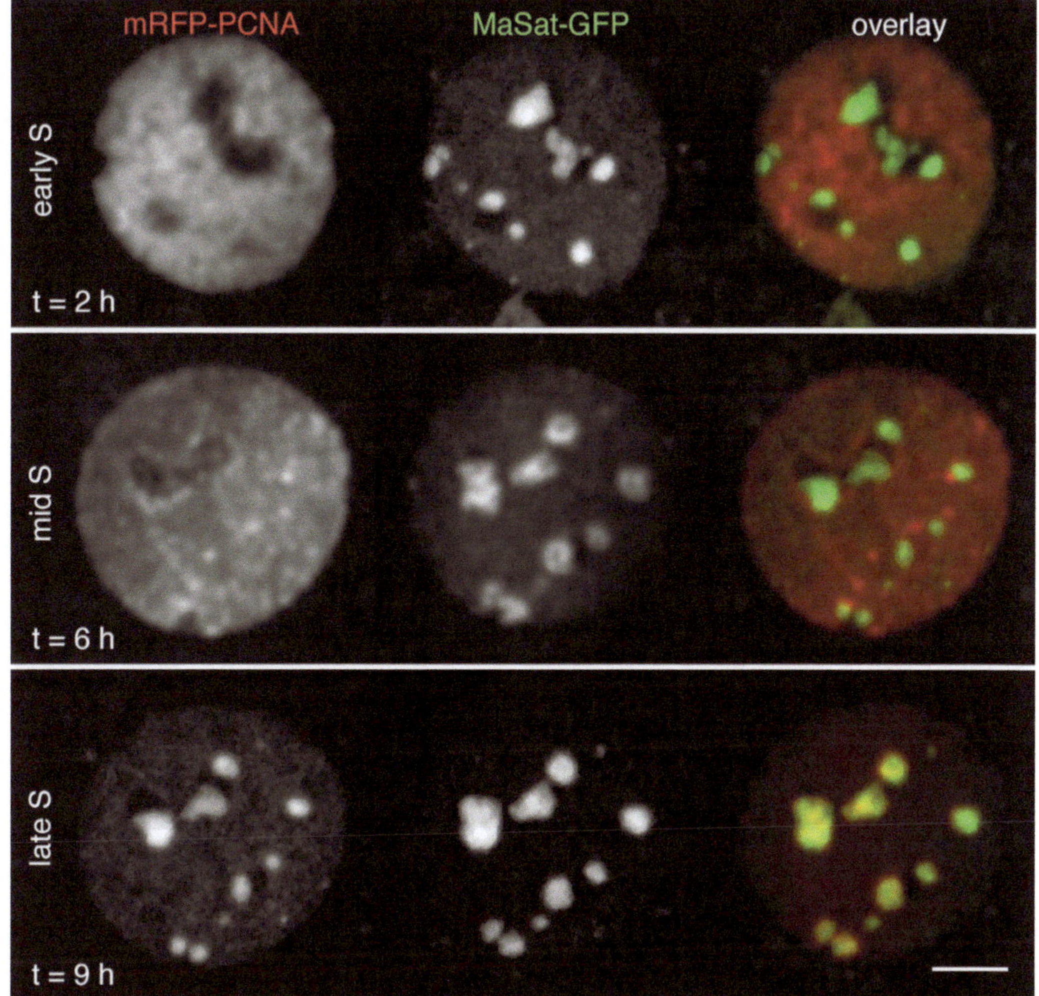

Fig. 1 Mouse embryonic fibroblast transiently co-expressing mRFP-PCNA, as a marker for sites of ongoing DNA replication and MaSat-GFP, as a marker for heterochromatin. Cells were imaged in a spinning disk microscope equipped with a climatization chamber so as to maintain constant temperature (37 °C), 5 % CO_2 and 60 % humidity. 3D confocal stacks were acquired at 1 h time intervals for a period of 12 h. Exemplary images of the same cell undergoing different S-phase stages and exhibiting the corresponding characteristic patterns. Scale bar: 5 μm

Different nuclear regions can be visualized specifically using various in vivo markers [23–25]. The protocol is adapted to image cells in a 35 mm glass-bottomed dish.

1. Pre-warm growing medium and PBS + EDTA to 37 °C and trypsin and nucleofection solution to room temperature. Prepare the dish where electroporated cells will be seeded by adding the final volume of growing medium (2 ml for a 35 mm dish) and keep it in an incubator so that the medium reaches 37 °C and CO_2 diffuses into it.
2. Use a $0.5–1 \times 10^6$ adherently growing cells (*see* **Note 6**) on, e.g., a 10 cm diameter plate. Remove growing medium and wash carefully with 5 ml PBS EDTA so as not to detach cells from the surface. Add 0.5 ml trypsin and incubate at 37 °C for 2–5 min. Monitor cell detachment under a microscope. When most cells have detached from the growing substrate and are now single cells, stop the reaction by addition of 4.5 ml growing medium. If cells clump, carefully pipette the cell suspension up and down a couple of times before stopping the trypsin reaction. Centrifuge the cells for 7 min at $300 \times g$.
3. Prepare 100 μl of nucleofection solution with 2 μg total plasmid DNA.
4. Once the cells are pelleted, discard the supernatant and carefully resuspend them in 100 μl nucleofection solution with plasmid DNA. Transfer the cell suspension into an appropriate cuvette avoiding air bubbles. Immediately perform the electroporation using the appropriate program for your cells (*see* **Note 7**). Take the previously prepared dish from the incubator, pipette approximately 500 μl medium from the dish into the cuvette, and carefully resuspend the cells. Transfer cell suspension into the dish with the rest of the medium; carefully shake the dish and return it to the incubator. Incubate overnight.
5. On the next day, remove the medium, carefully wash once or twice with pre-warmed medium to remove dead cells and debris, and add new medium.
6. Before bringing the cells to the microscope, make sure that the incubation chamber is already at 37 °C, 5 % CO_2, and >40 % humidity level.
7. Place the dish with the transfected cells on the microscope. Allow the dish to acclimatize to the new conditions for some minutes before starting imaging. Slight changes in temperature can affect the material in such a way that the focal plane can change dramatically during the first 10–20 min.
8. Look for transfected cells using the longest wavelength possible and short exposure times to minimize phototoxicity (i.e., in the case of co-expression of GFP and mRFP tagged proteins,

look for transfected cells using the red channel and only quickly check to see whether the cells also express the GFP-tagged construct). Select cells (*see* **Note 8**) that express the minimal amount of fluorescent protein that can be imaged properly. Too high expression levels can lower the chances that transfected cells will pass normally through S-phase. Extreme over-expression can also cause apoptosis. In case the lid is to be removed/replaced, do so before selecting the cells for imaging, since the dish might otherwise be shifted (*see* **Note 9**).

9. Set up the imaging conditions finding a compromise between phototoxicity and undersampling. The ideal conditions depend strongly on the cell line, since some cells are more sensitive to transfection and phototoxicity. In general, acquiring *z*-stacks at a time interval of 10–30 min is usually enough to follow changes in replication patterns. The minimal time to acquire an entire S-phase/cell cycle depends on how fast the cells divide. Under optimal conditions, cells can be kept on the stage and imaged for over 2 days (*see* **Note 10**).

3.2 Polytemporal DNA Replication Staining

Here, we present the detailed protocol for imaging multiple replication time points in mammalian cells (*see* Fig. 2). Proliferating cells are supplied with thymidine analogs, which are incorporated in the replicating genomic DNA for the duration of the pulse. Incorporation of modified nucleotides is stopped by chasing with excess unlabeled thymidine (*see* **Note 11**). The protocol is adapted for multiple 16 mm diameter coverslips in a 60 mm cell culture dish. Pulse-chase length needs to be adapted to doubling time and S-phase duration of the respective cell line, as well as to the purpose of the experiment. In principle, any replicating, cultured cells can be used for this approach.

1. Replication pulse labeling with thymidine analogs (IdU, EdU, and CldU).

   Label, e.g., 20 min with a final concentration of 10 μM IdU in growth medium at 37 °C, rinse twice with pre-warmed growth medium, and incubate cells for, e.g., 60 min with 200 μM thymidine in growth medium at 37 °C. Rinse again twice to remove thymidine from growth medium.

   Label 20 min with a final concentration of 10 μM EdU in growth medium at 37 °C, rinse two times with growth medium, and incubate cells for 60 min with 200 μM thymidine in growth medium at 37 °C. Rinse again twice to remove thymidine from growth medium.

   Label 20 min with a final concentration of 10 μM CldU in growth medium at 37 °C, rinse two times with growth medium, and incubate cells for 60 min with 200 μM thymidine in growth medium at 37 °C (*see* **Note 12**).

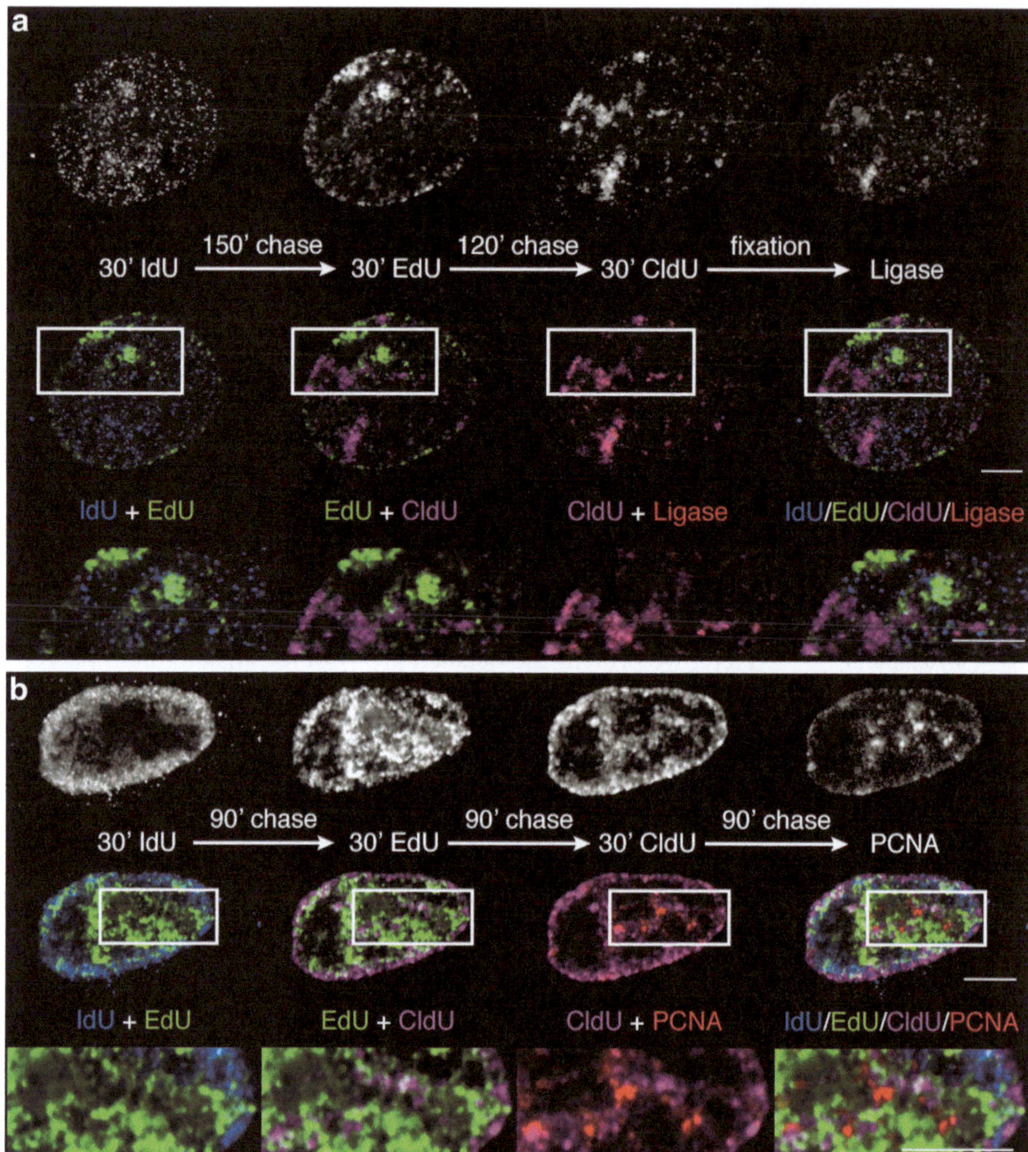

Fig. 2 Quadruple replication labeling in mammalian cells. Mid optical section of confocal images. (**a**) Six hour multiple pulse-chase replication staining in *M. cabrerae* fibroblast with immediate fixation after the CldU pulse. IdU pulse in *blue*, EdU pulse in *green*, CldU pulse in *magenta,* and DNA Ligase 1 in *red*, pulse and chase times as given in the image. (**b**) Six hour multiple pulse-chase replication staining in HeLa cells expressing mCherry-PCNA with fixation following CldU pulse and chase. IdU pulse in *blue*, EdU pulse in *green*, CldU pulse in *magenta,* and PCNA in *red*, pulse and chase times as given in the image. *Top row*: single replication stainings. *Middle row*: overlaid images of two sequential pulses followed by an overlay of all four pulses. *Bottom row*: selected ROI from the *middle row*. Scale bars: 5 μm

2. Fixation.
   Rinse cells once with PBS at room temperature to remove growth medium before fixing with 3.7 % formaldehyde in PBS for 10 min.
   Remove fixative and rinse two to three times with PBST.
   Cells can be stored at this point in PBS with 0.02 % Na-azide at 4 °C for several days.
3. Staining chamber.
   Prepare a lightproof chamber with a wet filter paper to maintain a humid environment and place parafilm on top of the filter [26]. Transfer coverslips to the chamber (*see* **Note 13**).
4. Permeabilization.
   To allow antibody penetration, permeabilize the cells with 0.5 % Triton X-100 in PBS for 10–20 min at room temperature. Wash the cells three times in PBST for 0, 5, and 10 min (hereafter referred to as triple wash) to remove the permeabilization reagent (*see* **Note 14**).
5. Detection of replication labeling (*see* **Note 15**).
   Use the ClickiT-EdU kit according to manufacturer's protocol to detect incorporated EdU followed by a triple wash (*see* **Note 16**).
   Block with 0.2 % fish skin in PBS for 45 min at room temperature.
   Mix 0.25 μl rat anti-BrdU Gentaur antibody solution (CldU detection) with 12.5 μl 2× DNase buffer and 12.5 μl 4 % BSA in PBS. Add 1 U DNaseI and incubate at 37 °C for 1 h, followed by a triple wash with PBSTE (*see* **Notes 17** and **18**).
   Mix 6.25 μl mouse anti-BrdU Becton Dickinson antibody solution (IdU detection) with 0.25 μl rabbit anti-DNA Ligase 1 antibody and 23.5 μl 4 % BSA in PBS and incubate at room temperature for 60 min, followed by a triple wash.
   Mix 0.06 μl anti-mouse IgG CF405M, 0.3 μl anti-rat IgG Cy5, 0.5 μl anti-rabbit IgG Cy3, and 24 μl 4 % BSA in PBS to perform secondary antibody incubation at room temperature for 60 min, followed by a triple wash.
6. Mounting.
   Rinse with ddH_2O to remove salts. Mount coverslips on microscope glass slide with Vectashield and seal with nail polish or mount with Moviol (*see* **Note 19**).

4 Notes

1. Factors to consider when choosing a cell line to perform live-cell microscopy include:
   - How well the cells can be transfected (transfection and expression rate)
   - How well the cells tolerate imaging-derived phototoxicity

- How much the cells move
- How fast they divide

2. While a good transfection rate is a factor to consider when choosing a transfection method, when following cell cycle progression at a single cell level, it is more important to achieve a moderate expression level. We recommend the nucleofection system from Amaxa (Lonza), although other methods can also be used.
3. Usually, the lid of microscopy dishes is of a thick plastic, which prevents good contrast images. For short-term imaging, it is possible to simply remove the lid.
4. If it is not possible to regulate the CO_2 and humidity levels, an alternative is to add 10 mM HEPES buffer to phenol-free medium and seal the dish with, for instance, paraffin to avoid evaporation of the medium.
5. Secondary antibodies indicated here can be substituted with your own favorite antibodies, as long as they are cross-absorbed to avoid cross-reaction with the other antibodies and the fluorophores conjugated provide enough spectral separation to be imaged by your microscope.
6. Cell density is a key factor for live-cell imaging of DNA replication. While a too high density can result in cell contact inhibition and therefore prevent cells from cycling, a too low density can result in cells moving more freely along the growing surface, making it extremely hard to keep them in frame over several hours. The optimal cell density depends on the cell line used: mouse fibroblasts and myoblasts tend to move, and therefore, a rather high density is recommended. On the other hand, HeLa cells, for instance, are less mobile and can therefore be plated at a lower density.
7. The program B-032 gives good results for mouse embryonic fibroblasts, C2C12 mouse myoblasts, HeLa cells, and *M. cabrerae* fibroblasts.
8. If the acquisition software allows the possibility of multipoint and mosaic imaging (stitching), the decision should be made depending on both the transfection rate and the mobility of the cell line used: a high transfection rate and cells that move fast are better imaged using the mosaic function, while sparsely transfected cells, or cells that barely move, are better imaged selecting separate points across the dish.
9. For time-lapse imaging over several hours, removing the lid would result in evaporation of the medium and possibly contamination of the sample. Therefore, in such cases it is better to replace the lid by a glass coverslip or keep the plastic lid.
10. In an asynchronous population at any given time point, the user will observe cells undergoing all of the different cell cycle substages.

Mitotic cells are easily recognized using phase contrast or DIC by their spherical morphology. At this stage of the cell cycle, PCNA diffuses throughout the cytoplasm. During G1 and G2, PCNA is exclusively nuclear but distributed homogenously. As S-phase begins, PCNA distribution becomes focal. During early S-phase the replication foci labeled by PCNA are distributed throughout the nucleus with the exception of the nuclear and nucleolar periphery. As cells progress into mid S-phase, the replication foci relocalize to these peripheral regions. During the second half, or late, S-phase, replication foci cluster at heterochromatic regions, giving the impression, at the confocal microscopy level, of fewer but larger foci (Fig. 1 and Movie S1).

11. With this pulse-chase-pulse labeling protocol, sequentially replicated regions of the genome are marked with differently modified nucleotides and visualized simultaneously within single cells. This protocol allows following the spatiotemporal dynamics of DNA replication progression in situ (Fig. 2).
12. We chose this specific chronological labeling order to facilitate visualization of potential cross-reaction between IdU and CldU antibodies. The additional intermediate EdU labeling time point allows a clear discrimination between both IdU and CldU labeling patterns. A high colocalization between both IdU and CldU antibodies signals indicates spurious cross-reaction of the primary antibodies with both nucleotides.
13. To reduce the required amount of antibodies, we use a “cell down protocol.” A 25 μl drop of staining reagent for each 16 mm diameter coverslip was placed on the parafilm in the humidified chamber. The coverslips were placed with cells facing downward onto the reagent carefully avoiding any air bubbles. If conventional methods are used instead of the cell down protocol, the respective amounts of reagents have to be adapted.
14. For PCNA detection, incubate cells for 10 min in ice-cold methanol and air-dry coverslips. Cell shape is usually affected by MeOH incubation and air-drying.
15. Nucleotide labeling and fluorescently coupled proteins are readily combinable with other (immuno) fluorescent labeling methods. Chromosome painting, methylation labeling, and chromocenter labeling are only a few among many possible permutations.
16. We observed intensity decrease in other fluorochromes when the ClickiT-EdU kit was applied as the final detection step.
17. Primary antibodies rat anti-BrdU Gentaur and mouse anti-BrdU Becton Dickinson were detected sequentially to minimize cross-reaction. A high salt wash in between is not essential but improves the quality of the staining by decreasing cross-reactions.

18. Alternatively to DNaseI treatment, acid denaturation can be performed after permeabilization and before the first detection step to expose the modified nucleotides for antibody recognition. For acid denaturation, incubate with 4 N HCl solution (*see* Subheading 2) for 10 min at room temperature. While acid denaturation can cause artifacts and partially damage protein epitopes, such as the ones recognized by the anti-DNA Ligase 1 antibodies, nucleotide recognition is generally improved.
19. Each mounting agent has different advantages. Vectashield allows later unmounting and restaining, whereas Moviol is a permanent mounting agent.

Acknowledgements

We thank Juan Alberto Marchal (University of Jaen, Spain) for the *Microtus cabrerae* fibroblasts and all present and past members of the laboratory for their contributions over the years. The laboratory of M. Cristina Cardoso is supported by grants of the German Research Foundation (DFG) and the Federal Ministry of Education and Research (BMBF).

References

1. Berezney R, Dubey DD, Huberman JA (2000) Heterogeneity of eukaryotic replicons, replicon clusters, and replication foci. Chromosoma 108:471–484
2. Nakamura H, Morita T, Sato C (1986) Structural organizations of replicon domains during DNA synthetic phase in the mammalian nucleus. Exp Cell Res 165:291–297
3. Nakayasu H, Berezney R (1989) Mapping replicational sites in the eucaryotic cell nucleus. J Cell Biol 108:1–11
4. O'Keefe RT, Henderson SC, Spector DL (1992) Dynamic organization of DNA replication in mammalian cell nuclei: spatially and temporally defined replication of chromosome-specific alpha-satellite DNA sequences. J Cell Biol 116:1095–1110
5. Huberman JA, Riggs AD (1966) Autoradiography of chromosomal DNA fibers from Chinese hamster cells. Proc Natl Acad Sci USA 55:599–606
6. Huberman JA, Tsai A, Deich RA (1973) DNA replication sites within nuclei of mammalian cells. Nature 241:32–36
7. Gratzner HG (1982) Monoclonal antibody to 5-bromo- and 5-iododeoxyuridine: a new reagent for detection of DNA replication. Science 218:474–475
8. Bravo R, Macdonald-Bravo H (1985) Changes in the nuclear distribution of cyclin (PCNA) but not its synthesis depend on DNA replication. EMBO J 4:655–661
9. Cardoso MC, Leonhardt H, Nadal-Ginard B (1993) Reversal of terminal differentiation and control of DNA replication: cyclin A and Cdk2 specifically localize at subnuclear sites of DNA replication. Cell 74:979–992
10. Aten JA, Bakker PJ, Stap J, Boschman GA, Veenhof CH (1992) DNA double labelling with IdUrd and CldUrd for spatial and temporal analysis of cell proliferation and DNA replication. Histochem J 24:251–259
11. Manders E, Stap J, Brakenhoff G, Van Driel R, Aten J (1992) Dynamics of three-dimensional replication patterns during the S-phase, analysed by double labelling of DNA and confocal microscopy. J Cell Sci 103:857–862
12. Schermelleh L, Solovei I, Zink D, Cremer T (2001) Two-color fluorescence labeling of early and mid-to-late replicating chromatin in living cells. Chromosome Res 9:77–80
13. Chalfie M, Tu Y, Euskirchen G, Ward WW, Prasher DC (1994) Green fluorescent protein as a marker for gene expression. Science 263:802–805
14. Cardoso MC, Joseph C, Rahn HP, Reusch R, Nadal-Ginard B, Leonhardt H (1997) Mapping

and use of a sequence that targets DNA ligase I to sites of DNA replication in vivo. J Cell Biol 139:579–587

15. Leonhardt H, Rahn H, Weinzierl P, Sporbert A, Cremer T, Zink D, Cardoso M (2000) Dynamics of DNA replication factories in living cells. J Cell Biol 149:271–280
16. Sporbert A, Gahl A, Ankerhold R, Leonhardt H, Cardoso MC (2002) DNA polymerase clamp shows little turnover at established replication sites but sequential de novo assembly at adjacent origin clusters. Mol Cell 10:1355–1365
17. Chagin VO, Stear JH, Cardoso MC (2010) Organization of DNA Replication. Cold Spring Harb Perspect Biol 2
18. Casas-Delucchi CS, Brero A, Rahn H-P, Solovei I, Wutz A, Cremer T, Leonhardt H, Cardoso MC (2011) Histone acetylation controls the inactive X chromosome replication dynamics. Nat commun 2:222
19. Heisenberg W (1927) Über den anschaulichen Inhalt der quantentheoretischen Kinematik und Mechanik. Zeitschrift für Physik 43:172–198
20. Shaner NC, Steinbach PA, Tsien RY (2005) A guide to choosing fluorescent proteins. Nat Methods 2:905–909
21. Sporbert A, Domaing P, Leonhardt H, Cardoso MC (2005) PCNA acts as a stationary loading platform for transiently interacting Okazaki fragment maturation proteins. Nucleic Acids Res 33:3521–3528
22. Bullejos M, Burgos M, Jimenez R, Sanchez A (1996) Distribution of sister-chromatid exchanges in different types of chromatin in the X-chromosome of Microtus cabrerae. Experientia 52:511–515
23. Lindhout BI, Fransz P, Tessadori F, Meckel T, Hooykaas PJJ, van der Zaal BJ (2007) Live cell imaging of repetitive DNA sequences via GFP-tagged polydactyl zinc finger proteins. Nucleic Acids Res 35:e107
24. Casas-Delucchi CS, van Bemmel JG, Haase S, Herce HD, Nowak D, Meilinger D, Stear JH, Leonhardt H, Cardoso MC (2012) Histone hypoacetylation is required to maintain late replication timing of constitutive heterochromatin. Nucleic Acids Res 40:159–169
25. Weidtkamp-Peters S, Rahn H-P, Cardoso MC, Hemmerich P (2006) Replication of centromeric heterochromatin in mouse fibroblasts takes place in early, middle, and late S phase. Histochem Cell Biol 125:91–102
26. Cardoso MC, Leonhardt H (1995). Immunofluorescence techniques in cell cycle studies. In: Pagano, M (ed) Cell cycle: materials and methods. Heidelberg, Springer-Verlag, pp 15–28

Chapter 16

The Dynamics of DNA Damage Repair and Transcription

Niraj M. Shanbhag and Roger A. Greenberg

Abstract

Recent advances have led to several systems to study transcription from defined loci in living cells. It has now become possible to address long-standing questions regarding the interplay between the processes of DNA damage repair and transcription—two disparate processes that can occur on the same stretch of chromatin and which both lead to extensive chromatin change. Here we describe the development of a system to create enzymatically induced DNA double-strand breaks (DSBs) at a site of inducible transcription and methods to study the interplay between these processes.

Key words Double-strand breaks, Transcription, ATM, Ubiquitin

1 Introduction

The repair of damaged DNA requires the timely and coordinated recruitment of multiple DNA damage repair (DDR) proteins to the lesion. These events have been especially well studied at DNA double-strand breaks (DSBs). The signal amplification that leads to the recruitment of many molecules of a given DDR protein to a site of damage results in a spreading of these repair factors along chromatin near a break and also makes this recruitment amenable to visualization by immunofluorescence microscopy. A common method of inducing damage to initiate these recruitment events is the use of ionizing radiation, which leads to the formation of multiple ionizing radiation-induced foci (IRIF) of repair factors throughout the irradiated nuclear volume. Alternatively, an ultraviolet laser can be used to create a stripe across the nucleus of a cell that has been first sensitized to DSB induction by treatment with bromodeoxyuridine. Subsequent laser microirradiation leads to the creation of a large and easily visualized swath of DNA DSBs across the nucleus [1, 2]. Methods such as these have been useful in the study of multiple repair factors and chromatin modifications that occur at DSBs, as well as the kinetics and relative requirements for accumulation of many of these marks at sites of damage [2–4].

Yaron Shav-Tal (ed.), *Imaging Gene Expression: Methods and Protocols*, Methods in Molecular Biology, vol. 1042, DOI 10.1007/978-1-62703-526-2_16,

In addition to these methods, which create damage at somewhat random sites within the genome, several groups have created methods for enzymatically induced breaks at defined locations in the genome. Some of these involve the use of restriction enzymes which target endogenous loci [5–7], while others have integrated exogenous restriction enzyme target sites at defined genomic loci to study the DDR [8–10].

A long-standing question in the field of DNA damage repair has been the interplay between the creation and repair of DSBs, and transcription occurring on nearby chromatin. Though a variety of DNA lesions are known to interfere with transcription [11, 12], the question has remained largely unanswered for DSBs. To address this question, varying groups have used ionizing radiation, followed by nuclear run-on assays [13, 14], as well as enzyme-induced DSBs coupled with chromatin immunoprecipitation and quantitative reverse-transcription PCR at nearby endogenous or reporter genes [6, 7, 15].

In this chapter, we describe the experimental procedures to efficiently create multiple nuclease-induced DSBs at a known distance away from an inducible and visualizable transcriptional reporter and methods to study both the DDR and its effects on local transcription. To do this, we utilize a previously described genomically integrated transcriptional reporter system in which genomic DNA, transcribed RNA, and translated protein can be visualized simultaneously in real-time or in fixed cells [16, 17]. The genomic locus where the reporter is integrated is visualized by the tandem lac operator present in the reporter. Expression of a mCherry-fluorescent-tagged lac-repressor protein (mCherry-LacI) leads to a collection of fluorescent proteins at the reporter, which is easily visualized by fluorescence microscopy. Depending on the specific cell line used, transcription can be activated by expressing a "tet-off" transactivator that translocates to the nucleus upon treatment with 4-hydroxytamoxifen (2-6-3 cells) [16, 17], or by doxycycline treatment of a variant cell line that stably expresses YFP-MS2, as well as a "tet-on" transactivator (2-6-3 rtTA + YFP-MS2 cells) [16]. The 2-6-3 cells do not stably express any fluorescent proteins and therefore are amenable to the study of fluorescence-tagged or immunostained repair factors of interest. The 2-6-3 rtTA + YFP-MS2 cells stably express a YFP-MS2 fusion protein and are therefore useful in studying transcription from the reporter. Transcription is visualized due to the 24 repeats of the MS2 sequence, which are integrated in the 3′-UTR of the reporter gene. The sequence, from the MS2 bacteriophage, forms a stem–loop structure in the transcribed RNA, which is then bound strongly and specifically by dimers of the YFP-tagged MS2 coat protein [17–19]. Finally, after induction of transcription, the reporter protein—a CFP-tagged peroxisome-targeting SKL peptide—can be visualized collecting in the cytoplasmic puncta of peroxisomes. RNA and protein

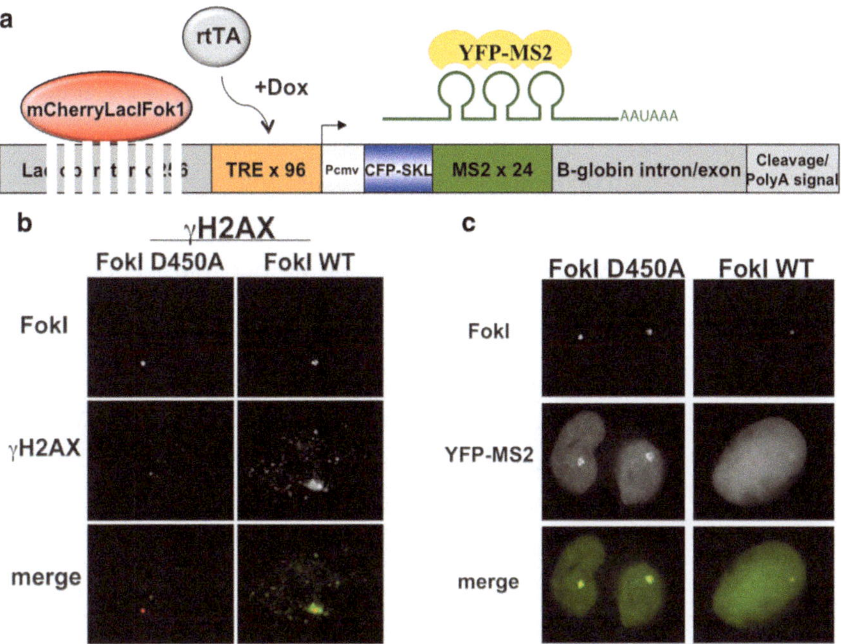

Fig. 1 System for monitoring communication between DNA damage responses and local transcription. (**a**) Schematic of reporter. The mCherry-LacI-Fok1 fusion protein creates DSBs within 256 lac operator repeats. Transcription is inducible upon doxycycline (Dox) addition, which activates a CFP gene. YFP-MS2 focally accumulates at the locus by binding to stem–loop structures within the 3′-UTR of the nascent transcript. (**b**) Expression of wild-type mCherry-LacI-Fok1 (FokI WT), but not nuclease-inactive FokI D450A, induces DSBs as indicated by gH2AX focus formation. (**c**) YFP-MS2 accumulates at the locus in FokI D450A-expressing cells indicating the presence of nascent transcription. This does not occur in FokI WT cells indicating that DSB responses silence transcription

production in this system can also be measured through qRT-PCR and immunoblotting, respectively.

To create DSBs in this system, we employ the nonspecific nuclease domain of the FokI nuclease [20, 21], fused to the mCherryLacI protein (FokI-mCherry-LacI) (Fig. 1). Upon expression of this construct, the lac-repressor domain targets the nuclease domain to a site approximately 5 kb upstream of the inducible reporter gene, where it can dimerize and create DSBs. As a control, we use a catalytically ineffective mutant of FokI (FokI-D450A) [20]. In Subheadings 2.1 and 2.2, we describe the induction of damage at the reporter and subsequent detection of DDR proteins through immunostaining or expression of fluorophore-tagged proteins. In Subheadings 2.3 and 2.4, we describe the quantification of transcription from the reporter by immunofluorescence and by qRT-PCR, respectively. The system entails the first example of a method that enables simultaneous, real-time analysis of DNA damage responses and their impact on local transcriptional events occurring on a contiguous stretch of chromatin. The DSBs induce chromatin changes that influence transcription by preventing elongation of RNA polymerase and subsequent transcription.

2 Materials

2.1 Expression of FokI-mCherry-LacI in Reporter Cells and Induction of Transcription

1. 2-6-3 and 2-6-3 rtTA+YFP-MS2 reporter cell lines [15, 16].
2. 50 mg/ml Hygromycin sterile solution.
3. 100 mg/ml G418 filter sterilized.
4. FokI-mCherry-LacI plasmid: use 200 ng per transfection [15].
5. ER-TA plasmid: use 200 ng per transfection; only for 2-6-3 cells [16].
6. LipoD-293 transfection reagent (Signagen Laboratories, Rockville, MD, USA).
7. Dulbecco's Modified Eagle's Medium (DMEM) + Glutamax.
8. Tet-system-approved fetal bovine serum.
9. Glass optical cover slips.
10. PBS for washing.
11. Formaldehyde fixative solution: 3 % paraformaldehyde/2 % sucrose in PBS.
    - Dissolve 15 g paraformaldehyde and 10 g sucrose in 500 ml PBS and place in a 65 °C water bath shaking occasionally until almost the PFA is in solution.
    - Adjust pH to 7.0–7.4.
    - Allow solution to cool slightly, then filter through Whatman paper and store at −20° C (*see* **Note 1**).
12. 24-Well tissue-culture-treated plates.
13. 2 mg/ml doxycycline in DMSO.
14. 1 mM 4-hydroxytamoxifen in EtOH.

2.2 Immuno-fluorescent Detection of DNA Damage Response

1. Fixed coverslips of transfected cells in a 24-well plate (*see* Subheading 2.1).
2. Ice-cold PBS.
3. Ice-cold PBS with 0.1 % Tween (PBST).
4. Ice-cold permeabilization buffer: PBS with 0.5 % Tween.
5. Parafilm.
6. Humidifying chamber: a small plastic lidded container big enough to hold the lid of a 24-well plate, with a damp paper towel or cheesecloth at the bottom.
7. 37-Degree incubator.
8. IF-validated antibodies to targets of choice.
9. Fluorescent-conjugated secondary antibodies.
10. Glass optical cover slips.
11. Fluorescence mounting medium with DAPI.

2.3 Quantification of DDR Foci/YFP-MS2 with Imaging Software

1. Fluorescence microscope (wide-field or confocal) of choice with camera attachment. We use a QImaging RETIGA-SRV camera connected to a Nikon Eclipse 80i microscope driven by ImagePro 6.2 software.
2. Computer equipped with ImageJ software (NIH).

2.4 Quantification of Transcript with qRT-PCR

1. 6-Well tissue-culture-treated dishes.
2. Trizol reagent or other RNA isolation reagent of choice.
3. High-capacity cDNA reverse transcriptase kit.
4. Control human primers. We use GAPDH.
   hGAPDH_F: CTCAAGATCATCAGCAATGCC
   hGAPDH_R: CATCACGCCACAGTTTCCC
5. Reporter transcript qPCR primers. We use the following two sets of primers, both of which amplify regions of the ECFP reporter gene:
   ECFP_1F: GACGTAAACGGCCACAAGTT
   ECFP_1R: GAACTTCAGGGTCAGCTTGC
   ECFP_2F: CGACCACTACCAGCAGAACA
   ECFP_2R: GAACTCCAGCAGGACCATGT
6. SYBR Green 2× master mix.
7. 384-Well qPCR plates and seals.
8. Conventional PCR block.
9. qPCR machine with analysis software. We use the 7900HT real-time PCR machine from Applied Biosystems.

3 Methods

3.1 Expression of FokI-mCherry-LacI into Reporter Cells and Induction of Transcription

1. Grow reporter cells in DMEM with 10 % Tet-system-approved FBS and 1 % Penicillin-Streptomycin. 2-6-3 cells are also grown in the presence of 100 μg/ml hygromycin, and 2-6-3 rtTA + YFP-MS2 cells in the presence of 100 μg/ml hygromycin, as well as 400 μg/ml G418, both of which are added fresh at each passage (*see* **Note 2**).
2. 1 day prior to transfection, seed approximately 5×10^4 cells per well in a 24-well dish, onto sterilized glass coverslips, without antibiotics. The cells should be 60–80 % confluent the day of transfection (*see* **Note 3**).
3. The next day, change medium on the cells before transfection.
4. We transfect the cells using the LipoD-293 protocol (Signagen), with some modifications. For each transfection, place 200 ng total DNA to be transfected in 1 sterilized 1.5 ml microcentrifuge tube with 10 μL serum-free DMEM. For instance, if cotransfecting 2 plasmids, add 100 ng each. Tap to mix.

5. After the DNA for each transfection has been added to the tubes, create a master mix of transfection reagent by adding 0.5 μL transfection reagent per transfection to 10 μL serum-free DMEM per transfection, with some extra to account for some loss upon pipetting. Tap to mix.
6. Incubate the transfection reactions for 15 min at room temperature.
7. Immediately after the 15 min incubation, add 20 μL transfection reagent/DNA mixture to each well and swirl gently to mix.
8. 18–24 h after transfection, wash cells 1× in ice-cold PBS.
   (a) If using 2-6-3 rtTA+YFP-MS2 cells and transcription activation is desired, first treat cells with 1 μg/ml doxycycline in complete growth medium for 3 h before washing and fixing (*see* **Note 4**).
   (b) If using 2-6-3 cells which have been transfected with the ER-TA, a fusion of TA with estradiol receptor transactivator, and transcription is desired, first treat cells with 1 μM 4-hydroxytamoxifen in complete growth medium for 3 h before washing and fixing.
9. Fix in formaldehyde fixative solution for 10 min at room temperature.
10. Wash cells 2× in ice-cold PBS and proceed to immunofluorescence (Subheading 3.2). At this step, cells can be kept at 4 °C in the dark before staining, but mCherry fluorescence will be slowly lost over this time.

3.2 Immunofluorescent Detection of DNA Damage Response

1. Aspirate PBS and add 0.5 ml ice-cold permeabilization buffer to each well.
2. Incubate on ice 5 min.
3. Aspirate permeabilization buffer and wash 3× with ice-cold PBST, leaving the coverslips in PBST after the third wash.
4. Using tweezers, place coverslips atop a piece of parafilm that has been pressed onto the lid of a 24-well dish.
5. Dilute primary antibodies of choice in ice-cold PBST. For instance, if staining with rabbit anti-protein X and mouse anti-protein Y, both can be diluted in the same PBST.
6. Add 25–30 μL primary antibody solution to the top of each coverslip.
7. Incubate, in the humidifying chamber, at 37 °C × 20 min.
8. Place coverslips back in wells with cold PBST (this is the first wash).
9. Aspirate PBST and wash 3× with cold PBST (do not aspirate after final wash).

10. Remove coverslips and place on fresh piece of laboratory film pressed onto lid of 24-well plate.
11. Add 25–30 μL diluted secondary antibody to each coverslip (we use secondary antibodies from Jackson Immunoresearch, each diluted 1:200).
12. Incubate, in the humidifying chamber, at 37 °C × 20 min.
13. Place coverslips back in wells with cold PBST, for the first wash.
14. Aspirate PBST and wash 4× with cold PBST. Do not aspirate after final wash.
15. With fine tweezers, remove coverslips, dab edge on paper towel to remove excess liquid, and place facedown onto 4 μL mounting medium on glass microscopy slides.
16. Pat dry gently with paper towels or laboratory tissue.
17. Seal edges of coverslips with clear coverslip sealant or clear nail polish. Let dry in dark.
18. Store slides at −20 °C, protected from light, until ready for microscopy.

3.3 Quantification of DDR Foci/YFP-MS2 with Imaging Software

1. Using the microscopy settings of your choice, obtain images of your cells of interest. We obtain pictures at 60× oil-immersion magnification.
2. When taking pictures, all pictures for a given experiment should be taken on the same day. When comparing conditions, the same exposure settings should be used for all comparisons.
3. Open images of interest in the ImageJ program.
4. For each cell measured, measure a region of interest (e.g., a focus of 53BP1 accumulation), as well as a background fluorescence image elsewhere in the cell, using the elliptical ROI tool. Note that adjusting brightness and contrast in ImageJ will alter the appearance of the image, but not the values obtained, and can therefore be helpful in measurements. For images in which a given signal is absent (e.g., no 53BP1 accumulation at the reporter locus in a cell expressing the FokI nuclease-deficient mutant), overlay the mCherry image to locate the locus and draw an ROI at that point.
5. Measure 50–100 cells per experiment, then subtract each background from each signal, and calculate the mean fluorescence intensity (MFI) for a given signal.

3.4 Quantification of Transcript with qRT-PCR

1. We use a modified version of the RNA isolation protocol from Trizol (Invitrogen).

   You can use any RNA isolation reagent/protocol. Be sure all reagents/equipment have been cleaned and are free of any RNAse/DNAse contamination.

2. After isolating RNA, measure the concentration and convert 1–2 μg to cDNA using the high-capacity cDNA kit or another cDNA kit of your choice.
3. After the reverse-transcription reaction is complete, dilute the cDNA with sterile water. 1 μg is diluted to 100 μl.
4. We use the following reaction mixture per well in a 384-well qPCR plate.

   5 μL 2× SYBR Green master mix.

   4 μL diluted cDNA.

   1 μL 9 μM primer mix (a mix of both forward and reverse primers for a given amplicon, with each primer being at a final concentration of 9 μM).

   Final volume—10 μL

   Run the qPCR with the following conditions.

   Stage 1 (×1): 50° × 2 min.

   Stage 2 (×1): 95° × 10 min.

   Stage 3 (×40): 95° × 15 s.

   60° × 1 min.
5. Analyze the data using the SDS and RQ manager programs (ABI), and the relative quantity method, normalizing to a given sample, and to human GAPDH within samples. For example, a given experiment may compare transcript levels after induction with doxycycline in the presence of FokI(WT)-mCherry-LacI, as compared to expression of the nuclease-deficient mutant FokI(D450A)-mCherry-LacI.

4 Notes

1. For formaldehyde fixative solution, perform all steps in a chemical hood to avoid fumes. Single-use aliquots can be stored at −20 °C (we use pop-cap culture tubes).
2. When growing and passaging the cells, we split the cells 1:2–1:4 every 2–3 days, once the cells reach 80–90 % confluence. It is also important to freeze down multiple vials of early passage cells in 10 % DMSO-containing freezing media, as later passage cells can lose the integrated reporter or inducibility.
3. We sometimes see problems with transfection toxicity if the cells are too sparse (less than approximately 50–60 % confluent) and problems with transgene inducibility if the cells are too dense (close to or completely confluent).
4. As the doxycycline is light sensitive, we store small aliquots of the 2 mg/ml stock in the dark at −20 °C.

References

1. Lukas C, Falck J, Bartkova J, Bartek J, Lukas J (2003) Distinct spatiotemporal dynamics of mammalian checkpoint regulators induced by DNA damage. Nat Cell Biol 5(3):255–260
2. Bekker-Jensen S, Lukas C, Kitagawa R, Melander F, Kastan MB, Bartek J et al (2006) Spatial organization of the mammalian genome surveillance machinery in response to DNA strand breaks. J Cell Biol 173(2):195
3. Lisby M, Barlow JH, Burgess RC, Rothstein R (2004) Choreography of the DNA damage response: spatiotemporal relationships among checkpoint and repair proteins. Cell 118(6): 699–713
4. Lisby M, Mortensen UH, Rothstein R (2003) Colocalization of multiple DNA double-strand breaks at a single Rad52 repair centre. Nat Cell Biol 5(6):572–577
5. Berkovich E, Monnat RJ, Kastan MB (2007) Roles of ATM and NBS1 in chromatin structure modulation and DNA double-strand break repair. Nat Cell Biol 9(6):683–690
6. Pankotai T, Bonhomme C, Chen D, Soutoglou E (2012) DNAPKcs-dependent arrest of RNA polymerase II transcription in the presence of DNA breaks. Nat Struct Mol Biol 19(3): 276–282
7. Iacovoni JS, Caron P, Lassadi I, Nicolas E, Massip L, Trouche D et al (2010) High-resolution profiling of ||[gamma]|H2AX around DNA double strand breaks in the mammalian genome. EMBO J 29(8):1446–1457
8. Soutoglou E, Dorn JF, Sengupta K, Jasin M, Nussenzweig A, Ried T et al (2007) Positional stability of single double-strand breaks in mammalian cells. Nat Cell Biol 9(6):675–682
9. Honma M, Izumi M, Sakuraba M, Tadokoro S, Sakamoto H, Wang W et al (2003) Deletion, rearrangement, and gene conversion; genetic consequences of chromosomal double-strand breaks in human cells. Environ Mol Mutagen 42(4):288–298
10. Rouet P, Smih F, Jasin M (1994) Introduction of double-strand breaks into the genome of mouse cells by expression of a rare-cutting endonuclease. Mol Cell Biol 14(12):8096–8106
11. Anindya R, Ayg\ün O, Svejstrup JQ (2007) Damage-induced ubiquitylation of human RNA polymerase II by the ubiquitin ligase Nedd4, but not Cockayne syndrome proteins or BRCA1. Mol Cell 28(3):386–397
12. Svejstrup JQ (2007) Contending with transcriptional arrest during RNAPII transcript elongation. Trends Biochem Sci 32(4): 165–171
13. Solovjeva LV, Svetlova MP, Chagin VO, Tomilin NV (2007) Inhibition of transcription at radiation-induced nuclear foci of phosphorylated histone H2AX in mammalian cells. Chromosome Res 15(6):787–797
14. Kruhlak M, Crouch EE, Orlov M, Montano C, Gorski SA, Nussenzweig A et al (2007) The ATM repair pathway inhibits RNA polymerase I transcription in response to chromosome breaks. Nature 447(7145):730–734
15. Shanbhag NM, Rafalska-Metcalf IU, Balane-Bolivar C, Janicki SM, Greenberg RA (2010) ATM-dependent chromatin changes silence transcription in cis to DNA double-strand breaks. Cell 141(6):970–981
16. Janicki SM, Tsukamoto T, Salghetti SE, Tansey WP, Sachidanandam R, Prasanth KV et al (2004) From silencing to gene expression: real-time analysis in single cells. Cell 116(5): 683–698
17. Rafalska-Metcalf IU, Janicki SM (2007) Show and tell: visualizing gene expression in living cells. J Cell Sci 120(14):2301–2308
18. Bertrand E, Chartrand P, Schaefer M, Shenoy SM, Singer RH, Long RM (1998) Localization of ASH1 mRNA particles in living yeast. Mol Cell 2(4):437–445
19. Yunger S, Shav-Tal Y (2011) Imaging mRNAs in living mammalian cells. Methods Mol Biol 714:249–263
20. Bitinaite J, Wah DA, Aggarwal AK, Schildkraut I (1998) FokI dimerization is required for DNA cleavage. Proc Natl Acad Sci USA 95(18):10570–10575
21. Wah DA, Bitinaite J, Schildkraut I, Aggarwal AK (1998) Structure of foki has implications for DNA cleavage. Proc Natl Acad Sci USA 95(18):10564–10569

Chapter 17

Fluorescence Microscopy-Based High-Throughput Screening for Factors Involved in Gene Silencing

Sebastian Bultmann and Heinrich Leonhardt

Abstract

Gene silencing in eukaryotes is a highly controlled process. It involves the concerted action of histone and DNA-modifying enzymes as well as transcription factors and chromatin-associated proteins. To understand how epigenetic gene silencing is regulated, it is important to identify the factors involved in this process. Here we describe an assay that allows high-throughput screening for factors involved in gene silencing. This assay exploits the susceptibility of the viral cytomegalovirus (CMV) promoter to epigenetic silencing in embryonic stem cells (ESCs) and uses reporter constructs with an optical readout to determine the gene silencing potential of candidate factors. This approach allows to study mechanisms and kinetics of gene silencing in living cells and to evaluate the role of DNA methyltransferases, histone-modifying enzymes, and other chromatin-associated factors during gene silencing.

Key words Epigenetic gene silencing, Automated microscopy, Embryonic stem cells, Cytomegalovirus promoter, Transgene silencing

1 Introduction

Gene silencing mechanisms play an essential role in cellular differentiation and development. In addition, gene silencing represents an effective defense mechanism against viral infection, retrotransposon activation, and nonhost DNA in general. Over the past decades numerous *cis*- and *trans*-acting factors contributing to eukaryotic gene silencing have been identified. However, a systematic screen and functional evaluation of factors involved in these repressive epigenetic pathways would contribute to our understanding of gene regulation in general and also help to develop vector systems and strategies for long-term, stable transgene expression.

Viral promoter silencing in embryonic stem cells (ESCs) represents a valuable model system that can be utilized to screen for novel factors involved in epigenetic gene silencing. Genetic engineering of ESCs by targeted homologous recombination and/or

Yaron Shav-Tal (ed.), *Imaging Gene Expression: Methods and Protocols*, Methods in Molecular Biology, vol. 1042,
DOI 10.1007/978-1-62703-526-2_17,

transient depletion of proteins using RNA interference (RNAi) allow large-scale analysis of epigenetic regulators [1–4].

In this chapter we describe an assay that allows screening for factors involved in epigenetic gene silencing by utilizing the different epigenetic stability of two promoters. In ESCs, viral promoters are rapidly silenced by epigenetic mechanisms as a protection against nonhost gene expression and to preserve genomic stability [5–8]. In line with these findings, we observed that upon transient transfection of ESCs, the viral cytomegalovirus (CMV) promoter is rapidly silenced [9]. In contrast, the chimeric CMV early enhancer/chicken β-actin (CAG) promoter yielded stable, long-term expression. Thus, cotransfection with reporter constructs expressing distinct fluorescent proteins under the control of the CMV and the CAG promoter allows monitoring gene silencing in vivo. By comparing the silencing kinetics of the CMV promoter in wild-type and mutant ESCs, we could identify cellular factors involved in epigenetic gene repression. In wild-type ESCs, the silencing gradually proceeded over the course of 10 days, while in ESCs lacking de novo methyltransferases *Dnmt3a* and *Dnmt3b* or the histone methyltransferase *G9a*, no silencing was observed [9].

Based on these findings, we developed an epigenetic silencing assay in which the distinct silencing kinetics of CAG- and CMV-driven fluorescent proteins are monitored over time. In combination with the available tools for the generation of knockout/knockdown cell lines, this assay allows high-throughput screening for factors involved in epigenetic gene silencing.

2 Materials

2.1 Cell Culture

1. Dulbecco's Modified Eagle's Medium (DMEM).
2. DMEM high glucose w/o phenol red for live cell imaging.
3. 1 M HEPES solution.
4. Fetal bovine serum (FBS).
5. Trypsin for detaching cells from tissue culture plates (0.25 % trypsin/EDTA solution).
6. 0.2 % Gelatin for plate coating.
7. PBS solution for washing cells.
8. 100× nonessential amino acids (NEAA).
9. 200 mM L-glutamine.
10. 100× Penicillin/Streptomycin (Pen/Strep).
11. 50 mM β-mercaptoethanol in PBS.
12. 15 mM CHIR99021 GSK3β inhibitor (*see* **Note 1**).
13. 5 mM PD0325901 Mek 1/2 inhibitor (*see* **Note 1**).
14. 10. 5×105 U/ml leukemia inhibitory factor (LIF) (*see* **Note 1**).

2.2 Media

1. Embryonic stem cell (ESCs) lines: wild-type (wt) and knockout/down cell lines for candidate genes.
2. ESC medium: 400 ml of DMEM with 80 ml FBS, 1 ml β-mercaptoethanol (final conc. 100 μM), 5 ml L-glutamine (final conc. 2 mM), 5 ml NEAA (final conc. 1×), 5 ml Pen/Strep (final conc. 1×), 100 μl PD0325901 (final conc. 1 μM), 100 μl CHIR99021 (final conc. 3 μM), and 1 ml LIF (final conc. 1,000 U/ml).
3. Live cell medium: 400 ml of DMEM without phenol red with 80 ml FBS, 1 ml β-mercaptoethanol (final conc. 100 μM), 5 ml L-glutamine (final conc. 2 mM), 5 ml NEAA (final conc. 1×), 5 ml Pen/Strep (final conc. 1×), 100 μl PD0325901 (final conc. 1 μM), 100 μl CHIR99021 (final conc. 3 μM), and 1 ml LIF (final conc. 1,000 U/ml).

2.3 Transfection

1. Lipofectamine™ 2000 transfection reagent.
2. Opti-MEM serum-free medium.
3. CAG-eGFP plasmid (Meilinger et al. [9]).
4. CAG-mRFP plasmid (Meilinger et al. [9]).
5. CMV-eGFP plasmid (Meilinger et al. [9]).
6. CMV-mRFP plasmid (Meilinger et al. [9]).

2.4 Microscopy

1. Automated fluorescence microscope (e.g., InCell Analyzer 2000, GE Healthcare; as used in our previous study [9]) or any standard fluorescence microscope. The microscope should be equipped with a CCD camera, a live cell chamber, and with filters for GFP and mRFP excitation/emission. For easy and fast image acquisition, the microscope should be equipped with a stage that holds multi-well plates. Best results in terms of maximal number of cells per image and sensitivity were achieved using a 20× air objective (NA = 0.45).
2. Image evaluation software (InCell Analyzer Workstation or ImageJ/Fuji).

3 Methods

3.1 Culture of Embryonic Stem Cell Lines

The epigenetic silencing assay depends on embryonic stem cell lines that are depleted or devoid of the factors of interest. Knockdown cell lines can be generated using standard RNAi techniques [1]. Knockout cell lines can either be generated by targeted homologous recombination or obtained from the International Knockout Mouse Consortium (IKMC) [2–4].

1. ESCs are cultured on gelatinized plates. For this, culture dishes are coated for 5 min using 0.2 % gelatin prior to seeding of the cells.

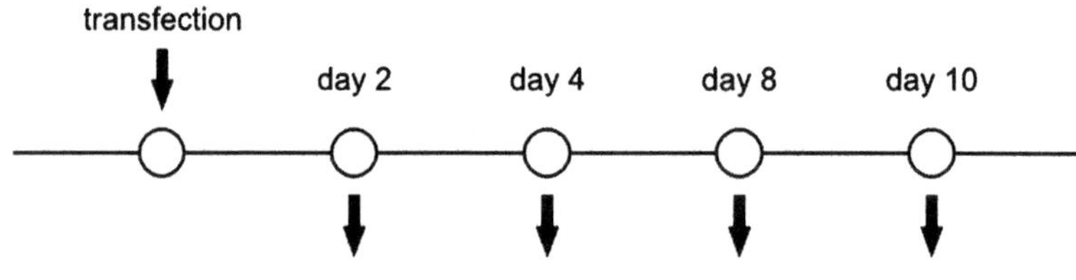

Fig. 1 Schematic representation of the time course of the epigenetic silencing assay. Cells are cotransfected with a CMV and CAG reporter construct. Every 2 days cells are split and images are acquired

2. Split ESCs every second day by washing with PBS and subsequent trypsinization for 5 min at 37 °C. After trypsinization carefully resuspend cells in fresh ESC medium (*see* **Note 1**).
3. Transfer the appropriate amount of the cell suspension onto a freshly gelatin-coated culture dish (*see* **Note 2**).
4. Exchange medium every day to keep ESCs in an undifferentiated, pluripotent state.

3.2 Transfection of Embryonic Stem Cells Using Lipofectamine™ 2000

1. In general, any transfection reagent can be used to transfect ESCs. However, optimal transfection conditions should be determined before assays are performed. Using Lipofectamine™ 2000 we achieved transfection efficiencies between 70 and 90 %. For the epigenetic silencing assay transfection, efficiency >30 % is recommended.

3.3 Epigenetic Silencing Assay

This assay utilizes the susceptibility to epigenetic silencing of the CMV promoter to screen for factors involved in epigenetic silencing. For this, ESCs are cotransfected with two distinct reporter constructs, one expressing mRFP under the control of the CMV promoter, the other expressing eGFP driven by the CAG promoter. To exclude artifacts produced by different stabilities of mRFP and eGFP proteins, it is recommended to confirm the obtained results by swapping eGFP and mRFP reporter sequences. A schematic representation of the time course is depicted in Fig. 1.

1. Of each knockout/down and corresponding wild-type ESC line, 1×10^6 cells are plated onto separate, gelatin-coated wells of a 6-well plate.
2. Per cell line analyzed, prepare transfection mixture by mixing 2 μg plasmid DNA (1 μg CMV and 1 μg CAG construct) with 250 μl Opti-MEM. In a separate reaction tube, mix 8 μl Lipofectamine™ 2000 with 250 μl Opti-MEM and incubate for 5 min at room temperature (*see* **Note 3**).
3. Combine the DNA and Lipofectamine premixes, mix gently, and incubate for 20 min (up to 6 h) at room temperature.
4. Add the transfection mixture to the wells containing the cells and medium. Mix gently by rocking the plate back and forth.
5. Change medium 8–12 h after transfection.

6. 48 h after transfection, split the ESCs (1:6–1:10 depending on cell density) onto gelatin-coated 6-well plates containing live cell medium.
7. 5–6 h after seeding, the cells have attached to the surface of the plate and can be imaged. It is important that exposure times and the objective are kept constant throughout all time points.
8. Record 90–150 images per time point and cell line.
9. After imaging place cells back into the incubator and change medium.
10. Repeat **steps 6–9** every 48 h for 10 days.

3.4 Image Evaluation

To quantify the silencing kinetics of the CMV promoter, the ratio between CAG-driven eGFP signals and the CMV-driven mRFP is calculated for each time point. In principle any image evaluation software capable of segmenting cells and quantifying signal intensity and/or signal area can be used. This protocol describes the evaluation using the Fiji (http://fiji.sc/wiki/index.php/Fiji) open-source software package. The evaluation can easily be automated by using a macro generated with the macro recording tool (Plugins > Macros > Record). A schematic overview of the image evaluation process is depicted in Fig. 2.

1. Open image stack of time point X cell line Y by clicking: File > Import > Image Sequence ….
2. For each image in each channel, apply Gaussian blur algorithm with radius (sigma) = 2 by clicking: Process > Filters > Gaussian Blur ….
3. For each image in each channel, apply a threshold so that the background is completely covered and minimal signal is lost by clicking: Image > Adjust > Threshold …; adjust threshold so that background is completely red. Click on "Set" and note the value for the lower and upper threshold. It is important to keep these values constant throughout the stack and the channel (*see* **Note 4**).
4. Convert Threshold into selection by clicking: Edit > Selection > Make Selection ….
5. Invert selection by clicking: Edit > Selection > Make Inverse ….

At this step only the signals should be converted into a selection. You can test this by clicking: Edit > Copy, followed by: File > New > Internal Clipboard ….

6. Set Measurements to "area" by clicking: Analyze > Set Measurements > area.
7. Measure the area covered by the signal in each channel for every image by clicking: Analyze > Measure ….
8. Calculate the ratio between the summarized signal area of the mRFP channel and the summarized signal area of the mRFP channel for each time point and cell line.

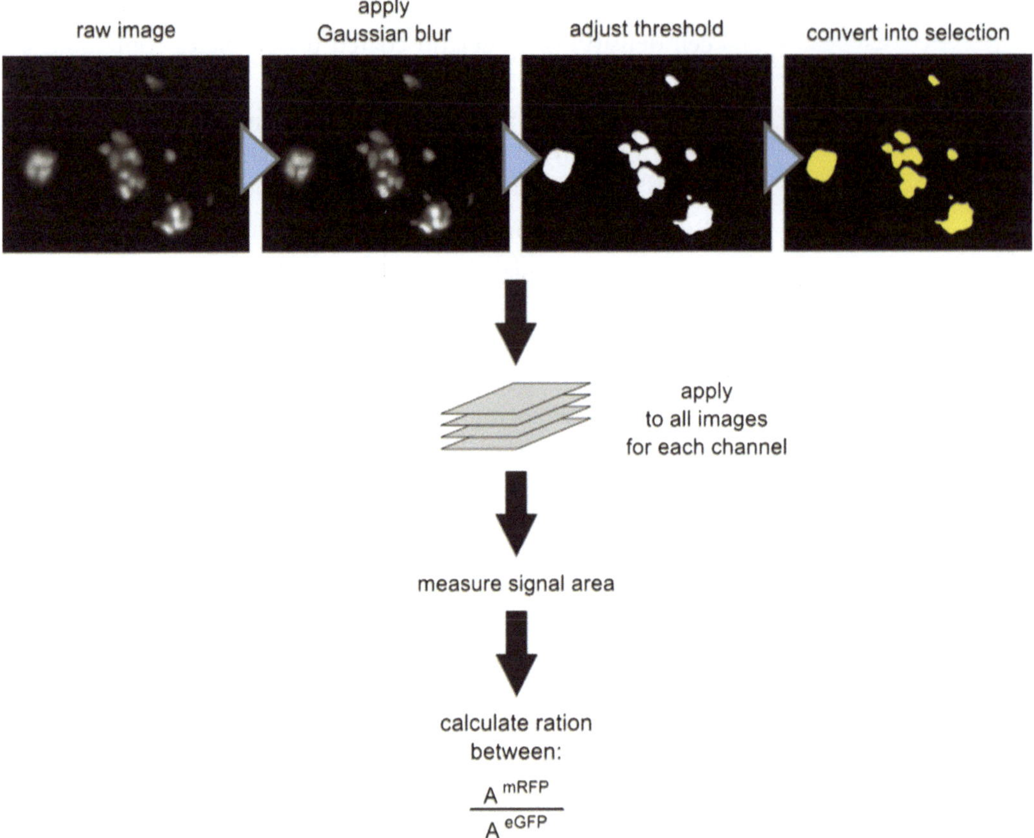

Fig. 2 Image evaluation using Fuji. Raw images are processed using a Gaussian blur filter. Subsequently, a threshold is applied to segment signal from background. This procedure is applied to all images in the stack and for mRFP and eGFP channels separately. The area covered by the signal is measured and the ratio of the mRFP total signal area over the GFP signal area is calculated for each stack

9. Repeat **steps 1–8** for every time point and every cell line assayed.
10. Plot ratio of mRFP versus eGFP signal area against time after transfection.

At this step wild-type cell lines should exhibit a rapid silencing of the mRFP (CMV) signal (for an example data set, *see* Fig. 3).

4 Notes

1. The two inhibitors (GSK3β inhibitor, Mek 1/2 inhibitor) and LIF are required to keep the ESCs in a pluripotent state.
2. It is important to resuspend the cells thoroughly generating a single-cell suspension to avoid plating of cell clumps that produce big colonies prone to differentiate.

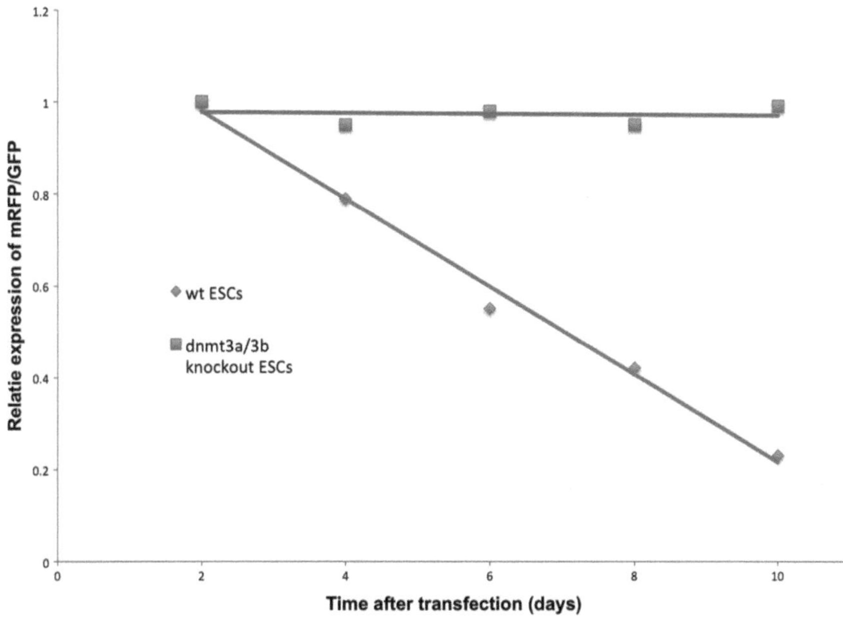

Fig. 3 Example data set. Relative expression of the CMV-driven mRFP construct is plotted against the time after transfection. Wild-type cell exhibit rapid silencing of mRFP expression, while cells lacking the de novo methyltransferases Dnmt3a and 3b do not silence the CMV promoter

3. Splitting ratios are dependent on the growth rate of the cell lines used. Wild-type cells are typically split in a ratio between 1:8 and 1:10; knockout/down cell lines may grow considerably slower.
4. Depending on the number of cell lines assayed, it is recommended to prepare a transfection master mix.
5. It is essential to keep the threshold settings constant throughout the channel and the stack of a given time point. Threshold setting usually requires some optimization as background levels can differ from image to image.

Acknowledgements

This study was supported by the Nanosystems Initiative Munich, by the BioImaging Network Munich, and by grants from the Deutsche Forschungsgemeinschaft (SPP1230 and SFB646) to H.L.

References

1. Montgomery MK (2004) RNA interference: historical overview and significance. Methods Mol Biol 265:3–21
2. Ullrich M, Schuh K (2009) Methods in molecular biology. Gene Trap: Knockout on the Fast Lane. In: Cartwright EJ (ed). Humana, Totowa, NJ

3. Wielders E, Dekker M, Riele HT (2009) Methods in molecular biology. Generation of Double-Knockout Embryonic Stem Cells. Wurst W, Kühn R (eds). Humana, Totowa, NJ
4. Hofker MH, Deursen J (eds) (2010) Methods in molecular biology. Humana, Totowa, NJ
5. Xia X, Zhang Y, Zieth CR, Zhang S-C (2007) Transgenes delivered by lentiviral vector are suppressed in human embryonic stem cells in a promoter-dependent manner. Stem Cells Dev 16:167–176
6. Karimi MM, Goyal P, Maksakova IA, Bilenky M, Leung D, Tang JX, Shinkai Y, Mager DL, Jones S, Hirst M et al (2011) DNA methylation and SETDB1/H3K9me3 regulate predominantly distinct sets of genes, retroelements, and chimeric transcripts in mESCs. Cell Stem Cell 8:676–687
7. Macfarlan TS, Gifford WD, Agarwal S, Driscoll S, Lettieri K, Wang J, Andrews SE, Franco L, Rosenfeld MG, Ren B, et al. Endogenous retroviruses and neighboring genes are coordinately repressed by LSD1/KDM1A. genesdev.cshlp.org
8. Magnusson T, Haase R, Schleef M, Wagner E, Ogris M (2011) Sustained, high transgene expression in liver with plasmid vectors using optimized promoter-enhancer combinations. J Gene Med 13:382–391
9. Meilinger D, Fellinger K, Bultmann S, Rothbauer U, Bonapace IM, Klinkert WEF, Spada F, Leonhardt H (2009) Np95 interacts with de novo DNA methyltransferases, Dnmt3a and Dnmt3b, and mediates epigenetic silencing of the viral CMV promoter in embryonic stem cells. EMBO Rep 10:1259–1264

Chapter 18

Actin as a Model for the Study of Nucleocytoplasmic Shuttling and Nuclear Dynamics

Kari-Pekka Skarp and Maria K. Vartiainen

Abstract

A great number of molecules are constantly being exchanged between the nucleus and the cytoplasm via nuclear pore complexes (NPCs). Importantly, this nucleocytoplasmic trafficking is used to transfer information between the two compartments, thereby permitting the manipulation of critical nuclear processes such as transcription. Constant shuttling of actin is an example of the versatility of this regulatory avenue, as this protein has the capability to drive the transcriptional activity of certain gene sets as well as influence transcription on a global scale. Nuclear import and export are extremely dynamic phenomena and require imaging tools capable of rapid sampling rates for proper quantitative observation. Here we describe live-cell imaging assays based on fluorescence recovery after photobleaching (FRAP) and fluorescence loss in photobleaching (FLIP) for monitoring both import and export of fluorescently labelled molecules. Our assays are performed with GFP-actin, but the same principle is applicable to most proteins shuttling between the nucleus and the cytoplasm. Furthermore, these assays may also expose novel qualities of the intranuclear dynamics of a protein, which can polymerize or partake in complexes, because such behavior is mirrored in the nuclear retention of the protein detectable by both import and export assays.

Key words Actin, Nucleus, Nuclear transport, Actin polymerization, Nucleocytoplasmic shuttling

1 Introduction

Nuclear pore complexes (NPCs) are large macromolecular complexes permeating the nuclear envelope (NE) and responsible for the exchange of particles between the nucleus and the cytoplasm [1]. Transcriptional regulation is an example of a fundamental nuclear process, which requires constant flux of molecules in and out of the nucleus. For example, signaling molecules are transported into the nucleus to transmit information to the transcription machinery [2, 3] and the mRNA molecules produced after transcriptional activation are exported from the nucleus to the cytoplasmic ribosomes [4]. Recent studies have clearly demonstrated a role for actin, a traditional component of the cytoskeleton, in the nucleus especially in transcription [5–7]. Here we describe a method to study nuclear import and export by using confocal microscopy.

Yaron Shav-Tal (ed.), *Imaging Gene Expression: Methods and Protocols*, Methods in Molecular Biology, vol. 1042, DOI 10.1007/978-1-62703-526-2_18, © Springer Science+Business Media, LLC 2013

In addition, by using actin as an example, we demonstrate how these assays can also yield precious information about other dynamic properties of the molecule under study.

Imaging of fluorescently labelled molecules with LSCM (laser scanning confocal microscope) makes it possible to capture and record in real time rapid cellular events. To study the exchange of molecules between the nucleus and the cytoplasm, we use photobleaching where fluorescent chromophores are photolytically destroyed and the resulting change in fluorescent signal is harnessed to measure the dynamics of various cellular events, in this case nuclear import and export. An alternative approach is photoactivation, which permits local highlighting of specific photoactivatable molecules and then monitoring either their local dispersion or accumulation somewhere else [8].

For these microscopy experiments, we most often use tagging with the enhanced green fluorescent protein (EGFP) to visualize the protein of interest. EGFP is suitable for our assays due to the inherent resistance to bleaching at lower laser power suitable for imaging, while still retaining the ability to be comprehensively bleached at higher laser power [9]. Also, the 488 nm line used to excite EGFP is commonly the most powerful laser available in imaging systems. When using EGFP, it is always important to determine that the fusion does not perturb protein function. In addition, the size of EGFP introduces an aspect to be considered in experiments involving the passage of fluorescent molecules through NPCs. These channels allow the passive diffusion of small molecules and the active transport of larger molecules capable of utilizing karyopherins—the proteins responsible for the energy-consuming Ran-dependent traffic between the nucleus and the cytoplasm. The NPC passive diffusion limit is close to 40 kDa [10]. The size of GFP is 27 kDa, which may easily be enough to complicate the natural flow of any small protein or peptide. However, via karyopherins the active machinery can facilitate the transport of large complexes sizing hundreds of kilodaltons such as the continuous export of mRNAs in mRNP particles [4]. Here, such a small fluorescent tag will hardly interfere with transport kinetics although the usual care must be taken to place it in the least interfering position regarding the molecular machinery present. In our study, we used N-terminally tagged actin, which has been shown to polymerize together with the wild-type protein into functional filaments and is routinely used to study the kinetics of this protein in various organisms [11, 12]. If there is doubt that size of EGFP may influence the results, alternative approaches should be considered. In our case, actin is globular and 42 kDa in molecular weight, which is close to the NPC exclusion limit. Therefore, we routinely also use for our assays in vitro fluorescently labelled actin, which is microinjected into the cells. Importantly, as we found no significant difference in import rates with Alexa Fluor

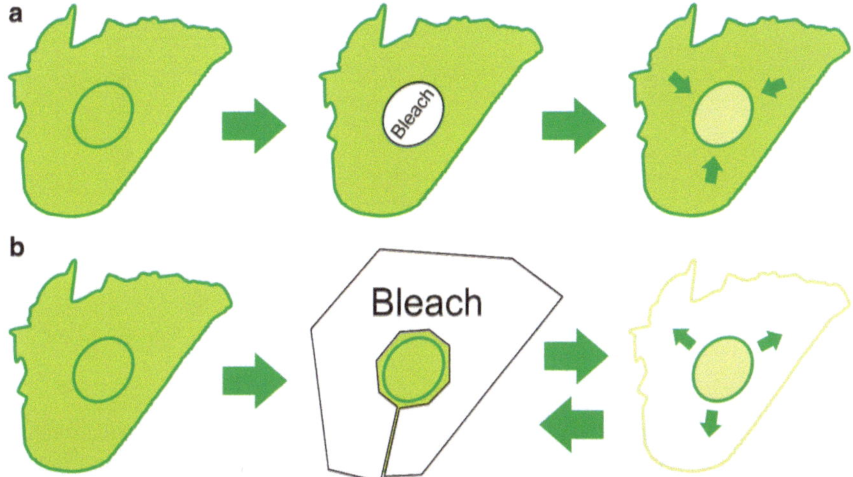

Fig. 1 Schematic of the bleaching strategy. In the import assay, the nucleus is bleached once and increase of nuclear fluorescence is measured (**a**). In the export assay, the whole cytoplasm is repeatedly bleached and loss of nuclear fluorescence is recorded (**b**)

488 (AF488)-labelled actin, which is very close to wt-actin size, GFP-actin nor 2GFP-actin, this implicates an active transport mechanism is at work because passive transport slows down as size increases [6].

Our nuclear import FRAP assay is performed accordingly: draw a region of interest (ROI) corresponding to the nucleus and bleach it with full laser power (Fig. 1a). After bleaching, follow the recovery of fluorescence in the nucleus, which represents import of unbleached fluorescent molecules from the cytoplasm. In FRAP experiments, the recovery should always be recorded until steady state is reached. However, since only the beginning of the FRAP curve properly represents import due to export starting to have an influence on the experiment, this may not be always necessary in import assays but may yield additional information. For example, with GFP-actin, full recovery of nuclear fluorescence takes more than half an hour due to reasons specific to actin discussed below. We describe here a fast 1 min FRAP experiment ($FRAP_1$) and also a longer ~40 min FRAP experiment ($FRAP_2$), where the recovery is followed until the signal is stable. For GFP-actin, the latter experiment shows the presence of three pools of nuclear actin, with different recovery rates. Traditionally, any cytoplasmic actin FRAP curve contains two phases. The first and fastest recovery phase corresponds to monomeric actin, which can rapidly diffuse to the area of bleaching. The second phase represents filamentous actin, which can only recover at a rate dependent on treadmilling. These two phases, monomeric and polymeric, have earlier been shown to be present also in the nucleus [13]. However, on top of the two classical actin phases, the $FRAP_2$ assay shows the presence of an additional

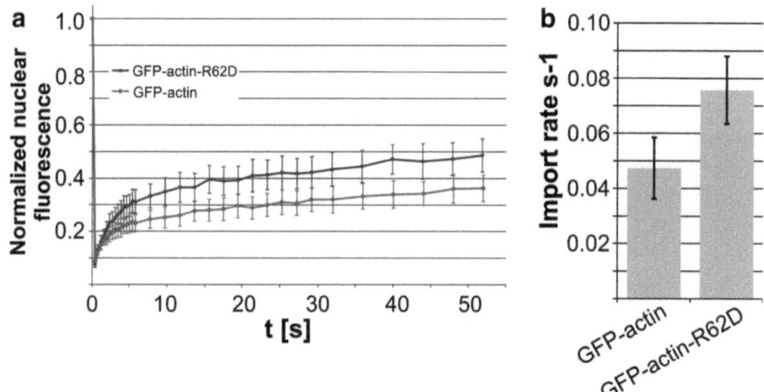

Fig. 2 Import curves of GFP fusions of actin (*red*) and an unpolymerizable mutant actin-R62D (*blue*) obtained by FRAP (**a**). In the experiment, prebleach images are first obtained and their average nuclear fluorescence intensity is set as 1. The nucleus is then photobleached resulting in a significant drop in nuclear fluorescence. Immediately after bleaching, the fluorescence recovery in the bleached area, which can only occur if unbleached particles are imported from the cytoplasm, is recorded. As the assay proceeds, export begins to affect nuclear fluorescence and therefore later data points reflect a combination of both nuclear import and export processes. Therefore, slopes calculated from initial data points represent the import rate (**b**). This allows the quantitative comparison of import rates of different constructs and/or conditions. In the case of actin, the mutant R62D recovers faster and is therefore imported into the nucleus faster than the wild-type protein. This data thus shows that actin monomer levels limit the nuclear import rate

phase, which in comparison recovers very slowly. This probably represents actin bound to various nuclear machineries and interestingly was also at least partially present for an actin mutant incapable of polymerizing [6].

Our nuclear export FLIP assay is performed accordingly: draw a region of interest (ROI) covering the entire cytoplasm of the cell while simultaneously carefully excluding the nucleus from this area (Fig. 1b). The cytoplasmic ROI is then bleached and the nucleus imaged repeatedly. Bleaching the cytoplasm results in gradual decrease of nuclear fluorescence, because during the assay, unbleached particles from the nucleus are exported to the cytoplasm and exposed to bleaching. The assay assumes unbleached particles exported from the nucleus are not reimported before they can be bleached. Loss of fluorescence from the nucleus is then used as a measure of export.

In import, the data is shown as a fluorescence recovery curve (Fig. 2) and, in export, as a fluorescence loss curve (Fig. 3). For both assays, the apparent import/export rates can be most reliably derived from the very beginning of the curve, where a straight line can be approximated and the slope determined. This is especially important in the import assay, where export soon begins to undermine the

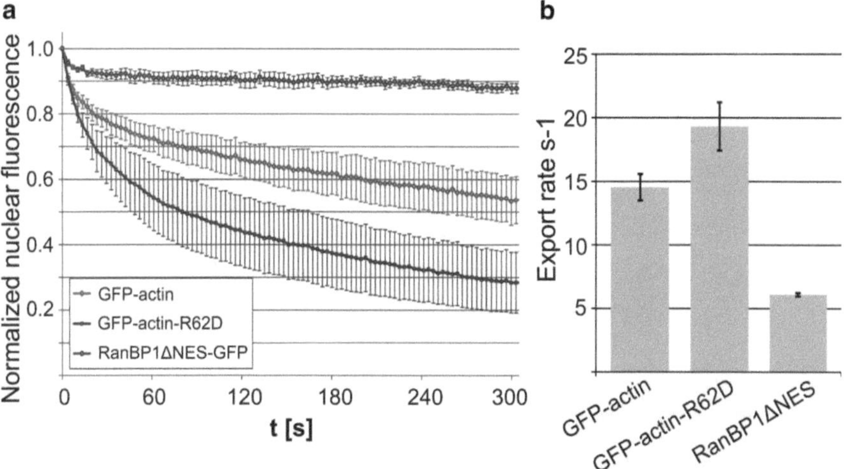

Fig. 3 Export curves of GFP fusions of actin (*red*), an unpolymerizable mutant actin-R62D (*blue*) and RanBP1ΔNES (*purple*) obtained by FLIP (**a**). In the experiment, the average fluorescence of prebleach images is set as 1. Then, a cycle of cytoplasmic photobleaching + imaging is initiated. This aims to immediately bleach cytoplasmic fluorescent particles as soon as they are exported from the nucleus. While this continues, nuclear fluorescence is slowly drained and the loss is recorded. RanBP1ΔNES serves as a negative control to demonstrate how a construct, which does not shuttle, behaves in this assay. Slopes calculated from initial data points represent the export rate (**b**). Similar to the situation in Fig. 2, the monomeric mutant is exported faster from the nucleus than the wild-type actin, while the export of RanBP1ΔNES is negligible (Color figure online)

fluorescence recovery making the proper determination of import rates unreliable with later data points. Ultimately, the exact manner of transport rate quantitation depends on the import and export rates of the protein of interest and has to be decided individually. Both assays can also give insight into the relative stability and internal dynamics of the nuclear pool of the protein of interest. For GFP-actin, in the import assay, the relative stability manifests as the third phase of slower turnover rate mentioned above. In the export assay, the relative stability is portrayed by the portion of molecules still remaining in the nucleus at the end of the assay. While the nuclear fluorescence of wild-type actin has been cut in half after the export assay, the actin mutant incapable of polymerizing has only one quarter left. This demonstrates that actin needs to be monomeric to be transport competent and that this mutant cannot enter all the nuclear complexes that the wild-type actin can.

2 Materials

1. Cell culture materials for culturing the cells of interest.
2. 35 mm plates for live-cell imaging. We routinely use Falcon 353001 Easy-Grip™ cell culture dishes with an upright Leica SP5 LSCM. For inverted microscopes, Mattek P35G-1.5-14-C glass bottom microwell dishes are recommended.

3. Transfection reagent (we use jetPRIME® from Polyplus).
4. Plasmid encoding the protein of interest fused to a fluorescent protein-encoding sequence. We use pEGFP-C1 vector (Clontech), which places the insert in the C-terminus of the GFP-encoding sequence.
5. LSCM with 60×/63× objective and 37 °C/5 % CO_2 incubation chamber.
6. Software capable of reading data in spreadsheet format (e.g., OpenOffice, LibreOffice, or Microsoft Office). The output format in Leica LAS AF software is .csv (comma-separated values).

3 Methods

3.1 Cell and Microscope Setup

1. Two days before the experiment, plate the cells on 35 mm dishes. The number of cells to be plated depends mostly on the type of cells used, the transfection reagent, and the dish in use. We have found ~120,000 NIH 3T3 cells on a Falcon dish to work well. For a Mattek dish, ~90,000 NIH 3T3 cells would already suffice due to greater adherence to a glass surface than to plastic.
2. The following day, transfect the cells according to manufacturer's instructions using as little of transfecting DNA as possible. For 35 mm plates with 1.5 ml of medium, we use 1 μg of total DNA. Of this 1 μg, only 100 ng is actual GFP-actin DNA and the remaining 900 ng can be topped with an empty vector (*see* **Note 1**).
3. On the 3rd day, the cells are ready to be imaged after ~24 h of transfection to guarantee satisfactory expression levels.
4. Microscope systems from different manufacturers exhibit differential implementation and terminology in their software. The following imaging parameters were used:
   - Pinhole: 1 AU.
   - Resolution: 256 × 256 (for $FRAP_1$ and FLIP) or 1,024 × 1,024 ($FRAP_2$).
   - Pixel size (with Zoom 3 for all the assays): 321.8 × 321.8 nm with 256 × 256 and 80.1 × 80.1 nm with 1,024 × 1,024 (*see* **Note 2**).
   - Scanning speed 700 Hz with bidirectional scanning.
   - Line average 2.
   - Data for quantitation should always be collected with a minimum bit depth of 12.
5. Choose a normal healthy-looking non-mitotic cell for the assay, which should not fluoresce too brightly. Similarly, a cell of too low signal may not yield properly quantifiable data due

to defects in normalization, inaccuracy of low fluorescence intensity measurements, etc. If possible, include a control cell in the field of view, which will help exposing any instabilities in the imaging system (*see* **Note 3**).

6. Select the optimal *z*-axis level in order to visualize the nucleus (*see* **Note 4**). Take an image with high-quality settings (such as 1,024 × 1,024, scanning speed 400 Hz, line average 3 in the Leica system) to be sure where to draw the ROIs.

Next, we will describe in more detail how the import and export assays described above can be performed on a Leica SP5 LAS AF system v. 2.4.1.6384.

3.2 FRAP for Measuring Import

For the Leica system at our disposal, the system is set up accordingly:

1. Go to the menu on the upper left corner labelled Leica Microsystems LAS AF and select FRAP wizard. A selection of tabs appears, which enable configuring the assay: Setup, Bleach, Time Course, and Evaluation.
2. In the first tab, Setup, set the imaging preferences. Rapid image acquisition often requires compromising other parameters, and because import rates are best portrayed in the start of the curve, it is wise to cover the immediate moments after photobleaching with a high sampling rate. Thus, 256 × 256 pixels are used in $FRAP_1$ and 1,024 × 1,024 pixels in $FRAP_2$.
3. The next tab, Bleach, allows to define parameters regarding bleaching. Use the ellipse tool to create a ROI completely covering the nucleus (Fig. 1a). Set laser power to 100 % and select Zoom In. More robust fluorophores may require more bleaching (*see* **Note 5**).
4. In the Time Course tab, it is possible to choose how many prebleach, bleach, and post-bleach frames to record. Table 1 shows the number of frames we use in our assays. After the setup, click Run Experiment.
5. When the assay is complete, go to Evaluation tab, draw ROIs in the nucleus, background, and control cell, and save the data by right-clicking on the graph window. To place the measured import rates into context, it is highly recommended to perform the assay with control constructs (*see* **Note 5**).
6. Open the raw data files with a spreadsheet program and normalize the data accordingly: reduce background from all data points and set background as 0. Then, take an average of the two prebleach images and set them as 1. All data points will now fall between these and allow comparison between different cells of varying fluorescence intensities. Finally, produce a linear fit of the first data points to quantitate the import rate. As the fit degrades with further data points, care should be taken to include only the very first ones producing the best fit.

Table 1
Frame settings for import (FRAP) and export (FLIP) assays

Assay	Prebl	Bleach	Post-bl1	Post-bl2	Post-bl2	Total *t*
$FRAP_1$	2	1	75	15	–	
S/frame	0.388	0.388	0.388	2	–	60.3 s
$FRAP_2$	2	2	10	12	120	
S/frame	1.486	1.486	1.486	5	20	2480.8 s
FLIP	2	3	1	–	–	
S/frame	0.388	0.388	0.388	–	–	~300 s

3.3 FLIP for Export Measurements

1. Go to the menu on the upper left corner labelled Leica Microsystems LAS AF and select Live Data Mode. Configure the resolution and other hardware settings as described above for this assay. Unlike in the other two FRAP assays, summing up the frames will not yield the total time shown in Table 1. This is due to Leica's Live Data Mode introducing an unnecessary lag between frames, which almost doubles the experiment time. Thus, on another system, an export assay may only take ~155 s. Then, place additional tasks (or "Jobs") inside this experiment (or "Pattern") according to Table 1.
2. Set laser power to 100 % and cross the ROI checkbox to direct laser only to the ROI. Carefully draw the ROI following the edges of the cytoplasm making sure no nuclear area is included (Fig. 1b, *see* **Notes 4** and 7).
3. Finally, after configuring the post-bl tab, program the microscope to loop 100 times by selecting both bleach and post-bl tabs (keep Shift pressed) and right-clicking on them. When everything is set, start the experiment by pressing Start Pattern in the lower right corner (*see* **Note 8**).
4. In the Quantify tab, draw ROIs in the nucleus, cytoplasm, background, and control cell and right-click on the graph to save your data.
5. The data is then normalized as described in the above FRAP Subheading 3.2, **step 6**.

4 Notes

1. Transfection levels should be optimized to be as low as possible while still allowing proper imaging. Instead of transfection, a purified, in vitro fluorescently labelled protein of interest can also be microinjected into the cells.
2. In the assays designed for the measurements of import and export rates ($FRAP_I$ and FLIP, respectively), high scanning speeds were used at the expense of resolution to include many data points at the beginning of the assay, which is used for quantitation. Despite the oversampling, 1,024 × 1,024 resolution was used in the longer assay for optimal image quality (and to make a nice video!).
3. For any individual microscope bleaching experiment, it is advantageous to have an internal control, which can be used to exclude anomalous data. A neighboring non-bleached cell can function for this purpose to ensure the focus does not drift due to, for example, unstable temperature. Additionally, it will reveal whether the actual imaging bleaches the chromophore. This, of course, should not happen as it interferes with the quantification and one should always use low enough laser power during imaging to prevent this. For example, we have set our 270 mW optically pumped semiconductor laser to 90 % hardware power. During imaging, we use 0.5 % software laser power (as opposed to 100 % during bleaching). A control cell is easy to include in the import assay but may be difficult to use in the export assay, because a larger portion of the screen is bleached. As with all live-cell imaging, an experiment probing the stability of the imaging system should always be performed prior to the actual experiments to verify that the microscope temperature is steady at 37 °C and after changing the sample dish to ensure it has stabilized to the ambient temperature.
4. Care should be taken to select the optimal z-level. According to our experience, this is achieved relatively easily by choosing the plane where the nucleus has the largest area and where the nucleoli and/or nuclear envelope are/is in focus. If one descends under this level towards the bottom of the dish or raises the z-level, freely diffusing cytoplasmic particles below and above the nucleus can contaminate the data by making, for example, the import appear faster than it actually is. To minimize such issues, we always use pinhole 1.
5. Depending on the fluorophore, it may be necessary to bleach the ROI in a FRAP experiment for several times to remove enough fluorescence in order to perform the experiment. For example, at our usual levels of fluorescence, two frames were needed to properly bleach AF488-actin, while one was enough for GFP-actin despite both having similar intensities.

6. The import FRAP assay controls should be of variable size. These controls should be ignored by any active transport machinery so they can serve to establish the rate of passive transport. For example, GFP and 2GFP or microinjected fluorescently labelled dextrans can be used. However, once the construct size is large enough to result in unambiguous nuclear exclusion (such as 3GFP or 70 kDa dextrans), the import assay can no longer be performed due to lack of signal in the nucleus. It may also be useful to investigate how the fluorescent tag of choice influences the traffic of the protein of interest. GFP is huge compared to small <1 kDa chemical tags such as Alexa Fluor 488. However, the large size of GFP may permit extraction of information about the very nature of the transport—whether it is active or passive [6,10].
7. It is advisable to keep the edges of the ROI ~1 μm distant from the NE to avoid signal loss within the nucleus. One may need to define this distance experimentally, and for this purpose it is important to have a fluorescent construct, which cannot exit the nucleus, as a negative control. To this end, we used GFP-RanBP1ΔNES, which lacks the nuclear export signal and loses only ~10 % of the nuclear signal during 5 min of cytoplasmic bleaching. In the same timeframe, GFP-actin signal dropped to ~50 %, which is indicative of the constant export of actin out of the nucleus.
8. It is very important to realize that in the Leica system, by default, each "Job" also has an individual gain and z-level settings. This means that once these have been set, for example, in the prebl settings (probably "Job1"), one needs to set the identical values for bleach and post-bleach imaging.

Acknowledgements

The work in the laboratory of MKV is funded by Academy of Finland and Sigrid Juselius foundation. K-PS is funded by a fellowship from the Viikki Graduate School in Biosciences.

References

1. Gorlich D, Kutay U (1999) Transport between the cell nucleus and the cytoplasm. Annu Rev Cell Dev Biol 15:607–660
2. Chen X, Xu L (2011) Mechanism and regulation of nucleocytoplasmic trafficking of smad. Cell Biosci 1(1):40. doi:10.1186/2045-3701-1-40
3. Kumar S, Saradhi M, Chaturvedi NK, Tyagi RK (2006) Intracellular localization and nucleocytoplasmic trafficking of steroid receptors: an overview. Mol Cell Endocrinol 246(1–2): 147–156. doi:10.1016/j.mce.2005.11.028
4. Xing L, Bassell GJ (2012) mRNA localization: an orchestration of assembly, traffic and synthesis. Traffic. doi:10.1111/tra.12004
5. Vartiainen MK, Guettler S, Larijani B, Treisman R (2007) Nuclear actin regulates dynamic subcellular localization and activity of the SRF cofactor MAL. Science 316(5832):1749–1752

6. Dopie J, Skarp KP, Kaisa Rajakyla E, Tanhuanpaa K, Vartiainen MK (2012) Active maintenance of nuclear actin by importin 9 supports transcription. Proc Natl Acad Sci USA 109(9):E544–E552. doi:1118880109 [pii] 10.1073/pnas.1118880109
7. Skarp KP, Vartiainen MK (2010) Actin on DNA-an ancient and dynamic relationship. Cytoskeleton (Hoboken) 67(8):487–495. doi:10.1002/cm.20464
8. Lippincott-Schwartz J, Patterson GH (2003) Development and use of fluorescent protein markers in living cells. Science 300(5616): 87–91
9. Tsien RY (1998) The green fluorescent protein. Annu Rev Biochem 67:509–544. doi:10.1146/annurev.biochem.67.1.509
10. Keminer O, Peters R (1999) Permeability of single nuclear pores. Biophys J 77(1):217–228. doi:S0006-3495(99)76883-9 [pii] 10.1016/S0006-3495(99)76883-9
11. Choidas A, Jungbluth A, Sechi A, Murphy J, Ullrich A, Marriott G (1998) The suitability and application of a GFP-actin fusion protein for long-term imaging of the organization and dynamics of the cytoskeleton in mammalian cells. Eur J Cell Biol 77(2):81–90
12. Verkhusha VV, Tsukita S, Oda H (1999) Actin dynamics in lamellipodia of migrating border cells in the Drosophila ovary revealed by a GFP-actin fusion protein. FEBS Lett 445(2–3): 395–401
13. McDonald D, Carrero G, Andrin C, de Vries G, Hendzel MJ (2006) Nucleoplasmic beta-actin exists in a dynamic equilibrium between low-mobility polymeric species and rapidly diffusing populations. J Cell Biol 172(4): 541–552

Chapter 19

Isochronal Visualization of Transcription and Proteasomal Proteolysis in Cell Culture or in the Model Organism, *Caenorhabditis elegans*

Anna von Mikecz and Andrea Scharf

Abstract

Investigation of differential gene regulation by protein degradation requires analysis of the spatial and temporal association between proteolysis and transcription. Here, we describe the isochronal visualization of proteasomal proteolysis and transcription in cell culture or in vivo in the model organism *Caenorhabditis elegans*. This includes localization of proteasome-dependent proteolysis by fluorescent degradation products of model and endogenous substrates of the proteasome in combination with immunolabelling of RNA polymerase II and transcription in situ run-on assays.

Key words Cell nucleus, In situ run-on assay, K48, Microinjection, Proteasome, Proteolysis, RNA polymerase II, Ubiquitination, Transcription

1 Introduction

1.1 Visualization of Transcription in Cell Culture

Time- and quantity-resolved transcription of a gene is a strictly choreographed interaction of hundreds of proteins [1]. Since proteasomes are known nuclear residents [2, 3] and proteasomal proteolysis has been shown to be an intrinsic function of the mammalian nucleus [4], the nuclear ubiquitin-proteasome system (nUPS) is suited to play a role in the regulation of transcription. It is now widely acknowledged that the 26S proteasome is a complex player in transcription regulation via proteolytic and non-proteolytic functions [5–8]. However, the understanding of the spatial and temporal organization of proteasomal proteolysis and transcription in response to different cellular and environmental stimuli is still incomplete. The nucleus has a functional architecture with defined microenvironments that support the different nuclear processes and that are characterized by their dynamic (ribo)nucleic acid and protein composition [9–12]. Proteasomal activity and proteins that are destined for degradation via labelling with a tetra-ubiquitin

Yaron Shav-Tal (ed.), *Imaging Gene Expression: Methods and Protocols*, Methods in Molecular Biology, vol. 1042, DOI 10.1007/978-1-62703-526-2_19, © Springer Science+Business Media, LLC 2013

chain are concentrated in focal domains in the euchromatin and represent proteolytic microenvironments in the nucleus [13]. Isochronal visualization of proteasomal proteolysis and transcription enables the understanding of the spatial organization and thereby the underlying mechanism of the interaction of these two active processes inside the functional architecture of the nucleus. It may represent an important determinant of regulation (a) whether active proteasomes are directly located at the transcription site to destroy proteins, which bears an inherent degradation risk for all proteins of the transcription machinery, or (b) whether a proteolytic microenvironment is located at a distance, and proteolysis-destined proteins have to shuttle for degradation. Investigations of the spatial and temporal association of proteasomal proteolysis and global transcription, as well as local transcription at an ectopic gene array, at ribosomal DNA genes, and at early transcribed genes in *Caenorhabditis elegans* embryos reveal that local proteasome-dependent proteolysis is not required for the gene expression of every gene [13]. However, under stress conditions local association of proteasomal proteolysis with transcription sites seems to be needed in order to remove dysfunctional proteins from stalled transcription machineries [13–16]. These data confine a model that proteasomes are generally needed at transcription initiation sites to promote transcriptional elongation via degradation of components of the initiation complex [8]. In line with this, inducible genes in eukaryotic cells show different mechanism and regulation strategies to form the pre-initiation complex on the promoter, to start transcription, and to proceed to productive elongation [1].

Transcription can be visualized on a global and on a local scale in cell culture. Localization of global transcription is based on the visualization of active transcription machineries in situ via transcription run-on assays or specific antibodies, whereas local transcription can be imaged with fluorescence in situ hybridization of RNA (RNA-FISH) or gene arrays.

In transcription run-on assays, cells are incubated in a medium containing nucleotide analogues, e.g., fluorouridine (FU), that integrate into nascent RNAs [17, 18]. After cell fixation the incorporated analogue can be labelled with antibodies and analyzed by fluorescence imaging. The incorporation time has to be short (<10 min), because otherwise, nascent RNAs are allowed to move away from the transcription machinery. Longer incubation times are tempting, because they result in intensified labelling; however, under these conditions RNA trafficking inside the nucleus is imaged, rather than transcription.

Specific antibodies represent an alternative to label transcription sites in a cell. The C-terminal domain (CTD) of RNAPII that in mammals comprises 52 tandem repeats of a heptapeptide is modified during the transcription cycle [19, 20]. The CTD is hypophosphorylated in its inactive form, while during initiation, the serines in position 5 are phosphorylated and in the elongation phase the serines in position 2 [21]. These differences in the phosphorylation status of the RNAPII have been used to produce specific antibodies to label the enzyme in different stages of the transcription cycle: The antibody 8WG16 recognizes hypophosphorylated CTDs (RNAPIIa), H5 recognizes CTDs with phosphorylation at serines in position 2, and H14 binds at CTDs with phosphorylated serines in position 5 (RNAPIIo) [22, 23].

1.2 Visualization of Proteasomal Proteolysis in Cell Culture

The 26S proteasome is assembled by two regulatory 19S subunits that flank the 20S core particle responsible for proteolytic activity and can be described as a self-compartmentalizing protease [24]. Substrate proteins are labelled with a tetra-ubiquitin chain via interaction with the ubiquitin-activating enzyme (E1), a ubiquitin-conjugating enzyme (E2), and a ubiquitin ligase (E3) [25]. For subsequent proteolysis these ubiquitins are typically linked at lysines in position 48 (K48); alternative linkages are more relevant in other processes, like protein trafficking [26, 27]. The 19S regulator recognizes tetra-ubiquitinated proteins that consequently become unfolded and are channeled through an opened gateway into the proteolytic chamber. There, three proteolytic activities (trypsin-, chymotrypsin-, and caspase-like) degrade substrates into peptides that are subsequently released from the chamber [28, 29].

To analyze ongoing proteolysis at a transcription site, the following methods were developed:

1. *In nucleus* microinjection of an ectopic substrate fluoresces after degradation and unravels about 7–90 proteolytic foci throughout the euchromatin [13]. The ectopic substrate is overloaded with BODIPY FL dyes resulting in a strong quenching effect due to sterical hindering. After degradation into peptides, the fluorochromes are released and fluoresce after excitation [4].
2. A second method is to use antibodies that specifically label the K48 linkage of tetra-ubiquitinated proteins, the endogenous substrates of the proteasome [30]. Similar to the pattern of proteolytic foci after microinjection of ectopic substrates, K48-linked tetra-ubiquitinated proteins are focally concentrated throughout the nucleoplasm. A combination of both methods reveals a colocalization of 57 % of the proteolytic foci with K48-linked tetra-ubiquitins [13].

1.3 Visualization of Transcription in Caenorhabditis elegans

Caenorhabditis elegans (*C. elegans*) is a 1.2 mm free-living soil nematode that is easy to cultivate under laboratory conditions [31, 32]. The worms are maintained on agar plates with nematode growth medium between 15 and 25 °C and fed with the uracil auxotroph *Escherichia coli* strain OP50 [33]. *C. elegans* mainly

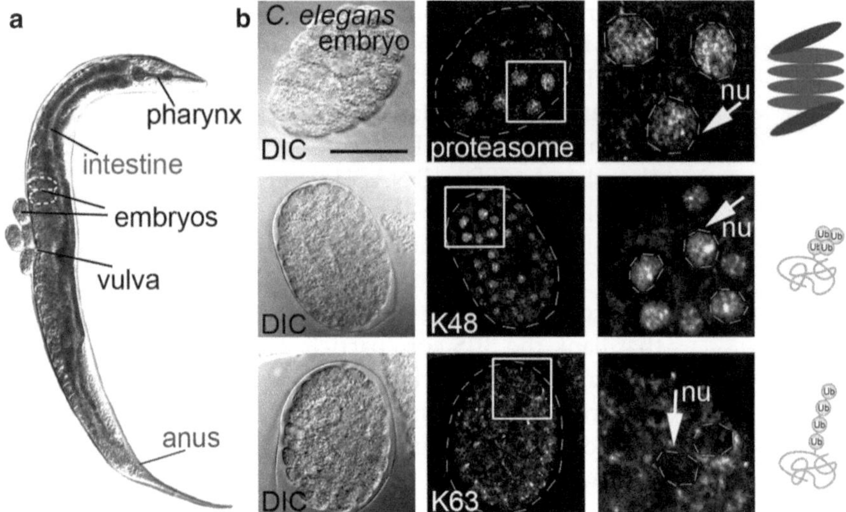

Fig. 1 The model organism *Caenorhabditis elegans*. (**a**) Anatomy of an adult hermaphrodite with fertilized eggs/embryos in the uterus. Note four recently laid embryos next to the worm. (**b**) Localization of proteasomes, K48-linked or K63-linked ubiquitin in *C. elegans* embryos. K48-linked tetra-ubiquitin and proteasomes are concentrated within nuclei, while K63-linked tetra-ubiquitin is exclusively localized in the cytoplasm. Bar, 25 μm; *nu* nucleus

occurs as a hermaphrodite that reproduces by self-fertilization. It has a generation time of 3 days under optimal conditions and a brood size of about 300 and lives about 15 days depending on environmental factors such as temperature [32]. *C. elegans* is ideal for imaging experiments, because the worm is transparent and easy to examine by microscopy. In contrast to cell culture, the worm develops from a zygote with one nucleus to a multicellular organism with 959 somatic nuclei and functionally different cells like neurons, intestinal cells, or muscle cells (Fig. 1a) [34]. The embryo is well suited for transcription analysis in relation to proteasomal proteolysis, because (a) transcription stops during oogenesis, (b) the early embryo is primarily under the control of maternal factors, and (c) transcription of the early embryonic genes restarts during the first cleavages representing a defined gene program [35–37]. The *C. elegans* RNAPII shares high sequence similarities with other species comprising a CTD with 42 heptad repeats that allows usage of phosphorylation status-specific antibodies [38]. Although the basic transcription machinery shows similarities to the mammalian system, including TATA-binding protein (TBP), *C. elegans* exhibits some particularities, such as trans-splicing and operon assembly of genes [39].

1.4 Visualization of Proteasomal Proteolysis in *Caenorhabditis elegans*

Proteasomal proteolysis plays an important role in the development and cellular physiology of *C. elegans*. Depletion of subunits of the 26S proteasome via RNAi leads to an arrest in embryonal development that causes lethality. Depletion of ubiquitin or proteasome subunits in the adult worm induces a reproductive defect that is characterized by an embryonal arrest in the one-cell stage [40]. Ubiquitin is highly conserved between the species, and the proteasome as well shares high homology with its yeast and human counterparts with 14 conserved 20S and 18 conserved 19S subunits [40–42]. This homology is helpful for the visualization of proteasomes and proteasomal activity, because it enables the use of the same tools, e.g., antibodies. Proteasomes and K48-linked tetra-ubiquitins can be conveniently visualized in the worm via a freeze-crack method [13]. The worms are fixed with formaldehyde between an objective slide and a coverslip. Then they are shortly frozen in liquid nitrogen and the coverslip is cracked away to break the worm's strong cuticle. This step is followed by an additional fixation and permeabilization step in ice-cold methanol and acetone. After applying indirect immunofluorescence, proteasomes and proteolytic activity can be localized prominently in the nucleus, while non-proteolytically linked ubiquitins, e.g., K63-linkage, are concentrated in the cytoplasm (Fig. 1b).

2 Materials

2.1 Isochronal Visualization of Proteasomal Proteolysis and Transcription in Human Cell Culture

2.1.1 Preparation of the Cells

1. Cell lines: HEp-2, human epithelioma type 2, CCL-23, U-2 OS, human epithelial osteosarcoma cell line, all obtained from ATCC, USA.
2. Cell culture medium: RPMI 1640, supplemented with 10 % bovine growth serum and 5 % supplement complex: 2 mM L-Glutamine, 1 mM sodium pyruvate, 1 % nonessential amino acids, 10 U/ml penicillin, and 10 μg/mL streptomycin and 0.25 μM 2-mercaptoethanol.
3. Trypsin-EDTA (1×): 0.05 % trypsin and 0.53 mM Na_2EDTA.
4. 100 % absolute ethanol, analytical grade.
5. 22×22 mm No. 1.5 glass coverslips.
6. 35 mm tissue culture dishes.

2.1.2 Microinjection-Based Visualization of Proteolysis in Cultured Cells

1. Proteasomal proteolysis substrates: DQ™ ovalbumin (DQ-ova, Molecular Probes, Life Technologies, CA, USA) or DQ™BSA (DQ bovine serum albumin, Molecular Probes). Prepare a 2 mg/ml stock solution: Add 0.5 ml distilled H_2O to 1 mg DQ-ova or DQ BSA.
2. Microcapillaries: Femtotips I (Eppendorf, Hamburg, Germany).
3. Microinjection apparatus: FemtoJet and InjectMan (Eppendorf).
4. Scalpel.

2.1.3 Transcription In Situ Run-On Assay

1. 5-Fluorouridine (MP Biomedicals, Solon, OH, USA).

2.1.4 Fixation of the Cells with 4 % Formaldehyde

1. Formaldehyde 10 % ultrapure (methanol-free), EM grade (Polysciences, Warrington, PA, USA): Prepare a 4 % formaldehyde solution in PBS freshly for each experiment (use in Subheadings 3.1.4 and 3.2.2) (*see* **Note 1**).

2. Triton X-100, molecular biology grade: Prepare a 0.25 % (v/v) Triton X-100 solution in PBS at least 60 min before use.
3. Phosphate buffered saline (PBS): Prepare a 10× stock solution with 1.37 M NaCl, 27 mM KCL, 43 mM Na_2HPO_4, and 14 mM KH_2PO_4. Prepare working solution by diluting the stock 1:10 with distilled H_2O.
4. Coplin jars.

2.1.5 Immunolabelling of Proteasomal Substrates, Proteasomal Proteolysis, and Transcription (Use in Subheadings 3.1.5 and 3.2.3)

1. Primary antibodies: polyclonal rabbit antibody PW8155 against the proteasome 20S core particle (Enzo Life Science, Farmingdale, NY, USA), monoclonal mouse antibody PW8195 (clone MCP231) against α-subunits (Enzo Life Science), mouse monoclonal antibody H5 against RNAPIIo (Covance, Princeton, NJ, USA), mouse monoclonal antibody H14 against RNAPIIo (Covance), mouse monoclonal antibody 8WG16 against RNAPIIa (Covance), human antibody Apu2.07 against K48-linked tetra-ubiquitin [30], human antibody Apu3.A8 against K63-linked tetra-ubiquitin [30], and rat antibody OBT0030 against BrdU (AbD Serotec, Kidlington, UK).
2. Secondary antibodies: goat anti-mouse IgG/IgM conjugated to Cy5 or fluorescein (FITC), goat anti-rabbit conjugated to Cy5 or FITC, and goat anti-rat conjugated to Cy5 or FITC.
3. Vectashield mounting medium (Vector Laboratories, Burlingame, CA, USA).
4. Nail polish.
5. Microscope slides.
6. Assembly of a humidified chamber: Take a plastic box, wet the bottom, press a piece of parafilm on the bottom, dry the surface of the parafilm with a tissue, and place moistened tissues around the parafilm at the edges of the box. Close the lid of the box to ensure a humidified atmosphere.
7. Quantification: MetaMorph image analysis software package (Molecular Devices, Sunnyvale; CA, USA).

2.2 Isochronal Visualization of Proteasomal Proteolysis and Transcription in *Caenorhabditis elegans*

2.2.1 C. elegans Culture

1. *Caenorhabditis elegans* N2 wild type (Caenorhabditis Genetics Center (CGC), University of Minnesota, USA).
2. Worm picker: Take a Pasteur pipette, flame the tip, and melt a 32-G platinum wire into the glass [33].
3. 9 cm Petri plates with cams.
4. *E. coli* strain OP50.
5. Nematode Growth Medium (NGM) agar: Mix 15 g BD Bacto Agar, 2.25 g NaCl, and 1.9 g BD Bacto Proteose No. 3 with

750 ml distilled H_2O in a 1 l lab flask. Sterilize by autoclaving, cool down to under 55 °C, and add 750 μl solution A (0.5 g cholesterol in 100 ml EtOH), 375 μl solution B (11.08 g $CaCl_2$ in 100 ml distilled H2O, autoclaved), 750 μl solution C (24.65 g $MgSO_4 \times 7H_2O$ in 100 ml distilled H_2O, autoclaved), and 18.75 ml solution D (108.3 g KH_2PO_4, 36 g K_2HPO_4 in 1 l distilled H_2O, autoclaved). Shake the flask well and pour the agar into the Petri dishes under sterile conditions. Let the plates dry. Add 500 μl of overnight culture of OP50 and use an inoculation spreader to spread the solution in the center of the plate (*see* **Note 2**).

6. M9: Dissolve 3 g KH_2PO_4, 6 g Na_2HPO_4, 0.5 g NaCl, and 1 g NH_4Cl in 800 ml distilled H_2O, mix it until everything has dissolved, and adjust with distilled H_2O to 1 l. Sterilize by autoclaving.
7. Bleach 10 ml: 5 ml distilled H_2O, 2 ml 4 M NaOH, and 3 ml 12 % NaClO (*see* **Note 3**).

2.2.2 Fixation of the C. elegans Eggs

1. Poly-L-lysine-coated objective slides: Wash uncoated slides by mounting approx. 12 mm diameter sample chambers (e.g., cytoslides, Thermo Fisher Scientific, Waltham, MA, USA) in 100 % EtOH and in distilled H_2O. After complete drying, apply approx. 100 μl of a 0.5 mg/ml poly-L-lysine solution (mol wt ≥300,000) to the sample chambers and incubate at room temperature for 10 min. Remove the poly-L-lysine solution with a pipette and wash the slides in distilled H_2O. Let the slides dry over night.
2. Formaldehyde 10 % ultrapure (methanol-free), EM grade (Polysciences, Warrington, PA, USA): Prepare a 4 % formaldehyde solution in PBS fresh for each experiment (use in Subheadings 3.1.4 and 3.2.2) (*see* **Note 1**).
3. Liquid nitrogen (*see* **Note 4**).
4. Methanol, extra pure grade (*see* **Note 1**).
5. Acetone, synthesis grade (*see* **Note 1**).
6. Phosphate buffered saline (PBS): Prepare a 10x stock solution with 1.37 M NaCl, 27 mM KCL, 43 mM Na_2HPO_4, and 14 mM KH_2PO_4. Prepare working solution by diluting the stock 1:10 with distilled H_2O.
7. Staining jars, Scalpel.

3 Methods

3.1 Isochronal Visualization of Proteasomal Proteolysis and Transcription in Human Cell Culture

3.1.1 Preparation of the Cells

Use adherent cells with nuclei of >5 μm diameter. Smaller nuclei are harder to microinject. HEp-2 or U-2 OS cells, for example, are suitable for intranuclear microinjection:

1. Flame 20 mm square coverslips and transfer them to a 35 mm Petri dish. Overlay the coverslip with 2–4 ml medium.
2. Add 1–2 droplets of 1×10^6/ml cells onto the coverslips and carefully spread them over the coverslip. Avoid clumping of the cells.
3. Incubate at 37 °C with 5 % CO_2 until the cells reach subconfluence of 50–70 % (*see* **Note 5**).

3.1.2 Microinjection-Based Visualization of Proteolysis in Cultured Cells [4, 43]

DQ ovalbumin (DQ-ova) is a precursor substrate loaded with sterically hindered fluorochromes that exclusively fluoresce after proteolytic cleavage. Thus, it can be used to localize active proteolysis after intracellular/intranuclear microinjection.

1. Mix 4 μl of DQ-ova stock solution (2 mg/ml) with 36 μl sterile PBS.
2. Centrifuge the solution for 10 min by ultracentrifugation (or similar) to clear the solution (*see* **Note 6**).
3. Put a Petri dish with a coverslip on the microscope and focus on the cells with a 40× objective for a longer working distance (*see* **Note 7**).
4. Load 5 μl DQ-ova solution (0.2 mg/ml) with a Microloader into a Femtotip. The Microloader must be inserted through the narrow opening up to the last third of the microcapillary and then release the DQ-ova solution, avoiding any air bubbles (*see* **Note 8**).
5. Carefully loosen the protection cap with your fingers by turning it and letting it drop down. Screw the microcapillary into its holder and connect it to the pressure tube. Then pressurize the system by pressing "menu" (compensation pressure: approx. 500 hPa) and position the microcapillary into the medium right over the cells. Try to move the microcapillary as near as possible over the cells without touching the cells or the coverslip. For correct positioning of the Femtotip, use the micromanipulator and the beam of light as orientation point. Try to focus rotationally between cells and microcapillary. Check, by pressing the "clean" button, that the Femtotip is not blocked and adjust the compensation pressure to 50–100 hPa. Now the system is ready to microinject cell nuclei.
6. Move the capillary over the first nucleus, then pierce this nucleus by moving the capillary down and directly up again. The nucleus should shortly inflate during a successful microinjection (*see* **Note 9**).

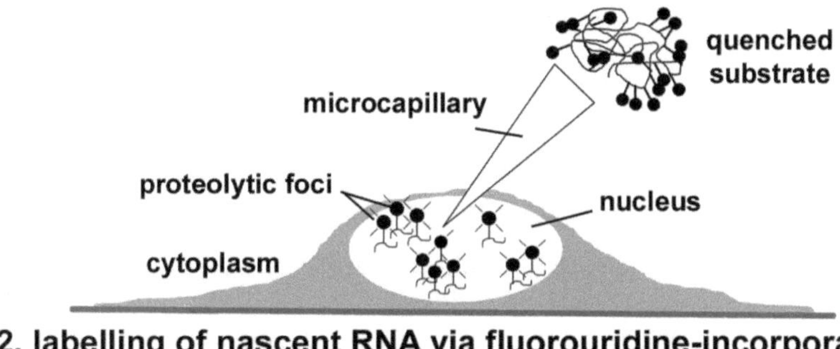

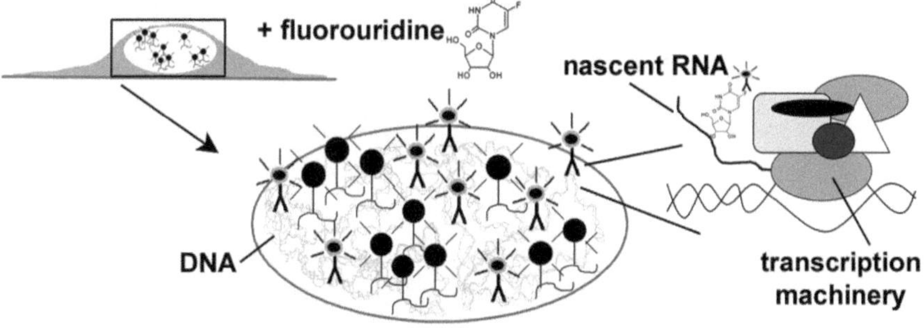

Fig. 2 Schematic of isochronal visualization of proteasomal proteolysis and transcription in cultured cells. (1) Microinjection of quenched substrates into the nucleus. Subsequent proteasomal degradation and release of the fluorochromes result in fluorescent proteolytic foci visible throughout the nucleoplasm. (2) The cells are incubated in a medium containing nucleotide analogues, e.g., fluorouridine. Transcription machineries incorporate nucleotide analogues into transcripts which enables antibody-based detection of nascent RNAs. Nascent RNAs are distributed as small foci throughout the nucleus

7. Continue to microinject approx. 100 nuclei in the same region.
8. After microinjection, incubate the Petri dish for 10–50 min at 37 °C and 5 % CO_2 (Fig. 2).

3.1.3 Transcription In Situ Run-On Assay [17, 18]

In order to localize transcription sites in cells that were microinjected with DQ-ova, nucleotide analogues, e.g., fluorouridine (FU), are incorporated into nascent transcripts by in situ run-on assays. Start with FU incorporation 10–50 min after microinjection.

1. Have Petri dishes with 37 °C prewarmed medium/0.02 M fluorouridine ready.
2. Transfer the microinjected coverslips with forceps into Petri dishes and incubate them for 5–10 min at 37 °C and 5 % CO_2.
3. Remove the coverslips for fixation (Fig. 2) (*see* **Note 10**).

3.1.4 Fixation of Cells with 4 % Formaldehyde [44]

1. Rinse the coverslips by placing them in a small Coplin jar containing PBS.
2. Quickly exchange the PBS for 4 % formaldehyde (diluted from a 10 % formaldehyde stock with PBS) and incubate for 10 min at room temperature.

3. Permeabilize with 0.25 % Triton X-100 in PBS for 3 min at room temperature.
4. Rinse fixed coverslips 6× quickly with PBS, then rinse 3 × 10 min with PBS by gently rocking on a shaker.

3.1.5 Immunolabelling of Nascent RNAs

Next, transcription sites are labelled with fluorochrome-tagged antibodies against nucleotide analogues:

1. Prepare a humidified chamber and place one objective slide for each coverslip inside.
2. Prepare the appropriate primary antibody dilution in PBS: antibodies against BrdU 1:50 (*see* **Note 11**).
3. Take a coverslip; drain the excess PBS quickly with a tissue. Avoid drying of the cells. Apply 33 μl antibody on the coverslip and convert the coverslip with cells down on an objective slide in the humidified chamber (*see* **Note 12**). Close the humidified chamber and incubate the antibody for 1 h.
4. Float the coverslip from the objective slide by adding PBS on the objective slide with a squirt bottle next to the coverslip. Put the coverslips into a Coplin jar and wash three times with PBS.
5. Dilute the secondary antibody 1:100 in PBS. For labelling of transcription sites, use a Cy5-conjugated secondary antibody to avoid cross talk with the fluorochromes that label proteolytic foci. BODIPY FL has an excitation maximum at 505 nm and fluoresces maximally at 513 nm, whereas Cy5-conjugated antibodies have an excitation maximum at 650 nm and fluoresce maximally at 670 nm. Fluorescent signals of both fluorochromes can be recorded simultaneously due to the wide separation of their emissions.
6. Repeat **step 3**, but reduce incubation time to 45 min. Repeat **step 4**.
7. Take a labelled objective slide, add two drops of vectashield, and immerse the coverslip with the cell layer downwards onto the vectashield (*see* **Note 13**). Drain excess vectashield by covering the slide with a tissue and let soak.
8. Seal the coverslip with nail polish. After drying of the nail polish, the slides can be directly analyzed via microscopy or stored at 4 °C in the dark (*see* **Note 14**).
9. Analyze the stained cells by fluorescence imaging. Use MetaMorph to quantify the colocalization of proteasomal proteolysis and transcription. Draw a line of interest through proteolytic and transcriptionally active microenvironments/foci and measure the fluorescence intensities of both fluorescent signals (proteolysis and transcription) along this line of interest [13]. Representative micrographs of the interaction between transcription and proteasomal proteolysis in human cells are shown in Fig. 3a.

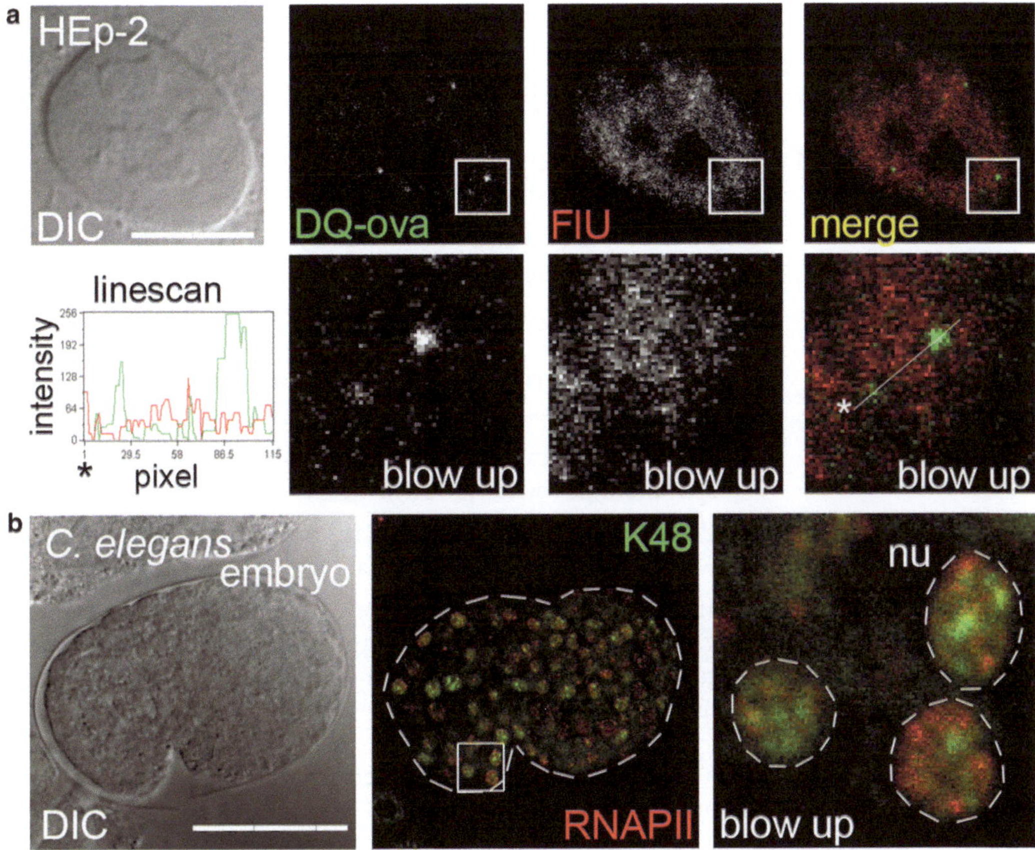

Fig. 3 Visualization of transcription and proteasomal proteolysis in cultured cells or *C. elegans* embryos. (**a**) Confocal fluorescence micrograph of a HEp-2 cell nucleus microinjected with DQ ovalbumin (*green*) followed by incubation with fluorouridine (FU) to label sites of active transcription (*red*). After fixation with 4 % formaldehyde, cells were immunostained with specific antibodies to visualize sites of fluorouridine incorporation. Bar, 5 μm. (**b**) Confocal fluorescence micrographs of a *C. elegans* embryo that was prefixed in 4 % formaldehyde, freeze cracked, refixed with methanol and acetone, and subjected to immunolabelling with antibodies against RNAPIIo to detect transcribed embryonal genes (*green*) and antibodies against K48-linked tetra-ubiquitin to localize intranuclear proteolytic centers (*red*). Bar, 25 μm

3.2 Isochronal Visualization of Proteasomal Proteolysis and Transcription in *Caenorhabditis elegans*

3.2.1 Culturing of *C. elegans*

C. elegans is cultured on agar plates seeded with a lawn of *E. coli* strain OP50. Synchronization of the worms is required for most of the experiments. A convenient method is to use bleach and isolate the eggs.

1. Pick approx. ten adult *C. elegans* with a worm picker and transfer them to a new agar plate with OP50.
2. Let the worms grow at 20 °C for approx. 4 days, e.g., until the plate contains between 200 and 500 adult *C. elegans* (*see* **Note 15**).
3. Wash the worms off the plate with M9 and collect them in a 1.5 ml tube.

4. Collect the worms by centrifugation at 500 × *g* in a tabletop centrifuge for 30 s, pour off the buffer, add 1 ml new M9, and resuspend the pelleted *C. elegans* by inverting the tube 3×. Repeat the washing steps 2×.
5. Add 1 ml bleach to the worms, mix by inverting the tube 3×, and spin them down at 3,000 × *g*.
6. Pour off the bleach, add 1 ml new bleach, and shake the worms by fast inverting of the tube (approx. 5×).
7. Transfer approx. 15 μl of the mixture on an objective slide (*see* **Note 16**) and observe the bleaching procedure. When the first eggs appear outside of the hermaphrodite, the worms are spun down at 3,000 × *g*.
8. Pour of the bleach, wash the eggs 3× with distilled H_2O, and spin them down at 3,000 × *g*.
9. Pour off the water after the last wash and transfer the pellet with the eggs to a new agar plate with OP50 (*see* **Note 17**).
10. Incubate the synchronized worms at 20 °C for at least 4 days. During this time period the eggs develop into adult hermaphrodites with eggs that in turn are ready to use in subsequent immunolabelling procedures.

3.2.2 Fixation of C. elegans Eggs

Eggs are extracted from adult hermaphrodites and prepared for immunolabelling.

1. Collect the synchronized adult worms by M9 wash off in a 1.5 ml tube and spin them down at 500 × *g*.
2. Wash the worms 3× with M9. After the last wash, leave the worms in approx. 100 μl M9.
3. Take a poly-L-lysine-coated objective slide; add 5 μl 8 % formaldehyde/PBS and 5 μl of the worm/M9 solution. Mix it with the pipette tip (*see* **Notes 1** and **18**).
4. Cover the worms with a coverslip and squeeze them gently by pressing the coverslip on the objective slide with forceps until the adults release their eggs (*see* **Note 19**).
5. Fix the worms for 5–30 min depending on the no. of coverslips and shock freeze in liquid nitrogen, e.g., dip the coverslip with forceps into a bowl with liquid nitrogen (*see* **Note 4**) for a few minutes.
6. Place the objective slide on a tissue, fix it with your left hand, and detach the coverslip with a scalpel. Press the scalpel under one side of the coverslip to loosen and lift the coverslip.
7. Transfer the objective slide directly to a Coplin jar containing −20 °C cold methanol to refix the worms and eggs (*see* **Note 1**). Incubate for 5 min at −20 °C (freezer).

8. Permeabilize the worms in −20 °C cold acetone for 2 min (*see* **Note 1**).
9. Gently wash the fixed eggs 3× in PBS. Use a new Coplin jar and new PBS in each step. Transfer objective slides with forceps.

3.2.3 Immunolabelling of Proteasomal Proteolysis and Transcription

1. Prepare the appropriate primary antibody dilution in PBS: antibody against K48-linked tetra-ubiquitin, Apu2.07, 1:50, and antibody against the RNAPIIo H5 1:50.
2. Take one objective slide, drain the excess PBS with a tissue, and place it into a humidified chamber.
3. Add 50–100 μl of the antibody solution. Close the humidified chamber and incubate the antibodies for 2 h.
4. Dip the objective slide in a beaker filled with PBS and transfer it into a Coplin jar containing PBS. Wash 3× with PBS (*see* **Note 20**).
5. Dilute the secondary antibody 1:100 in PBS. Use combinations of FITC- and Cy5-conjugated secondary antibodies to avoid cross talk between the fluorochromes during fluorescence microscopy-based analysis. FITC-conjugated antibodies have an excitation maximum at 494 nm and fluoresce maximally at 518 nm, whereas Cy5-conjugated antibodies have an excitation maximum at 650 nm and fluoresce maximally at 670 nm. Fluorescent signals of both fluorochromes can be recorded simultaneously, because of the wide separation of their emissions.
6. Repeat **step 3**, but incubate for 1.5 h.
7. Repeat **step 4**.
8. Take one objective slide, drain the excess PBS with a tissue and add two drops of vectashield to mount the eggs. Carefully place a coverslip on the vectashield (*see* **Note 13**). Drain excess vectashield by covering the slide with tissue and applying soft pressure.
9. Seal the coverslip with nail polish. After complete drying of the nail polish the slides can be directly analyzed via microscopy or stored at 4 °C in the dark.
10. Analyze the eggs by fluorescence microscopy. Representative micrographs of labelled *C. elegans* eggs are shown in Fig. 3b.

4 Notes

1. Formaldehyde, methanol, and acetone are toxic—protect yourself and discard the chemicals according to the respective institutional regulations.

2. Do not damage the surface of the agar; otherwise, worms will crawl into the agar. Keep the *E. coli* lawn in the center of the plate. Avoid spreading the lawn to the edges to prevent the worms from crawling off the plate.
3. Bleach is corrosive: Protect yourself.
4. It is important to familiarize yourself with safety methods of working with liquid nitrogen. Appropriate protection (thick gloves, safety goggles, lab coat) is required.
5. Let the cells grow for 2–3 days, because after 1 day, they are often not sufficiently adherent which results in wash off during microinjection.
6. It is important to lose all solid particles, e.g., clumps of proteins that are not dissolved, since these may clog the microcapillary.
7. We use an LCPlanFl 40×/0.60NA objective (Olympus, Shinjuku, Tokyo, Japan).
8. To get rid of air bubbles between the top of the microcapillary and the liquid, knock with one finger against the capillary without breaking it. During loading always leave the protection cap attached to the capillary.
9. Be careful—the nucleus should not burst. If that is the case, the compensation pressure is too high, and you will have to reduce it. In order to relocate cell nuclei, microinject them in the center of the coverslip. You may additionally scratch a cross into the cell layer with a scalpel and microinject nuclei in the four edges of the cross.
10. The protocol in this chapter suggests incorporation of nucleotide analogues as a method to label transcription; an alternative method is to combine the visualization of proteolytic foci with gene arrays [13]. (1) If you are interested in labelling the ribosomal DNA genes, the only endogenous gene array existing in mammalian cells [45], microinject the cells as described in Subheading 3.1.2, fix them and apply a silver stain as described in Mais et al. [46]. (2) If you want to use gene arrays as established by Janicki et al. and Darzacq et al. [47, 48], induce a transcription site as described and subsequently microinject nuclei with a proteolytic substrate. Labelling of induced transcription sites can be done by transfection of the appropriate fusion proteins or via fluorescence in situ hybridization (FISH) [13].
11. BrdU-antibodies likewise recognize incorporated FU.
12. Apply the antibody solution immediately to one corner of the coverslip to avoid drying of the cells. Always be aware of the coverslip orientation, e.g., the side with the cell lawn. If you are not sure, scratch with forceps over the coverslip. You will see the scratch exclusively in the cell lawn.

13. Avoid air bubbles.
14. The nail polish has to be completely dried before microscopic examination or storage at 4 °C.
15. Food should be abundant, because if the worms starve, they develop an internal hatch phenotype.
16. While the volume may vary, ensure that you observe around ten adults.
17. Use a pipette to transfer the eggs onto the new agar plate.
18. Pipet up and down until you have enough worms in the 5 μl volume. 5–20 worms are optimal.
19. Be careful! If the pressure is too high, the worms burst, if the pressure is too low, worms and eggs are lost during the fixation and staining procedures.
20. If you use different antibody combinations, always change the PBS in the beaker between dipping.

Acknowledgments

Work in the von Mikecz laboratory is supported by the German Science Foundation (DFG) through grants MI486/7-1 and GRK1033.

References

1. Weake VM, Workman JL (2010) Inducible gene expression: diverse regulatory mechanisms. Nat Rev Genet 11:426–437
2. Kleinschmidt JA, Hugle B, Grund C, Franke WW (1983) The 22 cylinder particles of *Xenopus laevis.* I. Biochemical and electron microscopic characterization. Eur J Cell Biol 32:143–156
3. Hugle B, Kleinschmidt JA, Franke WW (1983) The 22S cylinder particles of *Xenopus laevis.* II. Immunological characterization and localization of their proteins in tissues and cultured cells. Eur J Cell Biol 32:157–163
4. Rockel TD, Stuhlmann D, von Mikecz A (2005) Proteasomes degrade proteins in focal subdomains of the human cell nucleus. J Cell Sci 118:5231–5242
5. Muratani M, Tansey WP (2003) How the ubiquitin-proteasome system controls transcription. Nat Rev Mol Cell Biol 4:192–201
6. von Mikecz A (2006) The nuclear ubiquitin-proteasome system. J Cell Sci 119:1977–1984
7. Kodadek T (2010) No splicing, no dicing: non-proteolytic roles of the ubiquitin-proteasome system in transcription. J Biol Chem 285:2221–2226
8. Geng F, Wenzel S, Tansey WP (2012) Ubiquitin and proteasomes in transcription. Annu Rev Biochem 81:177–201
9. Wansink DG, Schul W, van der Kraan I et al (1993) Fluorescent labeling of nascent RNA reveals transcription by RNA polymerase II in domains scattered throughout the nucleus. J Cell Biol 122:283–293
10. Stein GS, Zaidi SK, Braastad CD et al (2003) Functional architecture of the nucleus: organizing the regulatory machinery for gene expression, replication and repair. Trends Cell Biol 13:584–592
11. Trinkle-Mulcahy L, Lamond AI (2007) Toward a high-resolution view of nuclear dynamics. Science 318:1402–1407
12. Misteli T (2007) Beyond the sequence: cellular organization of genome function. Cell 128:787–800

13. Scharf A, Grozdanov PN, Veith R et al (2011) Distant positioning of proteasomal proteolysis relative to actively transcribed genes. Nucleic Acids Res 39:4612–4627
14. Gillette TG, Gonzalez F, Delahodde A et al (2004) Physical and functional association of RNA polymerase II and the proteasome. Proc Natl Acad Sci U S A 101:5904–5909
15. Anindya R, Aygün O, Svejstrup JQ (2007) Damage-induced ubiquitylation of human RNA polymerase II by the ubiquitin ligase Nedd4, but not Cockayne syndrome proteins or BRCA1. Mol Cell 28:386–397
16. Selth LA, Sigurdsson S, Svejstrup JQ (2010) Transcript elongation by RNA polymerase II. Annu Rev Biochem 79:271–293
17. Boisvert FM, Hendzel MJ, Bazett-Jones DP (2000) Promyelocytic leukemia (PML) nuclear bodies are protein structures that do not accumulate RNA. J Cell Biol 148:283–292
18. Kießlich A, von Mikecz A, Hemmerich P (2002) Cell cycle-dependent association of PML bodies with sites of active transcription in nuclei of mammalian cells. J Struct Biol 140:167–179
19. Egloff S, Murphy S (2008) Cracking the RNA polymerase II CTD code. Trends Genet 24:280–288
20. Svejstrup JQ (2004) The RNA polymerase II transcription cycle: cycling through chromatin. Biochim Biophys Acta 1677:64–73
21. Phatnani HP, Greenleaf AL (2006) Phosphorylation and functions of the RNA polymerase II CTD. Genes Dev 20:2922–2936
22. Warren SL, Landolfi AS, Curtis C, Morrow JS (1992) Cytostellin: a novel, highly conserved protein that undergoes continuous redistribution during the cell cycle. J Cell Sci 103: 381–388
23. Patturajan M, Schulte RJ, Sefton BM et al (1998) Growth-related changes in phosphorylation of yeast RNA polymerase II. J Biol Chem 273:4689–4694
24. Baumeister W, Walz J, Zuhl F, Seemuller E (1998) The proteasome: paradigm of a self-compartmentalizing protease. Cell 92:367–380
25. Pickart CM, Eddins MJ (2004) Ubiquitin: structures, functions, mechanisms. Biochim Biophys Acta 1695:55–72
26. Thrower JS, Hoffman L, Rechsteiner M, Pickart CM (2000) Recognition of the polyubiquitin proteolytic signal. EMBO J 19:94–102
27. Pickart CM, Fushman D (2004) Polyubiquitin chains: polymeric protein signals. Curr Opin Chem Biol 8:610–616
28. Finley D (2002) Ubiquitin chained and cross-linked. Nat Cell Biol 4:E121–E123
29. Finley D (2009) Recognition and processing of ubiquitin-protein conjugates by the proteasome. Annu Rev Biochem 78:477–513
30. Newton K, Matsumoto ML, Wertz IE et al (2008) Ubiquitin chain editing revealed by polyubiquitin linkage-specific antibodies. Cell 134:668–678
31. Brenner S (1974) The genetics of *Caenorhabditis elegans*. Genetics 77:71–94
32. Hope IA (1999) Background on *Caenorhabditis elegans*. In: Hope IA (ed) *C. elegans*. A practical approach. Oxford University Press, Oxford, New York, pp 1–15
33. Stiernagel T (1999) Maintenance of *C. elegans*. In: Hope IA (ed) *C. elegans*. A practical approach. Oxford University Press, Oxford, New York, pp 51–67
34. Hall DH, Altun ZF (2008) Introduction to *C. elegans* anatomy. In: Hall DH, Altun ZF (eds) *C. elegans* Atlas. Cold Spring Harbor Laboratory Press, Cold Spring Harbor, New York, pp 1–15
35. Edgar LG, Wolf N, Wood WB (1994) Early transcription in *Caenorhabditis elegans* embryos. Development 120:443–451
36. Seydoux G, Mello CC, Pettit J et al (1996) Repression of gene expression in the embryonic germ lineage of *C. elegans*. Nature 382:713–716
37. Baugh LR, Hill AA, Slonim DK et al (2003) Composition and dynamics of the *Caenorhabditis elegans* early embryonic transcriptome. Development 130:889–900
38. Seydoux G, Dunn MA (1997) Transcriptionally repressed germ cells lack a subpopulation of phosphorylated RNA polymerase II in early embryos of *Caenorhabditis elegans* and *Drosophila melanogaster*. Development 124:2191–2201
39. McGhee JD, Krause MW (1997) Transcription factors and transcription regulation. In: Blumenthal T, Meyer BJ, Priess JR, Riddle DL (eds) *C. elegans* II. Cold Spring Harbor Laboratory Press, Cold Spring Harbor, New York, pp 147–184
40. Kipreos ET (2005) Ubiquitin-mediated pathways in *C. elegans*, In: The *C. elegans* Research Community (ed) WormBook, doi/10.1895/wormbook.1.36.1, http://www.wormbook.org
41. Hershko A, Ciechanover A (1998) The ubiquitin system. Annu Rev Biochem 67:425–479
42. Davy A, Bello P, Thierry-Mieg N et al (2001) A protein–protein interaction map of the *Caenorhabditis elegans* 26S proteasome. EMBO Rep 2:821–828

43. Scharf A, Rockel TD, von Mikecz A (2007) Localization of proteasomes and proteasomal proteolysis in the mammalian interphase cell nucleus by systematic application of immunocytochemistry. Histochem Cell Biol 127:591–601

44. von Mikecz A, Chen M, Rockel T, Scharf A (2008) The nuclear ubiquitin-proteasome system: visualization of proteasomes, protein aggregates, and proteolysis in the cell nucleus. Method Mol Biol 463:191–202

45. Darzacq X, Yao J, Larson DR et al (2009) Imaging transcription in living cells. Annu Rev Biophys 38:173–196

46. Mais C, Wright JE, Prieto JL et al (2005) UBF-binding site arrays form pseudo-NORs and sequester the RNA polymerase I transcription machinery. Genes Dev 19:50–64

47. Janicki SM, Tsukamoto T, Salghetti SE et al (2004) From silencing to gene expression: real-time analysis in single cells. Cell 116:683–698

48. Darzacq X, Kittur N, Roy S et al (2006) Stepwise RNP assembly at the site of H/ACA RNA transcription in human cells. J Cell Biol 173:207–218

Chapter 20

Considering Discrete Protein Pools when Measuring the Dynamics of Nuclear Membrane Proteins

Nikolaj Zuleger, David A. Kelly, and Eric C. Schirmer

Abstract

Measuring dynamics of nuclear proteins is complicated by the fact that many DNA- and chromatin-binding proteins have separate nucleoplasmic and nuclear membrane pools with distinct mobilities. Moreover, when measuring recoveries in FRAP experiments, it is important to be aware that the continuous transport of new protein through the nuclear pore complexes means that fluorescence recovery comes from both dynamic exchange of protein already within the nucleus and newly imported protein. Here we describe fluorescence recovery after photobleaching and photoactivation techniques designed to track nuclear membrane proteins and some methods we have developed that may help to distinguish these various pools. A combination of these approaches with standard FRAP approaches is necessary to understand the true dynamics of nuclear proteins.

Key words Nuclear envelope, Nuclear envelope transmembrane protein (NET), GFP, FRAP, Photoactivation, Nuclear transport

Abbreviations

ER	Endoplasmic reticulum
FRAP	Fluorescence recovery after photobleaching
INM	Inner nuclear membrane
NE	Nuclear envelope
NET	Nuclear envelope transmembrane protein
NPC	Nuclear pore complex
ONM	Outer nuclear membrane
ROI	Region of interest

1 Introduction

Live cell imaging of nuclear proteins is complicated by the existence of several distinct nuclear subcompartments (Fig. 1a). While standard FRAP approaches may apply for a subset of soluble nuclear

Yaron Shav-Tal (ed.), *Imaging Gene Expression: Methods and Protocols*, Methods in Molecular Biology, vol. 1042, DOI 10.1007/978-1-62703-526-2_20, © Springer Science+Business Media, LLC 2013

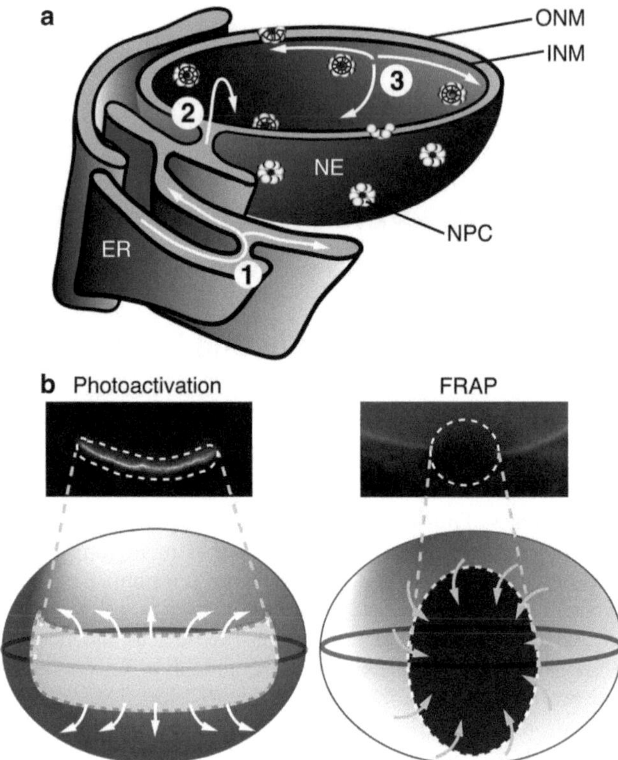

Fig. 1 The structure of the NE and different paths by which fluorescence recovery can occur. (**a**) The NE is a double membrane system typically in the shape of an ovoid. It is continuous with the endoplasmic reticulum (ER) and perforated by nuclear pore complexes (NPCs). Thus, membrane proteins could move (*1*) from the ER to the outer nuclear membrane, (*2*) from the outer nuclear membrane around the NPCs to the inner nuclear membrane, and (*3*) within the inner nuclear membrane. (**b**) Photoactivation (*left panel*) or photobleaching (*right panel*) of protein in the NE must be considered in context of the ovoid, i.e., activating at the edge means activating along an entire "wall" as well as the edge of both the top and bottom of the nucleus. This means in theory that activation of a carefully selected NE will result in a *rectangular activated shape* (*left panel*), while photobleaching of a circular area will result in an *ecliptic shape* (*right panel*)

proteins, many chromatin proteins have multiple populations based on both partners and localization. The nuclear subcompartment that provides the greatest stationary/stabilizing force is the nuclear envelope (NE). Most of the dense chromatin classically observed by electron microscopy accumulates at the nuclear periphery [1, 2] and modern fluorescence and biochemical approaches indicate that a large portion of the epigenetically silenced chromatin accumulates at the nuclear periphery, some of which is in direct contact with proteins of the NE [3–6]. Live cell imaging of NE proteins and associated proteins must consider that they are restricted in the two-dimensional context of the surface area of an ovoid (Fig. 1a, b) while proper imaging of soluble nucleoplasmic proteins often is complicated by there being two pools, one freely diffusing in the nucleoplasm and one restricted to the two-dimensional space of the NE by interaction with the intermediate filament lamin polymer or proteins embedded in the membrane. Examples of chromatin and chromatin regulatory/remodeling proteins that interact with NE proteins include the histone deacetylase HDAC3 [7], the chromatin cross-linking protein barrier-to-autointegration factor (BAF; [8]), histone H3 carrying trimethylation on lysine 9 [9], heterochromatin protein 1 [10], and several transcription factors and transcriptional repressors including such master regulators as pRb, Fos, and Smads [11–14]. As there are roughly 1,000 NE transmembrane proteins (NETs; [15–17] and most transcriptional and epigenetic regulators have not been directly tested for a NE pool, there are likely to be a great many more chromatin proteins for which the two populations are relevant.

The NE itself is a complex double membrane compartment that separates the nucleus from the cytoplasm. The outer nuclear membrane (ONM) makes connections to cytoplasmic filament systems, while the inner nuclear membrane (INM) forms connections to chromatin. These connections to chromatin are mediated both by the intermediate filament lamin proteins and membrane proteins resident in the INM. Like all membrane proteins, these INM proteins are synthesized in the ER. Because of the continuity between the ER and the NE, these proteins can be transported to the INM by diffusion in the membrane via the nuclear pore complexes (NPCs), which are large protein structures that connect the ONM with the INM (Fig. 1a) and which facilitate bidirectional nucleocytoplasmic transport of most cellular molecules. Experimental data suggests two possible transport routes for INM proteins through the NPC: (1) entirely through the peripheral channels of the NPC [18–20] and (2) simultaneously though both central and peripheral channels by cleaving through the NPCs [21]. Studying how these INM proteins are transported to and from the NE requires special approaches that take into account both the two-dimensional limitations of the environment of the membrane but also the three-dimensional structure of the ovoid surface of the nucleus. It also requires distinguishing movement within the INM from translocation of proteins through the NPCs and movement in the ER and ONM (Fig. 1a). Similarly, new protein translocated through the NPCs contributes to FRAP recovery for even soluble nuclear proteins that do not have NE pools.

This chapter will describe two live-cell imaging techniques, FRAP and GFP-photoactivation, which in combination serve as a powerful tool to study INM protein transport. While GFP-photoactivation is required to visualize the directionality of the activated material (e.g., protein activated in the ER can be monitored for accumulation in the NE; Fig. 2a) and thus distinguish the various protein pools, FRAP is more accurate at quantifying the mobility rate of the protein studied (Fig. 2b). When combining the data obtained from both methods, one can draw reasonably accurate

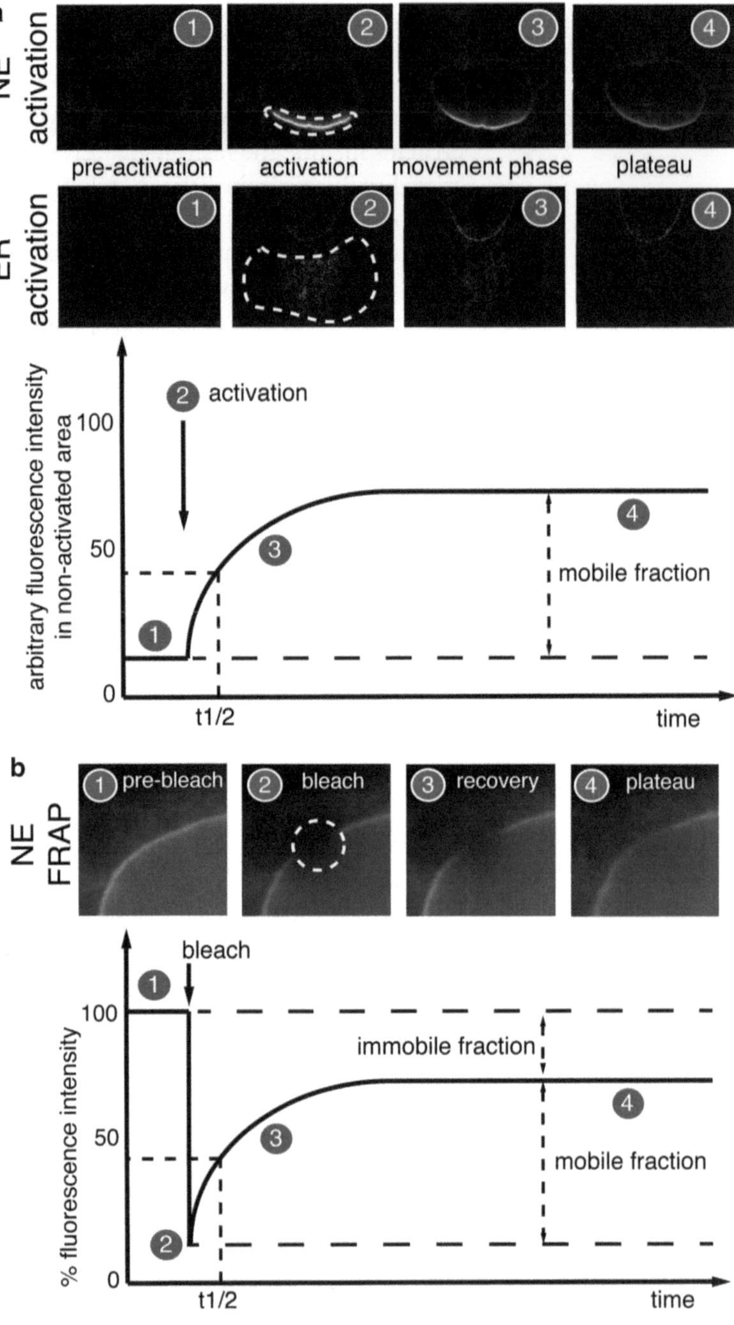

Fig. 2 Measurements in photoactivation and FRAP experiments. (**a**) A standard photoactivation experiment can be divided into the following phases: the pre-activation phase (*1*) where any background from illumination in the marker protein channel or due to activation from the lower intensity scans can be accounted for, the activation point (*2*) where a strong laser pulse activates the fluorophore in a discrete region of the cell (*dashed line*), the movement phase (*3*) where the mobile fraction of the protein moves out of the region of activation, and the plateau phase (*4*) where equilibrium is reached and the mobile population can be distinguished from the tethered population. (**b**) A standard FRAP experiment can be

conclusions about the nature of dynamics of nuclear membrane proteins. Applying the approaches used for nuclear membrane proteins together with standard approaches should also reveal insights into the presence of multiple protein pools for many chromatin proteins. We also provide an approach to limit transport of new soluble and some transmembrane proteins, and as many nuclear proteins require ATP hydrolysis for their dynamics, we describe an approach to rapidly deplete ATP prior to FRAP studies.

2 Materials

2.1 Preparation of Cells for Microscopy

1. Cells expressing protein of interest fused to GFP or photoactivatable GFP (*see* **Note 1**).
2. Coverslips: borosilicate glass, Thickness No. 1 of appropriate size for the live cell chamber used (e.g., 25 mm diameter round, VWR International).
3. 0.01 % poly-L-lysine solution (*see* **Notes 2** and **3**).
4. Tissue culture plates that fit coverslips and other tissue culture accessories to grow cells (e.g., 35 mm plates or 6 well plates for 25 mm coverslips).
5. 1 M HEPES-KOH: pH 7.4, sterile filtered (*see* **Note 4**).
6. Phenol-free medium supplemented with 25 mM HEPES-KOH: This should be the standard medium used for culturing the particular cell type (i.e., DMEM and RPMI), but using a phenol-free base and supplemented with HEPES, 10 % fetal bovine serum, 100 μg/ml penicillin and 100 μg/ml streptomycin sulfate, 25 mM HEPES-KOH (*see* **Notes 5** and **6**).
7. Reagents for mammalian cell transfection: For example, Fugene HD or Lipofectamine.

2.2 Microscope and Sample Setup

1. Live-cell imaging chamber. There are many forms for such a chamber ranging from peltier systems costing tens of thousands of dollars to tissue culture dishes with coverslip bottoms

Fig. 2 (continued) divided into following phases: the pre-bleach phase (*1*) where the fluorescence intensity is 100 %; the bleach pulse (*2*), which depletes the fluorescence level in the bleached area to a fraction of the pre-bleach intensity; a recovery phase (*3*), in which mobile non-bleached mobile molecules move into the bleached area to facilitate a regain of the fluorescence; and a plateau phase (*4*), at which the fluorescence is not significantly recovering. The recovered fluorescence often does not reach pre-bleach levels (100 %) and this phenomenon is due to immobile molecules within the bleached area (immobile fraction). The difference between the pre-bleach level intensity and the immobile fraction constitutes the mobile fraction

in an environmental chamber to a Dvorak-Stottler chamber that can be purchased for several hundred dollars and can be heated with a hair dryer. In general what is needed is to be able to maintain the cells in medium on a coverslip at 37 °C. One should ideally use a system that has been purchased with the microscope to be used, but a variety of imaging chambers are available that can be used with most systems. Examples of good non-peltier imaging chambers that can be used inside a temperature-controlled environmental chamber include for 25 mm coverslips the Attofluor Cell Chamber, Invitrogen, A-7816, or Chamlide CM-B25-1 (*see* **Note 7**).

2. Microscope: Photoactivation and FRAP experiments can be performed on both confocal and widefield systems; however, both FRAP and photoactivation are best performed on a confocal microscope. Photoactivation experiments will require a 405 nm laser which is often not present due to the cost and extra optics required to integrate it. FRAP can be performed using standard light sources but it tends to be slow and unwieldy and is therefore much better when performed with a laser tuned to the excitation wavelength of the fluorophore to be bleached (e.g., 488 nm for GFP). For effective and quick photobleaching in FRAP experiments, the laser should be capable of photobleaching the specimen to around 80 % of starting value in one short pulse (less than diffusion time of test protein).
3. Appropriate objective: The choice of objective for FRAP is dependent on two factors, the magnification required and the light throughput of the objective because one of the primary requirements for effective FRAP experiments is gathering the maximum amount of light. For FRAP experiments both factors generally come down to a choice between ×60 or ×100 objective. A ×100 objective is usually used for FRAP on very small structures such as the nuclear membrane as it provides a larger area for accurate placement and measurement of the photobleached spot. The light gathering capability of a particular objective can be calculated from the equation

$$10^4 \times (\mathrm{NA}^4 / M^2)$$

where NA is the numerical aperture of the objective and M is the magnification. From Table 1, the Plan Apo ×63 is the best compromise in terms of light gathering, resolution, and field of view. However, if available, the ×100 objective should be used when the GFP labeling is bright enough that light gathering is not a problem because the extra magnification of the ×100 makes positioning the bleach spot or photoactivated area much easier and more accurate.

Table 1
General characteristics of different objectives

Objective type	Light gathering	Resolution (nm)
Plan Apo ×40 NA 1.25 oil immersion	15.25	253
Plan Apo ×63 NA 1.40 oil immersion	9.67	226
Plan Apo ×100 NA 1.40 oil immersion	3.84	226

4. Control samples to calibrate laser/bleaching: A concentrated solution of free dye (e.g., 25 % w/v FITC or Alexa488 for GFP calibration) placed on a microscope slide under a coverslip will serve for this.

2.3 Photoactivation Experiments

1. Fusions of your protein of interest to photoactivatable GFP [22] (*see* **Notes 8** and **9**).
2. Fusion to mRFP (*see* **Note 10**) of a marker protein for the NE, e.g., emerin or NET20 (*see* **Note 11**).

2.4 FRAP Experiments

1. Fusions of your protein of interest to E-GFP or AcGFP (both Clontech; *see* **Notes 8, 12,** and **13**).

2.5 Data Analysis

1. Software platform for image analysis (e.g., Image-Pro, Volocity, or Metamorph).
2. Macros were designed based on flow diagrams in Fig. 3.

2.6 NPC Transport Inhibition and ATP Depletion

2.6.1 NPC Transport Inhibition

1. Plasmid expressing untagged Ran Q69L mutant under control of the CMV promoter and separately expressing RFP (e.g., [20]; *see* **Notes 14** and **15**).
2. Ran antibody that works for both immunohistochemistry and Western blot (e.g., Becton-Dickinson, 610341).
3. Antibody for loading control (beta-actin or tubulin) when testing the expression of the Ran mutant by Western blotting.

2.6.2 ATP Depletion

1. Normal Glucose-containing medium.
2. Glucose-free medium base (e.g., Gibco glucose-free DMEM).
3. 1 M stock 2-deoxy-D-glucose in molecular biology grade water.
4. 1 M stock sodium azide in molecular biology grade water.
5. Freshly prepared glucose-free growth medium containing normal supplements (i.e., serum and antibiotics) and 6 mM 2-deoxy-D-glucose, 10 mM sodium azide, 25 mM HEPES-KOH.

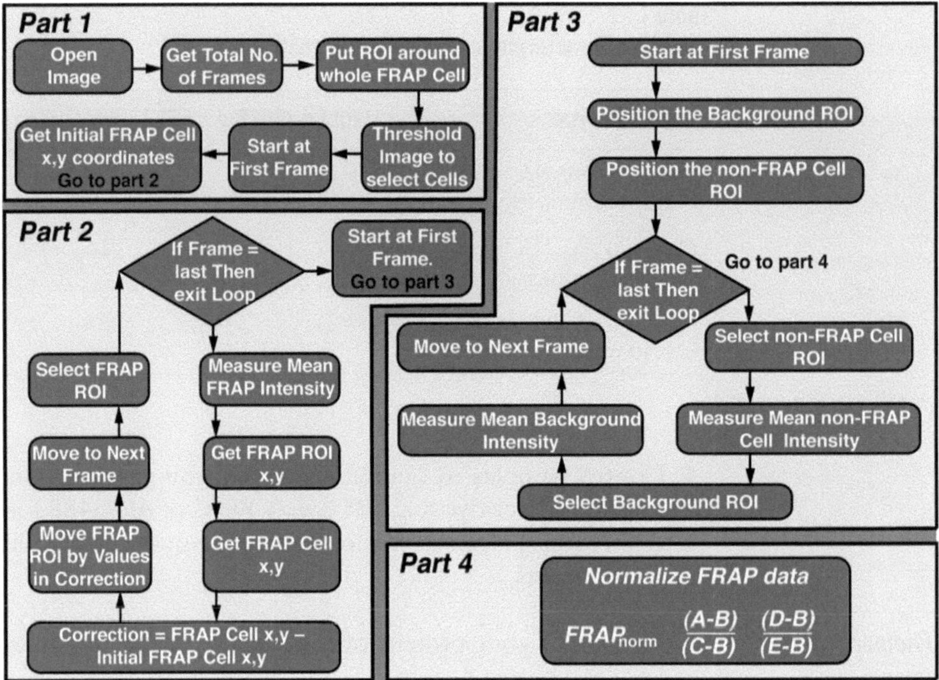

Fig. 3 Flow chart describing the steps important to include in macros to facilitate data analysis for FRAP experiments. There are four basic parts to the process: (1) open image and select FRAP region, (2) measure intensity and correct for movement in FRAP ROI, (3) background correction, and (4) normalization. The same macros were used with logical modifications for the regions of interest being investigated in photoactivation experiments. *A* Pre-bleach mean intensity from non-FRAP cell, *B* background fluorescence at time (t), *C* non-FRAP cell at time (t), *D* bleach ROI at time (t), *E* pre-bleach mean intensity from bleach ROI

3 Methods

From the time the cells are ready, a typical experiment comparing two samples and taking ten movies each should take roughly 3 h for a FRAP experiment and 5 h for a photoactivation experiment.

3.1 Preparation of Cells for Microscopy

1. Incubate 25 mm coverslips (or other system component for cell growth that can be used with your live-cell imaging setup) with 100 μg/ml poly-L-lysine for 10 min (*see* **Note 16**).
2. Wash coverslips (or other system component that can be used with your live-cell imaging setup) extensively with water (at least 100 volumes) to remove all traces of unattached poly-L-lysine and dry (*see* **Note 17**).
3. Plate cells onto the poly-L-lysine-treated coverslips (or other system component that can be used with your live-cell imaging setup) at ~15 % density to prevent them from becoming confluent at the time of the experiment (*see* **Note 18**). Transiently transfected cells can be used as well as stably transfected cells.

If stably transfected cells are to be used then skip to **step 5** provided that all necessary constructs are stably expressed (both RFP and GFP constructs are needed for photoactivation experiments).

4. Transfect cells using standard procedures with constructs for photoactivation or FRAP experiments 12–24 h after plating and time this for use in experiments typically at 16–24 h post-transfection. For FRAP experiments cells are transfected with only the GFP fusion proteins, but for photoactivation experiments they need to be transfected with both the photoactivatable GFP fusions and the RFP-NET reporter to visualize the nuclear envelope (*see* **Note 19**).
5. Plate and transfect a duplicate set of cells to fix and perform control for GFP photophysics (*see* Subheading 3.4, **step 8**).
6. Plate and transfect one extra slide of the main fusion protein to be characterized to be used in optimizing settings for the microscope (Subheadings 3.2.3 and 3.2.4).

3.2 Microscope and Sample Setup

3.2.1 Bleach Spot Calibration for Lasers with Widefield Systems

1. Align the laser so that it is tightly focused and perfectly central within the objective to be used as per manufacturer's guidelines (*see* **Note 20**).
2. Place a slide with a concentrated solution of free dye (e.g., FITC for GFP calibration) onto the microscope stage (*see* **Note 21**).
3. Open an image histogram window if it is not already part of the acquisition/alignment software and set the system to continuously acquire. While it is acquiring slowly adjust the focus until the histogram reaches its peak intensity and is steady. This should place the focus in the center of the reference slide.
4. Perform one test bleach using the same laser power (100 % usually) and only one pulse. It is important that only one laser pulse is used as this removes the possibility of protein recovery during any subsequent pulses, which can be a problem for proteins with fast recovery mobilities (*see* **Note 22**).
5. The laser should be aligned each time before a set of FRAP experiments are undertaken and the test bleach image should be kept because it is needed to adjust the data if it is to be modeled.

3.2.2 Placing Cells into Heated Chamber System

Details here will vary depending on the nature of the chamber setup associated with a particular microscope. Peltier systems tend to be completely enclosed and are very complex to set up and so should follow the detailed manufacturer's instructions. Here in order to highlight a few relevant considerations, we give details for using a standard open non-peltier imaging chamber, e.g., Attofluor Cell Chamber, Invitrogen, A-7816 inside an environmental chamber.

1. At least 2 h before beginning imaging, start the gas flow (5 % CO_2 and 95 % air) and pre-warm the environmental chamber together with the cell imaging chamber to 37 °C. If a peltier system is being used, the chamber and objective heater should be similarly heated prior to commencing imaging (*see* **Note 23**).
2. Use forceps to gently place the coverslip containing the cells to be imaged into the chamber and seal the chamber bottom properly (*see* **Note 24**).
3. Immediately add imaging medium containing 25 mM HEPES-KOH to maintain pH during the course of experiments.
4. Gently wipe the coverslip bottom that will face the microscope objective with a tissue to clean of medium before making contact with the objective. Before placing the chamber on the microscope, wait at least 10 s and place a clean tissue on the bottom to check for leaks and further tighten the thread or start over if any medium has leaked out.
5. Immediately transfer the chamber with the cells into the pre-warmed environmental chamber. Allow the chamber with the cells to equilibrate to 37 °C for about 10 min. Maintain the cells in the same gas temperature conditions throughout the entire experiment.

3.2.3 Microscope Settings for Photoactivation Experiments

Although a widefield system could be used for simple photoactivation experiments, a laser scanning confocal is recommended for photoactivation experiments analyzing the NE pool of a protein. This is because it is necessary to photoactivate shapes at the NE rather than just a spot in order to perform these experiments optimally, and the confocal is capable of generating such shapes because the laser scans each pixel separately:

1. Place the extra control transfection cells in the system.
2. Open the pinhole fully to gather as much light as possible.
3. Use the transfected RFP-NET to identify the NE and ER in a cell.
4. Use the region of interest function to highlight an area comprising both part of the nucleus and part of the ER.
5. Photoactivate this region of interest using a 0.5 s laser pulse at 10 % laser power for a 100 mW 405 nm laser.
6. If both NE and ER signals are bright, then these conditions can be used. If one or the other is quite weak (less than 25 % above background), then increase the laser power (*see* **Note 25**).
7. The scan speed should be set to support light collection according to the sample being analyzed, i.e., if the sample fluorescence in a control test is very low when photoactivated, then the scan speed should be set to slow in order to maximize the

light collection, while if the sample fluorescence is very bright in a control test, then a faster scan speed can often be used (*see* **Note 26**). However, it should be noted that if there are smaller pools where the activated protein might move, then they may only be seen using the slow scan speed.

3.2.4 Microscope Settings for FRAP Experiments

1. Open the pinhole completely if using a confocal (*see* **Note 27**).
2. Ensure image depth is 12 or 16 bit as this is important for measuring subtle intensity differences.
3. Test several cells from the control transfected sample to determine optimal bleach times for your protein of interest. A recommended starting point is a 0.5 s pulse at 100 % laser power. Ideally, conditions will be identified that generally result in only 20 % of the pre-bleach intensity for a whole-cell image taken at 1 % laser power within 2 s of the bleach (*see* **Notes 28** and **29**).
4. Acquisition speed. As for the photoactivation procedure described above, this must be determined experimentally using the extra control transfection cells. The speed of acquisition should be fast enough to capture the dynamic recovery of the protein under study but not too fast since illumination of the specimen causes excessive photobleaching. For example, if most recovery is observed in 30 s, then to obtain a good FRAP curve will require imaging every 1–2 s. If the laser/UV intensities are low, then this frequency of imaging should not too badly photobleach the sample in a 2–3 min experiment, whereas this frequency of imaging would photobleach a sample if the experiment transpired over 10 min for a slower recovering protein. In addition to the relationship between the general recovery time and frequency of imaging, the speed of acquisition for each image is also sample dependent as well as very dependent on whether a widefield or confocal microscope system is used.
5. Determining acquisition speed using a laser scanning confocal system. The laser scanning confocal tends to be the most widely used system for FRAP as the majority of manufacturers have incorporated software wizards to help users, and the fact that the lasers scan point by point across the specimen allows a variety of bleach spot shapes and positions to be used. There is a tradeoff between speed of imaging and signal-to-noise ratio where the faster the scan speed, the less time the detectors have to collect photons, and so image quality is compromised. Enabling bidirectional scanning (not usually enabled) is an easy way to double the speed of acquisition without compromising the signal-to-noise ratio, although this can introduce phase shifts that need to be corrected. To ascertain the fastest possible scan speed, take the following steps:

(a) Image a set of control cells over the same length of time that the experiment will run.

(b) Set scan speed to the fastest setting.

(c) Set laser power to a low setting (typically 2–5 %).

(d) Set detector gain to midrange (not too noisy).

(e) Run sample.

(f) Repeat runs, gradually increasing the laser power to improve image quality until bleaching begins to occur. Alternatively, if absolute speed is not essential, then reduce scan speed gradually until image quality is acceptable (but not too much, when the image capture is slower than the dynamic process being tested).

(g) Based on the above parameters, set a sequence for experiments as follows:

- At least five pre-bleach images to establish baseline.
- One bleach pulse set to ensure that the specimen bleaches between 70 and 80 %.
- Collect as many post-bleach events as necessary to cover the whole recovery process of your protein (i.e., when intensity reaches a stable plateau) (Fig. 2b).

6. Determining acquisition speed using a widefield microscope system. For absolute speed of acquisition, widefield FRAP systems are far superior to the confocal systems as they are more sensitive due to the use of EMCCD (electron-multiplying charge-coupled device) cameras and there are no pinholes to reduce light throughput. Resolution tends not to be as good as the confocals because of the large pixel size of the EMCCD cameras but that is not usually a problem for FRAP.

(a) Sample bleaching is determined by running control cells with different camera exposure times to find an exposure time that does not bleach the sample, but remember to set the hardware to close the excitation shutter between acquisitions as leaving the cells permanently exposed to the excitation light will rapidly photobleach/kill the cells.

(b) If significant bleaching is occurring, neutral density filters can be placed in front of the light source to reduce the light intensity.

(c) For very fast events (e.g., recovery rates of less than 20 s), the camera readout can be binned, where, for example, in 2×2 binning, the readout from four pixels is summed as one thus producing a brighter image which is faster for the camera to process (*see* **Note 30**).

(d) Set up experimental sequence as in 5(g).

3.3 Photoactivation Experiments

A standard photoactivation experiment follows the steps below:

1. First identify a region of interest in which to activate a pool of the protein of interest using a separately transfected marker protein.
2. Measure pre-activation fluorescence.
3. Photoactivate a population of the protein of interest.
4. Measure both the loss of fluorescence from this region and the accumulation of activated protein in areas outside of the region of interest where it was initially activated (Fig. 2a).

3.3.1 Photoactivation Experiment for Determining if Protein Is Accumulating in the Nucleus/at the NE after Activation in the ER

1. Find a cell that expresses the RFP-fused reporter NET (e.g., NET20-RFP).
2. Based on the distribution of the RFP reporter, select a region of interest for the photoactivation that encompasses a region of the ER close to but several microns away from the nucleus (*see* **Note 31**).
3. Take 5–10 pre-activation images of the whole cell at low laser power (e.g., 1 % laser power, Argon laser 488 nm line on a Leica SP5 confocal; *see* **Note 32**).
4. Photoactivate the region of interest using a single high laser power pulse as determined in Subheading 3.2.3 (e.g., ~10 % laser power using a UV laser at 405 nm on Leica SP5 confocal for 0.5 s).
5. Take subsequent interval images at low laser power (same intensity as in **step 3**) to monitor the distribution of the activated material over time (*see* **Note 33**).
6. Save experiment.

3.3.2 Photoactivation Experiment for Determining Mobility of Protein Activated at the NE

1. Perform experiment as in Subheading 3.3.1 except change the region of interest for activation. For NE activation, 30 % of the circumference of the nuclear rim is outlined and activated, while trying to avoid co-activation of the ER. Thus the width of the photoactivation area should be restricted to ideally under 1 μm and drawn carefully at the nuclear edge based on the NET-RFP marker.
2. There are several pathways for movement of a protein at the NE. For membrane proteins these are typically (1) movement within the two-dimensional space of the nuclear surface and (2) movement from the NE to the ER (Fig. 1). For soluble proteins associating with the NE in addition to exchanging binding partners at the nuclear surface, they could move into the nucleoplasm. Thus while the taking of interval images to follow movement of the activated material should be done at least initially by taking whole-cell images to determine if the protein generally fails to move, it is often beneficial to take rapid scans of a different region of interest. For example, in the

case of soluble proteins that may have a pool associated with the NE, it may be necessary to make the pinhole smaller as used for confocal mode to measure accumulation of the activated material in the nucleoplasm as distinct from measuring the nuclear surface and the nucleoplasm when the pinhole is open. This is not relevant for transmembrane proteins as, their being physically restricted to the two-dimensional environment of the membrane, any accumulation that would appear to be in the nucleoplasm would be from movement to both the top and bottom of the nucleus.

3.4 FRAP Experiments

A standard FRAP experiment follows the formula of first measuring pre-bleach fluorescence, then bleaching a population of the protein of interest, and finally measuring the recovery of unbleached protein into the bleached area (Fig. 2b).

1. Start a microscope program that assists FRAP (*see* **Note 34**).
2. Find a cell that expresses the NET-GFP and focus the nucleus to make sure the whole nucleus is captured by the imaging field (*see* **Note 35**).
3. Use the experimental sequence setup for **steps 4–6**.
4. Take five pre-bleach images
5. Bleach the area of interest at appropriate laser intensity determined in Subheading 3.2.4. So that ideally ~20 % of the fluorescence intensity is retained.
6. Take post-bleach images until no obvious fluorescence recovery is observed anymore (*see* **Notes 36** and **37**).
7. Save experiment.
8. Use the fixed coverslip cells, going through the same FRAP process, to test for GFP photophysics. If recovery in the fixed sample is observed, then this background will need to be subtracted from live cell results (*see* **Note 38**).

3.5 Data Analysis

The amount of information that can be extracted from these experiments is considerable. We give below the core methods to get information about the mobility of the different fractions. However, it is worth noting that, if additional data can be obtained by other methods that go beyond this chapter, more detailed analyses can be done. For example, in the case of membrane proteins, because neither FRAP nor photoactivation could distinguish protein in the ONM from the INM, the higher resolution information from immunoelectron microscopy was required to enable us to model the kinetics of different phases of movement [20].

3.5.1 Analysis of Photoactivation Experiments

1. Although image data is typically collected using a wizard with the microscope software, we recommend analyzing the data with a separate software package such as Image-Pro Plus 7

software (Media Cybernetics), Metamorph (Molecular Devices), or Volocity (Perkin Elmer).

2. Measure the pixel intensities in the photoactivated area in the different images taken over the time course of the experiment for each experiment. When measuring the pixel intensities in the photoactivated area, movement correction should be manual rather than automatic. This is because the larger irregular-shaped area is very difficult to correct automatically as rotational shifts can become a problem.
3. Measure the pixel intensities in various regions of interest (ROI) outside the photoactivated area from the images taken over the time course of the experiment for each experiment. For example, after activation of either the NE or the ER area, the mean fluorescence intensity of the non-activated NE portion would be measured for each frame (*see* **Note 39**).
4. The initial value of the non-activated NE part should be set to 100 and the intensities of the subsequent frames adjusted accordingly.
5. The equation for normalizing the data is

$$\text{Photoactivated}_{\text{Norm}} = \frac{A}{B-C} \times \frac{D-C}{E}$$

A = Mean non-photoactivated cell pre-activation

B = Non-photoactivated cell timepoint (t)

C = Background timepoint (t)

D = Photoactivated region timepoint (t)

E = Mean photoactivated cell pre-activation

6. Normalized intensities for each frame are averaged from different experiments and plotted to measure the rate of loss of the photoactivated material from the photoactivation region of interest and the percentage of the total photoactivated material accumulating in the different cellular compartments as well as their rate of accumulation (*see* **Note 40**).
7. Calculate $t½$s based on the curves generated for the rates of loss and accumulation (Fig. 2a; *see* **Note 41**). The $t½$ is the timepoint in the curve where one half of the total recovered fluorescence has been achieved. This is based on the point where a plateau in the recovery curve is achieved as opposed to the starting fluorescence. The $t½$ is a simple measure by which to compare multiple proteins or conditions with shorter $t½$s indicating faster recoveries.
8. Although it is possible to calculate diffusion coefficients from this data, because of the complexity of the shapes used for the regions of interest, this is not simply done and requires

complex mathematics. Therefore, we recommend this being done on a case-by-case basis depending on the local expertise and do not provide instructions here.

3.5.2 Analysis of FRAP Experiments

1. Intensity measurements and *t*½ calculations were made using a macro written in Visual Basic within Image-Pro according to [23]; *see* Subheading 2.5. This involves the following steps (Fig. 3):
   (a) A region of interest (ROI) was applied to the non-FRAP cell (a control cell in the same image field that is not photobleached), the bleach area, and the background.
   (b) The region of interest was applied to each frame in the FRAP sequence.
   (c) Each image was automatically corrected for photobleaching based on the equation below:

$$\text{FRAP}_{\text{norm}} = \frac{(A-B)}{(C-B)} \times \frac{(D-B)}{(E-B)}$$

   A = Pre-bleach mean intensity from non-FRAP cell

   B = Background fluorescence at time (t)

   C = Non-FRAP cell at time (t)

   D = Bleach ROI at time (t)

   E = Pre-bleach mean intensity from bleach ROI (FRAP ROI)

   (d) Movement of the bleach area because of cell movement was corrected for by applying any change in the center xy position of the whole FRAP cell to the position of the bleach ROI (*see* **Note 42**).
   (e) The *t*½s were calculated using normalized fluorescence values, setting the immediate post-bleach value to zero and the mean of the last 10 points of the recovery curves to 100.
2. It is essential when calculating diffusion coefficients to use an equation that is adjusted to account for the size and shape of the bleach spot; thus this should be done on a case-by-case basis and is beyond the scope of this chapter (*see* **Note 43**).

3.6 NPC Transport Inhibition and ATP Depletion

3.6.1 Inhibition of Nuclear Transport by Ran Inhibition

In all FRAP or photoactivation experiments, there is a likelihood that a component of the dynamics measured will have been contributed by translocation of newly synthesized protein through the NPCs. As a typical mammalian cell has 2,000–3,000 NPCs per nucleus, there are many potential entry points all throughout the nuclear periphery. There are two ways to control for this, both of which have the potential to generate pleiotropic effects that could affect the outcome of measurements. The most drastic for pleiotropic effects is cycloheximide treatment to inhibit protein synthesis.

We present here a less drastic approach where nuclear transport is inhibited by expressing the dominant-negative Q69L mutant of the transport factor Ran. It is important to note, however, that timing is very important for these experiments because Ran inhibition will also yield pleiotropic effects over time and so it is important to perform experiments shortly after the transport inhibition begins.

1. It is only worth doing Ran inhibition if a significant percentage of the recovery in FRAP is due to transport of new protein into the nucleus. To test for this, one can photoactivate the protein of interest in the ER and then measure if a reasonable fraction of the activated material is entering the NE or the nucleoplasm. If the amount of photoactivatable GFP signal measured in the nucleus is less than 10 % the fluorescence intensity measured during the recovery phase of standard FRAP experiments, then it is probably not necessary to control for this with Ran inhibition.
2. Transfect the plasmid expressing the Ran Q69L mutant either into cells already expressing the GFP fusion protein to be monitored or together with the GFP fusion protein (*see* **Notes 44** and **14**). Regardless, the RanQ69L mutant should be transfected precisely 24 h before the start of experiments (*see* **Note 45**).
3. Make certain to transfect duplicate coverslips so that one set can be used to confirm an increase in Ran expression in the RFP expressing cells by antibody staining.
4. Engage FRAP experiments 24 h after transfection of the Ran Q69L mutant construct as in Subheadings 3.4 and 3.5, selecting cells expressing both the GFP fusion protein and the mRFP for experiments (*see* **Note 46**).
5. Confirm that the Q69L mutant was being expressed by immunofluorescence staining of the same or equally transfected coverslips using the Ran antibody and/or Western blot using Ran and loading control antibodies.

3.6.2 Depletion of ATP

Some NETs require ATP for some step in their translocation and retention in the nucleus. Similarly the dynamics of many chromatin proteins depend on ATP. To gain more information about the dynamics of the protein being investigated, FRAP experiments can also be engaged in cells depleted for ATP.

1. Bearing in mind that the cells can only be used in a short time window, accordingly transfect multiple coverslips for each condition so that at least 10 separate FRAP experiments can be concluded for each condition.
2. 10 min before engaging FRAP experiments, change the medium from regular growth medium to the glucose-free medium containing 10 mM sodium azide, 6 mM 2-deoxy-D-glucose, and 25 mM Hepes-KOH.

3. Engage FRAP experiments as in Subheadings 3.4 and 3.5 (*see* **Note 47**).
4. End FRAP experiments within 40 min and discard cells (*see* **Note 48**).
5. Engage equivalent experiments on regular non-depleted cells to compare.

3.7 Final Notes

Though in most cases the goal of FRAP and photoactivation experiments is to generate numbers that can be compared to numbers obtained for other nuclear proteins, it is important to bear in mind that the *t*½s and diffusion coefficients calculated can vary based on differences between the bleach spot size, the bleaching parameters, the health of the cells during the experiments, as well as between different cell types and species. Moreover, the existence of the various pools that this chapter is focused on distinguishing between will also impact on the interpretation of results.

4 Notes

1. Adherent cells should be used in order to limit cell movement during the course of experiments.
2. Polylysine treatment of coverslips is not necessary, but as cells tend to be more spread out and less mobile on this substratum, it aids in the microscopy.
3. Longer poly-L-lysine chains (70–150 kDa) adhere better and are preferred because shorter chains sometimes release from the surface, in which case they can act like signaling molecules.
4. It is preferable to use a system that perfuses CO_2 into the medium, but in case one is not available, it is important to maintain the pH of the medium during experiments with the addition of the HEPES.
5. We give this as the default because some have reported that phenol-containing medium increases the background fluorescence levels so that it interferes with readings; however, we and several others have noted that this effect appears to be cell type or microscope specific as we have not observed any significant effect in our system using HeLa cells. Thus, if checked first and an increase in background is not observed, then regular phenol red-containing medium can be used.
6. It is often fine to do experiments in phenol-containing medium as the increased background is not a problem. However, if the fluorescence signal is particularly weak, it is recommended to use phenol-free medium.
7. A lighter chamber such as the CM-B25-1 made of aluminum is recommended as some of the heavier chambers can contribute

to drift during experiments, particularly when used in conjunction with z-galvo stages. It is always a good idea to check with the z-stage manufacturer what is the weight limit of the z-galvo drive, as placing too great a load onto it can cause damage.

8. Unless tags on one end of the protein completely perturb its localization, it is wise to perform all tests in parallel with constructs that have the GFP tag on both ends. This is because the binding sites for two different partners might be at both ends and the behavior observed with each partner would then depend on the construct being free at this binding site. In the case of transmembrane proteins, it is useful to first test the membrane topology (*see* refs. 24, 25 for a protocol for this) as placing the tag in the NE lumen should not interfere with chromatin/nucleoplasmic interactions.
9. There is a strong advantage to using cell lines stably expressing the photoactivatable GFP fusion protein because the experiment is set up using the distribution of a separately transfected mRFP fusion as a guide to identify the cell region to activate. When both constructs are transiently transfected, it is possible to set up an experiment based on an mRFP-transfected cell, only to often find that the cell does not co-express the photoactivatable GFP construct.
10. mRFP can be replaced by any fluorescent fusion protein that the microscope setup can visualize as long as it is not visualized with an emission wavelength that could partially activate the photoactivatable GFP.
11. In theory any NET could be used, but some NETs such as LAP2ß have very little protein accumulating in the ER when overexpressed while some others accumulate large aggregates in the ER. The ideal NET for this purpose is one that gives a crisp nuclear rim staining while also accumulating enough in the ER to distinguish the various regions of the cell as is the case for both emerin and NET20.
12. The original versions of GFP have a strong tendency to dimerize, which can result in inaccurate FRAP measurements. The E-GFP version has been mutated to reduce this tendency and can be used for FRAP experiments, but the AcGFP version is the best in this regard. For membrane proteins the two-dimensional restriction increases the likelihood of dimers forming with even the E-GFP versions so the AcGFP is highly recommended.
13. There are clear advantages to using cell lines stably expressing the GFP fusion protein in that both uniform and the typically lower levels of fluorescence compared to transient transfections result in more reproducible results and lower error bars. However, it is important to recognize that the generation of stable lines can also select for properties that might affect the

functioning of the protein being investigated. Therefore data may be more accurate despite the higher error bars when using transient transfections and it is recommended when using stable lines to test multiple clones.

14. Ran function and targeting is strongly inhibited by the addition of tags; therefore it is important to express the Ran independently of tags, such as in the bicistronic vector mentioned with the RFP expressed separately. The RFP is needed to identify cells transfected with the Ran construct.
15. Though theoretically any construct for mammalian expression of the Ran Q69L mutant should work, we have optimized this in the timing for expression to do the FRAP before the cells become sick from global inhibition of nucleocytoplasmic transport based on using the CMV promoter.
16. This step will improve cell adherence to minimize cell movements that can interfere with FRAP and photoactivation experiments.
17. Small amounts of soluble poly-L-lysine (i.e., that are not attached to the coverslip) can potentially affect signaling mechanisms in the cells and make them behave aberrantly.
18. It is important for cells to be well spread and not altered in shape by touching adjacent cells. The recommended density of 15 % refers to adherent HeLa cells to be imaged between 30 and 48 h post-transfection. As the degree of confluency at the time of experiments depends on both the plating density and the division rate of the cells used, it therefore needs to be titrated for each cell line individually.
19. When co-transfecting for photoactivation experiments, the ratio between the photoactivatable construct and the RFP reporter should be at least 2:1 to increase the probability that cells expressing the RFP-reporter NET also express the photoactivatable construct.
20. See if the acquisition software can put a crosshair centrally in the field of view. This helps to align the laser by gradually adjusting it until the laser spot is overlapping with the central crosshair.
21. A plastic reference slide will also work although not as well because it does not show the Gaussian nature of the beam profile that will be needed if modeling the data.
22. Computational modeling requires as many of the experimental variables to be fixed as possible. One such variable is the size of the bleach spot. On widefield systems this spot needs to be measured as they usually do not have the same vendor calibrations as the confocal systems.
23. It is crucial to pre-warm the environmental chamber/objective at least 2 h before the experiment. Otherwise the cooler

microscope will serve as a heat sink and will rapidly cool the imaging chamber and cells, thus slowing kinetics. Moreover, when using an environmental chamber, the heating will also affect the oil in the stage and the control of stage movement will vary over the course of experiments unless the whole system has been equilibrated to the temperature of the chamber prior to starting experiments.

24. This must be done quickly to prevent cells from drying out before medium can be added to the chamber.
25. It is important to find a good balance in the photoactivation intensity, as too weak activation will impair image quality and too strong activation will result in the photobleaching of the fluorophore.
26. The sensitivity of the detectors associated with the microscope system also enters into the decision of scan speeds. For example, standard confocal photon multiplier tubes are around 20 % quantum efficient and should generally be used with the slowest scan speed for these experiments whereas GaSP detectors are around 45–50 % quantum efficient and can generally be used with fast scan speeds.
27. When measuring protein dynamics with FRAP, fine confocal sectioning is not usually required. The primary objective is to gather as much light as possible from the specimen under scrutiny. For this reason the confocal pinhole is usually set to its fully open position or as far open as possible without allowing an excessive amount of background fluorescence from the surrounding media/cell organelles.
28. It is important to find the right balance for this as overbleaching often results in cellular damage/whole-scale protein aggregation that would interfere with recoveries while too little bleaching makes it difficult to distinguish the recovering protein.
29. This value fits more easily with the equations laid out by [26] for calculating diffusion coefficients.
30. There is a downside to this in that it seriously reduces the image resolution.
31. It is crucial not to cross-activate the NE while activating the ER as this will invalidate any measurement of NE accumulation due to translocation from the ER.
32. Ten pre-activation images are preferred, but five are acceptable in cases where the proteins are very weakly expressed.
33. Photoactivated GFP is less photostable than E-GFP or AcGFP and so photobleaches much faster. Therefore, if planning acquisition intervals based on FRAP experiments, the parameters should be titrated to roughly half the acquisition speed and number of intervals to minimize photobleaching.

34. Most microscope-controlling software contains FRAP assistance programs.
35. It is advisable to choose cells that are not massively overexpressing the protein of interest as this tends to result in larger populations from the ER contributing to recoveries in the nucleus.
36. This could take as little as 5 min for very mobile proteins and up to an hour or even longer for highly immobile NETs.
37. Because of the tendency to photobleach, it is important to set the time intervals based on the speed of recovery, i.e., for NETs with fast fluorescence recovery (<2 min), ~3 s time intervals will avoid photobleaching for the duration of the experiment, while for intermediate recovery times (~2–10 min), 10 s intervals would work, and for very long recoveries (>10 min), the time frame intervals could increase to 60 s. *See* Subheading 3.2.4.
38. This is a control that is sometimes asked for by reviewers.
39. For many proteins it will be informative to break the cell up based on the different regions where the protein could move, i.e., for an NE protein photoactivated in the NE to measure separately the accumulation in the ER and the accumulation in other regions of the NE.
40. In an ideal situation, the amount of photoactivated protein accumulated in other cellular areas should match the loss of the photoactivated protein in the activated region of interest. Although in practice this is rarely the case, if the differential is greater than 30 %, it indicates that there are significant problems with photobleaching that invalidate the experiment.
41. If these are similar to those from FRAP experiments, then the system is simple and its dynamics are accounted for without multiple phases of movement.
42. Any cell exhibiting rotational changes of position needs to be discarded or measured manually.
43. Diffusion coefficients may vary when using different cell types due to different binding partners present or absent, differences in nuclear size, differences in nuclear pore number and density, and other parameters.
44. If the GFP fusion protein expresses strongly by 24 h, then it is fine to co-transfect, but if longer times are needed to build up expression of the GFP fusion, then it should be transfected prior to the RanQ69L mutant.
45. This timing is based on our experimental determination that transport was beginning to be inhibited at this time, but pleiotropic effects of transport inhibition had not yet become apparent. This was determined for HeLa cells and may vary somewhat from one cell type to another. If you observe the cells appearing

sickly before initiating experiments (e.g., rounding up, losing refraction, aberrant nuclear shapes), then it might be necessary to engage experiments earlier than 24 h after transfection of the Ran dominant-negative construct.

46. There should be a window from 24 to 36 h where the Ran inhibition has taken effect to reduce transport, but general cell health remains fine.
47. Experiments should not be started until at least 10 min after addition of the ATP depletion medium in order for enough ATP to be depleted at the start of experiments.
48. 45–60 min (cell type dependent) after initiating the ATP depletion, the cells become sufficiently sick that we recommend performing no further experiments on the cells at this time.

Acknowledgments

This work was supported by Wellcome Trust Senior Research Fellowship 095209 to E.C.S. and Wellcome Trust Centre for Cell Biology core funding 092076.

References

1. Drings P, Sonnemann E (1974) [Phytohemagglutinin-induced increase of euchromatin contents in human lymphocytes]. Res Exp Med (Berl) 164:63–76
2. Hirschhorn R, Decsy MI, Troll W (1971) The effect of PHA stimulation of human peripheral blood lymphocytes upon cellular content of euchromatin and heterochromatin. Cell Immunol 2:696–701
3. Brown CR, Kennedy CJ, Delmar VA, Forbes DJ, Silver PA (2008) Global histone acetylation induces functional genomic reorganization at mammalian nuclear pore complexes. Genes Dev 22:627–639
4. Makatsori D, Kourmouli N, Polioudaki H, Shultz LD, McLean K et al (2004) The inner nuclear membrane protein lamin B receptor forms distinct microdomains and links epigenetically marked chromatin to the nuclear envelope. J Biol Chem 279: 25567–25573
5. Pickersgill H, Kalverda B, de Wit E, Talhout W, Fornerod M et al (2006) Characterization of the Drosophila melanogaster genome at the nuclear lamina. Nat Genet 38:1005–1014
6. Pindyurin AV, Moorman C, de Wit E, Belyakin SN, Belyaeva ES et al (2007) SUUR joins separate subsets of PcG, HP1 and B-type lamin targets in Drosophila. J Cell Sci 120: 2344–2351
7. Somech R, Shaklai S, Geller O, Amariglio N, Simon AJ et al (2005) The nuclear-envelope protein and transcriptional repressor LAP2beta interacts with HDAC3 at the nuclear periphery, and induces histone H4 deacetylation. J Cell Sci 118:4017–4025
8. Furukawa K (1999) LAP2 binding protein 1 (L2BP1/BAF) is a candidate mediator of LAP2- chromatin interaction. J Cell Sci 112: 2485–2492
9. Polioudaki H, Kourmouli N, Drosou V, Bakou A, Theodoropoulos PA et al (2001) Histones H3/H4 form a tight complex with the inner nuclear membrane protein LBR and heterochromatin protein 1. EMBO Rep 2:920–925
10. Ye Q, Worman HJ (1996) Interaction between an integral protein of the nuclear envelope inner membrane and human chromodomain proteins homologous to Drosophila HP1. J Biol Chem 271:14653–14656
11. Ivorra C, Kubicek M, Gonzalez JM, Sanz-Gonzalez SM, Alvarez-Barrientos A et al (2006) A mechanism of AP-1 suppression through interaction of c-Fos with lamin A/C. Genes Dev 20:307–320
12. Osada S, Ohmori SY, Taira M (2003) XMAN1, an inner nuclear membrane protein, antagonizes BMP signaling by interacting with Smad1 in Xenopus embryos. Development 130:1783–1794

13. Ozaki T, Saijo M, Murakami K, Enomoto H, Taya Y et al (1994) Complex formation between lamin A and the retinoblastoma gene product: identification of the domain on lamin A required for its interaction. Oncogene 9: 2649–2653
14. Pan D, Estevez-Salmeron LD, Stroschein SL, Zhu X, He J et al (2005) The integral inner nuclear membrane protein MAN1 physically interacts with the R-Smad proteins to repress signaling by the transforming growth factor-{beta} superfamily of cytokines. J Biol Chem 280:15992–16001
15. Korfali N, Wilkie GS, Swanson SK, Srsen V, Batrakou DG et al (2010) The leukocyte nuclear envelope proteome varies with cell activation and contains novel transmembrane proteins that affect genome architecture. Mol Cell Proteomics 9:2571–2585
16. Schirmer EC, Florens L, Guan T, Yates JR, Gerace L (2003) Nuclear membrane proteins with potential disease links found by subtractive proteomics. Science 301:1380–1382
17. Wilkie GS, Korfali N, Swanson SK, Malik P, Srsen V et al (2011) Several novel nuclear envelope transmembrane proteins identified in skeletal muscle have cytoskeletal associations. Mol Cell Proteomics 10(M110):003129
18. Ohba T, Schirmer EC, Nishimoto T, Gerace L (2004) Energy- and temperature-dependent transport of integral proteins to the inner nuclear membrane via the nuclear pore. J Cell Biol 167:1051–1062
19. Soullam B, Worman HJ (1995) Signals and structural features involved in integral membrane protein targeting to the inner nuclear membrane. J Cell Biol 130:15–27
20. Zuleger N, Kelly DA, Richardson AC, Kerr AR, Goldberg MW et al (2011) System analysis shows distinct mechanisms and common principles of nuclear envelope protein dynamics. J Cell Biol 193:109–123
21. Meinema AC, Laba JK, Hapsari RA, Otten R, Mulder FA et al (2011) Long unfolded linkers facilitate membrane protein import through the nuclear pore complex. Science 333:90–93
22. Patterson GH, Lippincott-Schwartz J (2002) A photoactivatable GFP for selective photolabeling of proteins and cells. Science 297: 1873–1877
23. Phair RD, Misteli T (2001) Kinetic modelling approaches to in vivo imaging. Nat Rev Mol Cell Biol 2:898–907
24. Lorenz H, Hailey DW, Lippincott-Schwartz J (2008) Addressing membrane protein topology using the fluorescence protease protection (FPP) assay. Methods Mol Biol 440:227–233
25. Shepshelovich J, Hirschberg K (2007) Expression, localization, and topology of fluorescently tagged plasma membrane-targeted transmembrane proteins. Methods Mol Biol 390:393–403
26. Yguerabide J, Schmidt JA, Yguerabide EE (1982) Lateral mobility in membranes as detected by fluorescence recovery after photobleaching. Biophys J 40:69–75

Chapter 21

Correlative Microscopy of Individual Cells: Sequential Application of Microscopic Systems with Increasing Resolution to Study the Nuclear Landscape

Barbara Hübner, Thomas Cremer, and Jürgen Neumann

Abstract

The term correlative microscopy denotes the sequential visualization of one and the same cell using various microscopic techniques. Correlative microscopy provides a unique platform to combine the particular strength of each microscopic approach and compensate for its specific limitations. As an example, we report results of a correlative microscopic study exploring features of the nuclear landscape in HeLa cells. We present a detailed protocol to first investigate distinct structural features of a living cell in space and time (4D) using spinning disk laser scanning microscopy (SDLSM). Then, after fixation and staining of selected structures (e.g., by means of immunodetection), details of these structures are explored at increasingly higher resolution using three-dimensional (3D) confocal laser scanning microscopy (CLSM); super-resolution fluorescence microscopy, such as three-dimensional structured illumination microscopy (3D-SIM); and transmission electron microscopy (TEM). We discuss problems involved in the comparison of images of a given cell nucleus recorded with different microscopic approaches, which requires not only a compensation for different resolutions but also for various distortions.

Key words Correlative microscopy, Live cell microscopy, Super-resolution fluorescence microscopy, Transmission electron microscopy, Relocalization of cells, Nuclear architecture, Hypercondensed chromatin, HCC

Abbreviations

1 Mbp CDs	Megabase-sized chromatin domains
3D	Three-dimensional
3D-SIM	Three-dimensional structured illumination microscopy
4D	Four-dimensional
CC	Chromatin compartment
CLSM	Confocal laser scanning microscopy
CT–IC	Chromosome territory–interchromatin compartment
EM	Electron microscopy
FCS	Fetal calf serum
FIB	Focused ion beam

Yaron Shav-Tal (ed.), *Imaging Gene Expression: Methods and Protocols*, Methods in Molecular Biology, vol. 1042,
DOI 10.1007/978-1-62703-526-2_21, © Springer Science+Business Media, LLC 2013

FISH	Fluorescence in situ hybridization
FRAP	Fluorescence recovery after photobleaching
GFP	Green fluorescent protein
H2B-mRFP	Histone 2B tagged with red fluorescent protein
H3K4me3	Histone 3 tri-methylated at lysine 4
H3K9me3	Histone 3 tri-methylated at lysine 9
HCC	Hypercondensed chromatin
IC	Interchromatin compartment
OTF	Optical transfer function
PBS	Phosphate buffered saline
PBST	1× PBS with 0.02 % Tween
PFA	Paraformaldehyde
PR	Perichromatin region
PS	Penicillin/streptomycin
PSF	Point-spread function
RNPs	Ribonucleoproteins
ROI	Region of interest
SDLSM	Spinning disk laser scanning microscopy
SEM	Scanning electron microscopy
SIM	Structured illumination microscopy
TEM	Transmission electron microscopy
WF	Wide field

1 Introduction

A variety of microscopic systems described below can be used to study individual cells. Typically, these imaging procedures are applied to different cells. Correlative microscopy (e.g., [1–4]), by contrast, aims at the sequential visualization of one and the same cell and provides a unique platform to combine the particular strength of each microscopic approach and to compensate for its specific limitations. A central issue in microscopy is resolution: all optical devices have an intrinsically limited capability of separating two point-like objects. This Abbe limit of resolution was first described in 1893 by Ernst Abbe and implies that in the case of conventional light microscopy, two objects optically merge laterally (xy-plane) when they get closer than approximately half the wavelength of the illuminating light. In axial (z) direction, resolution is even more limited by a factor of 2–3. Many subcellular structures and macromolecular complexes have sizes below this limitation and therefore could not be analyzed by light microscopic approaches so far. We refer to this resolution as the classical Abbe limit with the understanding that the resolution of every light microscopic approach is constrained by its own Abbe limit, which may in principle strongly differ from the classical limit. However, several new approaches, summarized under the term super-resolution fluorescence microscopy, allow to overcome the classical Abbe limit of resolution ([5], see below). In this chapter, we describe procedures for correlative microscopy based on the sequential application of laser scanning microscopy of living and fixed cells, structured illumination microscopy (one of the new microscopic approaches of super-resolution fluorescence microscopy), and transmission electron microscopy (TEM). The flow chart for our setup of correlative microscopy is depicted in Fig. 1. As an example, we present results exploring features of the nuclear landscape of HeLa cells. In addition to a detailed protocol, we point out problems of correlative microscopy that need to be overcome for an optimized comparison of light optical and physical sections of a given nucleus sequentially obtained with microscopic techniques with increasing resolution.

1.1 Microscopic Techniques

1.1.1 Fluorescence Microscopy with Conventional Resolution

Fluorescence microscopy is one of the most important techniques in biological sciences and has been widely used for the visualization of all kinds of cellular structures in both living and fixed cells. Sophisticated optical wide field (WF) microscopes were developed, equipped, for example, with specialized fluorescence filter sets for the simultaneous detection of fluorophores with different excitation and/or emission spectra, and optimized light sources, such as halogen lamps, and amplifier cameras that minimize sample stress and make the devices suitable for long-term observations of living cells. Resolution of these microscopes is limited to about 200–300 nm laterally and 600–1,000 nm axially. Confocal laser scanning microscopy (CLSM) and spinning disk laser scanning microscopy (SDLSM) have substantially improved three- and four-dimensional (3D and 4D) imaging beyond the possibilities of classical WF microscopy, although the resolution is at best marginally improved to about 180–250 nm in *xy*- and about 500–700 nm in *z*-direction. The great advantage of CLSM and SDLSM stems from allowing light optical serial sectioning of cells by a major reduction of detected out-of-focus light and a drastic increase of contrast. This has greatly improved the generation of 3D reconstructions from the collected serial sections. Furthermore, both systems provide excellent possibilities for multicolor imaging, and CLSM has become the method of choice for 3D studies of fixed cells. Although it has also been successfully employed for space-time (4D) studies of living cells, SDLSM is much better suited for such experiments. The reason is evident, when considering the difference in image recording between the two approaches: With CLSM, a 3D image of a cell is obtained with a single focused laser beam, which scans the sample by recording the fluorescent output separately at each focal point. This approach is time-consuming and typically results in a high total light exposure of the recorded cell. In case of SDLSM, the laser light passes through a disk, which contains approximately 20,000 small pinholes arranged in such a way that rotation of the disk leads to full coverage of the sample. Rotation of the disk with high speed (in the range of

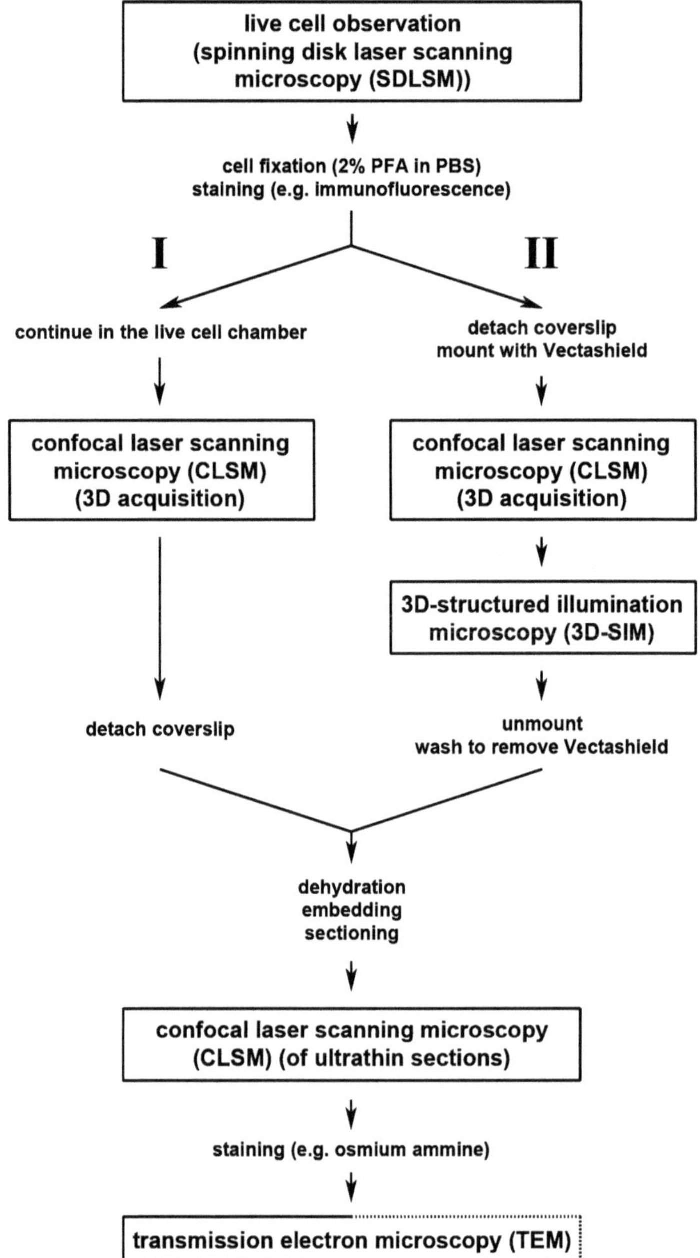

Fig. 1 Flow chart for our setup of correlative microscopy. Live cell observations are performed with spinning disk laser scanning microscopy (SDLSM). After cell fixation with 2 % PFA in PBS, to maintain the ultrastructure in the best possible way, and staining of the cells, e.g., by indirect immunofluorescence as in our case, two pathways (I or II) can be followed: (I) 3D acquisitions are acquired by scanning the sample with confocal laser scanning microscopy (CLSM) still in the live cell chamber in PBS. Subsequently the coverslip is detached from the live cell chamber. (II) This pathway involves 3D structured illumination microscopy (3D-SIM). For this purpose, the coverslip is detached during staining to allow the necessary mounting of the cells in the antifade Vectashield. After 3D scanning with

1,500–10,000 rpm) allows the illumination of one plane of the entire specimen within a few milliseconds. This setup leads to the simultaneous recording of fluorescence from numerous focal points and enhances the speed of image recording-up to 30 frames per second are possible-and therefore also reduces light stress on the sample. A downside of this technique is the so-called pinhole cross talk, i.e., light scattered or emitted by structures outside the focal plane can reach the detector through adjacent pinholes. As a consequence, the axial resolution is somewhat reduced in comparison to CLSM [6].

1.1.2 Super-Resolution Fluorescence Microscopy

During the last decade, a variety of so-called super-resolution fluorescence microscopic techniques have been developed. These approaches opened new ways to explore the details of structures beyond the classical Abbe limit of conventional light microscopy resolution [7]. Here, we focus on one of these new procedures, i.e., 3D structured illumination microscopy (3D-SIM). This method is based on increasing resolution by using an illumination pattern, in most setups [8, 9] generated by a diffraction grating: If a sample is illuminated with an incident pattern, the pattern heterodynes with the structure of the object and shifts higher frequency components-corresponding to subtle sample details-into the passband region of the optical transfer function (OTF) of the device. In extension, Gustafsson et al. [10] showed that a three-beam interference can be applied to produce a 3D (sinusoidal) illumination pattern within the sample. Most of the commercially available 3D-SIM microscopes still use the original (Gustafsson) setup with a classical diffraction grating to produce a sinusoidal 3D illumination pattern via three-beam interference of the −1, the 0, and the 1 diffraction orders. It can be shown that the reconstruction of a super-resolved image therefore requires five images per orientation, making it necessary to sequentially shift the illumination pattern on each image plane perpendicular to the stripes by one fifth of the grating's period. For an improvement of resolution in a second lateral direction, the orientation of the grating has to be rotated accordingly. In principle, the more orientations are

Fig. 1 (continued) CLSM and 3D-SIM, the coverslip is unmounted again and the cells are washed extensively to remove Vectashield. Subsequently both pathways are continued in the same way: the cells are dehydrated and embedded in resin, and ultrathin sections are cut. Fluorescence retained in the sections can be recorded again with CLSM. Next, staining for electron microscopy is performed, e.g., DNA-specific staining with osmium ammine as in our case, and images are collected with transmission electron microscopy (TEM). An example of correlative microscopy following pathway I is shown in Fig. 2 and Fig. 3 exemplifies pathway II. Note that pathway II in comparison to pathway I yields a considerable impaired TEM image quality of osmium ammine-stained DNA (compare Figs. 2C and 3D)

used, the more complete is the high-resolution coverage in the xy space. This advantage, however, is gained at the cost of prolonged acquisition time and additional light stress. Three to five equally spaced orientations are regarded as sufficient to completely enhance the resolution in xy space; usually only three orientations are used. The resolution increase in the axial direction follows the same principle and requires a precise z-positioning of the sample within the illumination pattern. In combination, the multiple shifts and rotations of the illumination pattern and the acquisition of images from many z-positions reveal previously inaccessible information that contributes to the image. This allows the reconstruction of a final image with a resolution that is increased in all spatial directions beyond the classical Abbe limit to about 100–130 nm laterally and 250–340 nm axially, representing an eightfold increase in 3D compared to CLSM [11]. The principle of this technique implies that the generation of a final 3D-SIM image requires a massive computational effort. The typical setup results in the acquisition of 15 images per z-position (three angles, five shifts each) and 960 images per color for the 3D stack (based on a typical z-volume of 8 μm and the standard z-spacing of 125 nm, resulting in 64 z-sections). Therefore, 3D-SIM requires samples with bright signal intensities, fluorochromes that are fairly resistant against bleaching and mounting in antifade to diminish bleaching during the recording of serial images. In addition, any fluorescent particles which move during image recording, can introduce imaging artifacts. A great advantage of SIM is its compatibility with standard sample preparation protocols and standard fluorophores (*see* also Subheading 3.2.5), which allows multicolor applications with optical sectioning with enhanced resolution in lateral and also axial direction in fixed cells. Whereas the current standard 3D-SIM setup can only be applied to the imaging of fixed cells-in particular due the inevitable long recording time-advanced SIM procedures have already yielded promising results in live cell applications [12, 13]. Other methods of super-resolution fluorescence microscopy allow imaging at even higher resolution than 3D-SIM and partly also show great potential for live cell imaging [5, 14, 15]. It seems realistic to expect that these methods will allow 3D and 4D imaging with a resolution of about 10 nm.

1.1.3 Electron Microscopy

Electron microscopic (EM) procedures are only applicable to fixed cells, and staining possibilities to depict several different proteins or nucleic acids simultaneously are much more limited compared to multicolor imaging with fluorescence microscopy procedures. TEM provides the highest possible resolution reaching the low nanometer or even the Angstrom scale. TEM images are recorded from ultrathin sections cut from fixed cells embedded in resin. In principle, serial ultrathin sections can be collected, imaged, and aligned to a 3D stack, but this procedure is very laborious and time demanding.

New approaches make use of the combination of scanning electron microscopy (SEM) with ultramicrotomy [16, 17] or focused ion beam (FIB) milling [18]. Here serial images of surfaces are recorded after the sequential removal of ultrathin slices (with a thickness of about 50 nm in case of ultramicrotomy and down to 10 nm in case of FIB). These procedures were already used to generate 3D reconstructions from nuclei [19, 20] and chromosomes [21]. Conventional EM staining procedures often employed in studies of the nucleus include uranyl acetate and lead citrate, which stains proteins, RNA, and DNA at the same time [22]. In contrast to that, several new methods have been developed in the last decades, which stain DNA in EM samples in a highly specific manner [23, 24]. Cryo-electron microscopy carries the promise to be the method with the best structural preservation [25]. However, this advantage is achieved at the cost of having to abstain from staining for specific substructures in such specimens [26].

1.2 Correlative Microscopy of Cell Nuclei

In our group, we established correlative microscopy as part of ongoing experimental tests of the chromosome territory–interchromatin compartment (CT–IC) model of nuclear architecture [27, 28]. This model globally distinguishes between a chromatin compartment (CC) and an interchromatin compartment (IC). The CC is composed of chromosome territories (CTs). Experimental evidence [29–32] as well as theoretical considerations and modeling [33, 34] argue that CTs are composed of interconnected megabase-sized chromatin domains (~1 Mbp CDs), which in turn are built up of a series of interconnected chromatin loop domains in the order of 100 kbp. Such ~1 Mbp CDs form larger chromatin clusters composed of a core of dense and a periphery of less condensed chromatin, called the perichromatin region (PR). Functionally, the PR serves as the nuclear compartment where transcription, pre-mRNA processing, DNA replication, and repair take place (for reviews, *see* refs. [28, 35]). Topographically, the PR forms a border zone between the CC and the IC. The IC represents a spatially contiguous, and both highly crowded and organized 3D network of IC channels and wider IC lacunas. Channels start at nuclear pores, pervade the peripheral nuclear layer of heterochromatin, and expand with larger and smaller branches between and within CTs. At numerous sites, the IC forms wider lacunas containing splicing speckles and nuclear bodies. It provides factors and factor complexes essential for functions carried out in the PR and may also serve as a channel system for the export of RNPs [36].

A previous study from our group showed that a change of the osmolarity of the cell culture medium from physiological (290 mOsm) to high salt (570 mOsm) conditions results in the rapid formation of hypercondensed chromatin (HCC) accompanied by a corresponding enlargement of the IC [37]. Both HCC and the concomitant widening of the IC are fully reversible after incubating the cells again in medium with normal osmolarity. The formation of HCC facilitates the spatial mapping of chromatin characterized by histone markers for transcriptionally competent and silent chromatin, respectively. The enlargement of the IC allows the indisputable allocation of structures to the IC, such as splicing speckles and nuclear bodies in IC lacunas.

Here we discuss this approach for the first time in a correlative microscopic study of HeLa cell nuclei. Imaging of nuclei in living cells was performed with SDLSM before and after HCC induction. In vivo labeling of chromatin by stable expression of histone 2B tagged with red fluorescent protein (H2B-mRFP) made it possible to record the dynamic structural changes during this process in the living cell in space and time (4D). As the formation of HCC is a very rapid process, a stepwise increase in osmolarity was used (340, 425, 570 mOsm, incubated for 5 min each) to facilitate imaging of the ongoing condensation (Figs. 2A and 3A). After fixation, nuclear DNA was stained with DAPI. For further insights into the functional nuclear landscape, transcriptionally competent chromatin was identified by immunodetection of H3K4me3, and the

Fig. 2 (continued) interchromatin compartment (IC). *Arrowheads* indicate the increasing width of the same IC lacunas. (**B**) CLSM of the same nucleus after fixation and staining. DAPI-stained DNA is shown in *gray*. Indirect immunofluorescence of splicing speckles and of transcriptionally competent chromatin was performed with anti-SC35 (*green*) and anti-H3K4me3 (*red*) primary antibodies, respectively. The sample was kept in PBS in the live cell dish and was not mounted in Vectashield (compare explanations in Figs. 1 and 3). (b1) shows approximately the same xy-midsection as (a1) with an overlay of DAPI, SC35, and H3K4me3 signals. (b2) shows only the DAPI channel. The two images shown on the *top left* represent a xz-light optical section, again with an overlay of DAPI, SC35, and H3K4me3 signals (*top*) and the DAPI signal only (*bottom*). *Arrowheads* exemplify the location of SC35-stained splicing speckles in IC lacunas, whereas the compacted chromatin reveals signals of transcriptionally competent chromatin marked by H3K4me3. (b3) DAPI image corresponding to (b2) recorded by CLSM from an ultrathin (100 nm) physical section obtained after resin embedding. Note that some shrinkage of the nucleus occurred during the embedding process (compare scale bars in b2 and b3). (**C**) TEM images of the same section shown in b3 with osmium ammine-stained DNA. Each boxed area is further magnified from left to right in the neighboring image. These images demonstrate the distinct separation between the highly compacted chromatin compartment (**CC**) and the expanded interchromatin compartment (IC) after HCC treatment. At highest magnification, CC and IC are marked by (*degree*) and (*asterisk*) symbols, respectively. HCC clusters are connected with each other by fiber-like bundles of less condensed chromatin (*arrow*). *Arrowheads* in **A**, **B**, and **C** point to examples of IC lacunas containing splicing speckles. (**D**) Overlays of images obtained with different microscopic methods. *Upper panel*: overlayed images of DAPI-stained (a1, b2, b3) and osmium ammine-stained DNA (c1) were corrected for nuclear rotation and size differences but not for distortions. (D1) overlay of a1 (*green*) and b2 (*red*); (D2) overlay of b2 (*red*) and b3 (*green*); (D3) overlay of b3 (*green*) and c1 (*gray*); (D4) overlay of b2 (*red*) and c1 (*gray*). *Arrows* indicate that overlayed images do not perfectly match with each other. *Lower panel*: corresponding overlayed images D1′–D4′ after the correction for distortions. D1′, D2′, and D3′ now include SC35-stained speckles (*blue*; compare b1). D3′: in the ultrathin section scanned with CLSM (b3), the DAPI-stained DNA could still be visualized, but not the fluorescence from SC35-stained splicing speckles. Scale bars: 3 μm, except for TEM blowups in (C) (*middle* 1 μm; *right* 0.25 μm)

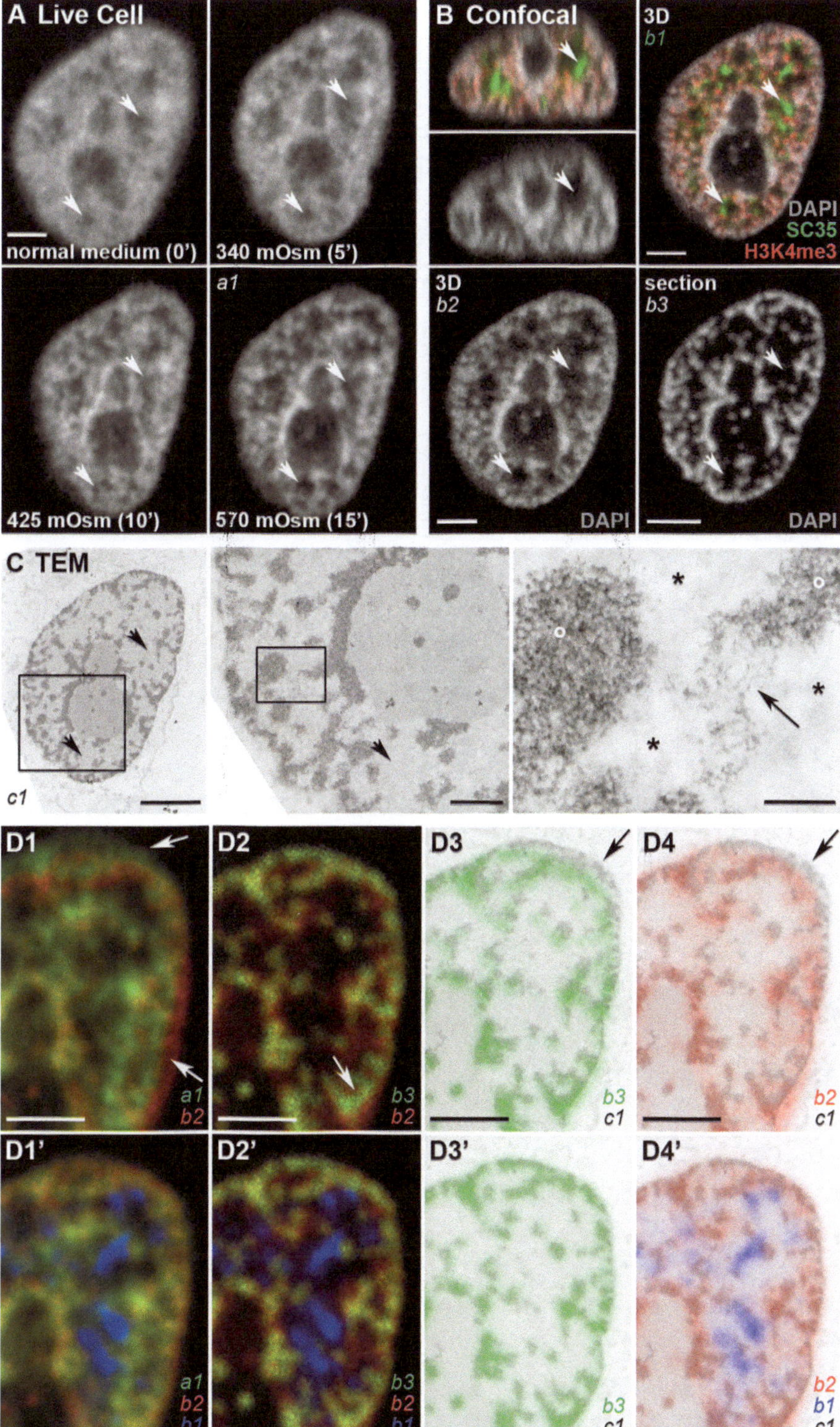

Fig. 2 Example of correlative microscopy of a HeLa cell nucleus according to pathway I (*see* Fig. 1). (**A**) Live cell imaging of H2B-mRFP (*gray*) in a HeLa cell nucleus was performed with SDLSM during the stepwise induction of hypercondensed chromatin (HCC) by increasing the medium osmolarity from 290 to 570 mOsm. Light optical xy-midsections are shown from the entire 3D-acquisition of light optical serial sections. This panel documents the increasing chromatin compaction correlated with increasing medium osmolarity. The image (a1) demonstrates the highest level of HCC obtained at 570 mOsm and the corresponding maximal enlargement of the

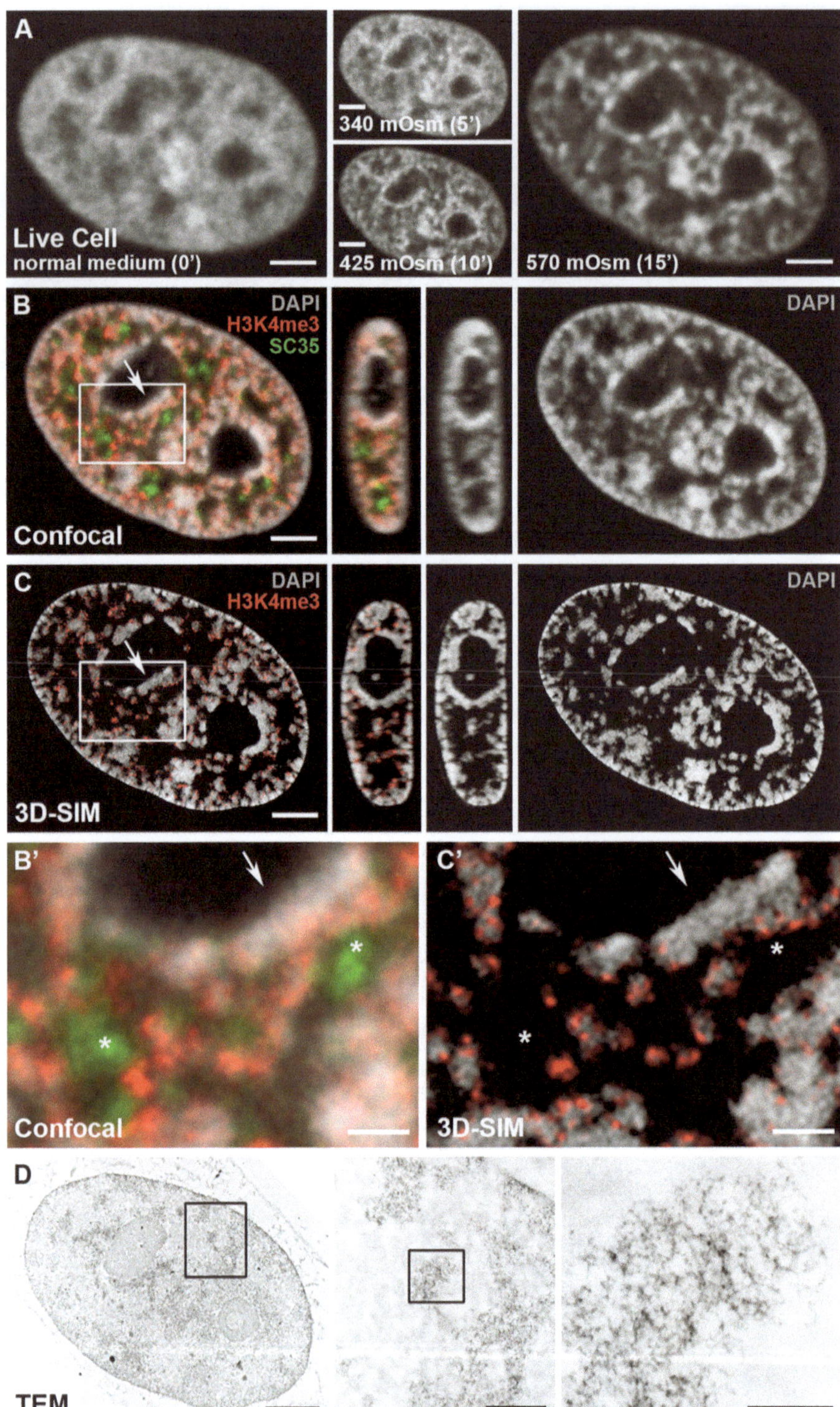

Fig. 3 Example of correlative microscopy of a HeLa cell nucleus according to pathway II (*see* Fig. 1). (**A**) SDLSM live cell imaging of H2B-mRFP (*gray*) in a HeLa cell nucleus during the stepwise induction of hypercondensed chromatin (HCC) by increasing the medium osmolarity from 290 to 570 mOsm (compare Fig. 2A). (**B**, **C**) CLSM (**B**) and 3D-SIM (**C**) of the same nucleus after fixation and immunodetection of splicing speckles (*green*) with anti-SC35 primary antibodies and transcriptionally competent chromatin (*red*) with anti-H3K4me3 primary antibodies, DAPI-stained DNA (*gray*). Shown are corresponding single midsections of the 3D acquisitions

location of splicing speckles was studied by immunodetection of SC35, a protein highly enriched in those. The same cells were relocalized and their nuclei were imaged again with conventional resolution using CLSM (Figs. 2B and 3B) and with increased resolution using 3D-SIM (Fig. 3C). Note that the use of an antifade is generally helpful in fluorescence microscopy to diminish bleaching during image acquisition. However, in case of CLSM, scanning of cells in PBS is often possible, while the increased amount of images collected with 3D-SIM (for details, *see* Subheading 1.1.2) requires the cells always to be mounted in an antifade containing embedding medium (preferably Vectashield). A comparison of correlative light optical sections sequentially obtained by CLSM and 3D-SIM (Fig. 3B, B′, C, C′) clearly demonstrates the advantage of the increased resolution of the super-resolution microscopic approach: With 3D-SIM, both the compaction of chromatin and the location of transcriptionally competent chromatin at the surface of HCC are much more obvious. The interior of many IC lacunas contains splicing speckles but lacks both DAPI-stained DNA and H3K4me3-stained chromatin. Notably, H3K9me3, a histone marker for transcriptionally silent chromatin, could be localized within HCC [38]. H3K4me3 was observed at the outer periphery of perinucleolar heterochromatin, but never at the inside (Fig. 3B, C, arrows). As SC35 was detected with a Cy5-coupled antibody, it could not be imaged with 3D-SIM, since the laser line necessary to excite this fluorophore is not available in our setup. Nevertheless, its localization inside the IC lacunas is obvious by the comparison with the corresponding light optical section obtained with CLSM (compare the asterisk in Fig. 3B′, C′). To explore features of higher-order genome organization with TEM, the samples were finally embedded in resin and ultrathin sections (100 nm thickness) were cut with an ultramicrotome. After relocalization of a given nucleus, its DAPI fluorescence in the ultrathin section was recorded with CLSM (Fig. 2B, b3; *see* also **Note 37**). Note that the nucleus

Fig. 3 (continued) (*xy*-view on the *right* and on the *left*, merge and DAPI only, respectively; *xz*-views in the *middle*). (**B′**, **C′**) Magnifications of the boxed areas in (**B**) and (**C**). The advantage of the improved resolution of 3D-SIM becomes obvious: it demonstrates much better the retraction of active chromatin marked by H3K4me3 together with chromatin in general and reveals that nucleoli and IC lacunas can clearly be distinguished by the lack of H3K4me3 staining at the inside of the former ones (*arrows*). Although SC35 was not imaged with 3D-SIM, its localization inside the IC lacunas is obvious by the comparison with the CLSM image (*asterisk*). (**D**) TEM images of the corresponding physical section after embedding and sectioning showing osmium ammine-stained DNA. Each boxed area is further magnified from left to right in the neighboring image. Note that the preceding imaging of this nucleus with 3D-SIM required the presence of the antifade Vectashield to diminish bleaching during the recording of serial SIM images. Although Vectashield was removed thereafter, the quality of the TEM preparation was impaired compared with nuclei scanned without mounting in Vectashield (compare Fig. 2C). Osmium ammine-stained HCC shows less contrast and appears somewhat less condensed. Scale bars: 3 μm, except for TEM blowups (D) (*middle* 1 μm; *right* 0.25 μm)

shows a certain degree of shrinkage after embedding (compare scale bars in Fig. 2B, b2, and b3). Subsequently, the sections were stained with osmium ammine [24]. This method includes RNA degradation by incubation in HCl and combines high sensitivity for DNA with TEM visibility. In contrast to post-embedding protocols for immunodetection of proteins or in situ hybridization of nucleic acids, osmium ammine staining is not restricted to the surfaces, since this compound can penetrate the whole section. The images of the same nucleus obtained by TEM (Figs. 2C and 3D) demonstrate the clear separation between HCC and IC. Occasionally clusters of chromatin fibers can be seen (Fig. 2C, arrow), which connect HCC clusters without a recognizable substructure. In addition, it seems possible that more resolved chromatin fibers may represent the periphery of a HCC cluster, which prior to cutting was present in a nuclear region above or below the section. In repeated TEM experiments we noted, that the previous mounting of the cells with Vectashield is correlated with a somewhat more dispersed structure of HCC and a weaker osmium ammine staining of the DNA (compare Figs. 3D with 2C). The antifade Vectashield is required to prevent the particularly strong bleaching effects involved in 3D-SIM (compare Subheading 1.1.2) as compared to CLSM. The reason(s) of this effect of Vectashield are not known, and it may be well possible to obtain better results with some modifications of the protocol. Arrowheads in Fig. 2A, B, and C point at IC lacunas containing splicing speckles. Live cell microscopy demonstrates that these IC lacunas were already present but more narrow in nuclei prior to HCC induction.

Figure 2D depicts overlays of images before and after the correction of distortions resulting from several problems that make it difficult to obtain perfect matches between images of the same biological structure sequentially obtained with different microscopic approaches. For a detailed comparison, one would like to view perfectly matched sections. For several reasons, however, possible solutions can only be an approximation to the ideal situation: (1) A detailed comparison of images requires corrections for resolution differences. At face value, this seems an easy task, but in practice, it is complicated by the fact that differences in the z-resolution of different types of microscopes result in the inclusion of more structures within a thicker optical section obtained with SDLSM or CLSM in comparison to the thinner corresponding section obtained with 3D-SIM or TEM. (2) Movements of structural components in nuclei of living cells, such as chromatin domains [39] or entire CTs [40], may result in positional changes between the living cell and the cell after fixation. The degree of the resulting matching problems depends on the speed with which a given structure may move and the time, which passes between the end of the live cell imaging and the time point where fixation is sufficiently advanced to inhibit any further movements. (3) Sectioning of resin-embedded cell nuclei with the ultramicrotome requires the alignment of the block containing the embedded cell parallel to the knife. This is done by eye, and although the well-trained user will achieve a pretty good alignment, it will never be perfect. Therefore, the plane of the sectioning will be slightly different from the imaging plane used for recording this cell with SDLSM, CLSM, or 3D-SIM. Furthermore, embedding of cells in resin causes some degree of shrinkage, and observation with TEM can lead to deformations caused by the electron beam.

1.3 Future Prospects of Correlative Microscopy as a Tool to Study Cellular Substructures in Space and Time at Nanometer Resolution

In summary, correlative microscopy, despite the difficulties pointed out above, opens new pathways for single-cell analyses. In contrast to high-throughput biochemical methods [41], this approach is suited to unravel the architecture of CTs and whole nuclei at the single-cell level. It enables the observation and investigation of dynamic processes first in the living cell, followed by high-resolution imaging of specifically stained subcellular structures in fixed cells. Although studies of living cells with resolution beyond the classical Abbe limit are about to be implemented, high-resolution imaging of fixed cells will surely be further needed as part of the imaging tool box, if only because of topographical studies of the nuclear landscape in living cells requiring even more complex strategies for the simultaneous visualization of several differently colored targets. Many multicolor strategies, such as multicolor FISH, currently can only be performed with fixed cells. One of the greatest problems we have to face in studies of fixed cells is the question to which extent the topography of structures may deviate from the original topography in their living counterparts. As soon as imaging of living cells with ultrastructural resolution becomes feasible, correlative microscopy will allow direct, detailed comparisons of nuclear topography between a given cell in its living and fixed state. However, multicolor live cell studies will have their own technical limitations and dangers of misleading artifacts. Yet, the possibility of correlative microscopy to compare super-resolution images obtained from a living cell with images from the same cell after fixation will give decisive new insights into the century-old problem to which extent a fixation procedure per se may or may not alter the topography of a given biological structure.

2 Materials

2.1 Cells

HeLa cells stably expressing histone 2B tagged with red fluorescent protein (H2B-mRFP).

2.2 Media, Buffers, and Solutions

1. Normal medium: DMEM supplemented with 10 % FCS, 1 % penicillin/streptomycin (PS), and 250 μM Trolox (6-hydroxy-2,5,7,8-tetramethylchroman-2-carboxylic acid) (*see* **Note 1**).

2. Live cell medium: DMEM without phenol red and sodium pyruvate containing 25 mM Hepes supplemented with 10 % FCS, 1 % PS, 25 mM additional Hepes, and 250 μM Trolox (*see* **Notes 1** and **2**).
3. HCC media (based on live cell medium): (a) 340 mOsm containing 1:100 20× PBS (e.g., 9.9 ml medium + 0.1 ml 20× PBS); (b) 425 mOsm containing 1:33 20× PBS (e.g., 16.5 ml medium + 0.5 ml 20× PBS); (c) 570 mOsm containing 1:20 20× PBS (e.g., 9.5 ml medium + 0.5 ml 20× PBS).
4. 20× PBS, pH 7.4: 2.8 M NaCl, 54 mM KCl, 130 mM Na_2HPO_4, 30 mM KH_2PO_4.
5. 2 % PFA and 4 % PFA dissolved in 1× PBS (*see* **Note 3**).
6. 1× PBS: dilute 20× PBS 1:20 with H_2O_{bidest}. This gives an osmolarity of 310 mOsm. If it is not used for the treatment of living cells, add 0.02–0.04 % sodium azide.
7. PBST: 1× PBS with 0.02 % Tween.
8. 20 mM glycine in 1× PBS.
9. 0.3 % Triton in 1× PBS.
10. Blocking buffer: 1× PBS with 0.02 % Tween, 2 % BSA, and 0.5 % fish skin gelatin (*see* **Note 4**).
11. 2.5 μg/ml DAPI: freshly diluted 1:200 in PBST from a 500 μg/ml stock prepared in H_2O (*see* **Note 5**).
12. 30, 50, 70, 90, 100 % absolute ethanol.
13. 0.25 N HCl.
14. 5 N HCl: freshly prepare from 12.5 N HCl stock using filtered H_2O_{bidest} (0.2 μm filter).
15. Osmium ammine staining solution: 0.2 % (or 2.8 mM) osmium ammine B (Polysciences Europe), 0.2 N HCl, 200 mM $Na_2S_2O_5$ using filtered H_2O_{bidest} (0.2 μm filter) (*see* **Note 6**).

2.3 Chemicals

1. Coverslip removal fluid (glass bottom fluid DCF OS 30) (MatTek).
2. Technical ethanol.
3. Vectashield (Vector Laboratories).
4. Absolute ethanol.
5. Chloroform.
6. LR white resin (resin only, medium, uncatalyzed) (ScienceServices); catalyzed according to company instructions.
7. Liquid nitrogen.
8. Acetone.
9. H_2O_{bidest}.

2.4 Antibodies

1. Primary antibodies: rabbit-anti-H3K4me3 (Abcam) (diluted 1:100), mouse-anti-SC35 (Sigma Aldrich) (1:2,000).
2. Secondary antibodies: donkey-anti-rabbit-Alexa594 (Invitrogen) (1:300), goat-anti-mouse-Cy5 (Dianova/Jackson Immuno Research) (1:100).

2.5 Other Laboratory Materials

1. Custom-made glass bottom dishes with 12 mm round gridded coverslips (Bellco Glass) attached (MatTek) (*see* **Note 7**).
2. Tweezers (for handling coverslips).
3. 12-well plates.
4. Plastic pasteur pipettes.
5. Parafilm.
6. Microscope slides.
7. Soft tissues.
8. Clear nail polish.
9. Precision cotton swabs with cone cotton tips (Critical Swab (TM)/VWR).
10. Embryo dishes with lids.
11. Box or plate.
12. Aluminum foil.
13. BEEM embedding capsules size "00" (ScienceServices).
14. Capsule rack.
15. Big tweezers.
16. 60 °C oven.
17. Liquid nitrogen.
18. Ultramicrotome.
19. Diamond knife.
20. Very sharp razor blades.
21. Petri dishes (preferably tightly closing).
22. Filter paper.
23. Grids: 300 square mesh gold grids (Plano) and 2 × 1 mm gold slot grids with a formvar/carbon membrane (ScienceServices) (*see* **Note 8**).
24. Tweezers (No. 2, 3, or 5 as you prefer) that are exclusively used for handling TEM grids.
25. Wooden sticks with a single eyelash attached (to handle ultra-thin sections).
26. Coverslips (e.g., 24 × 32 mm).
27. Fixogum (Marabu).
28. Scalpel.

2.6 Microscopic Equipment and Software

1. Spinning Disk microscope (Zeiss/PerkinElmer), operating software: Volocity (PerkinElmer), with a heatable live cell chamber (PeCon).
2. Confocal SP5 microscope (Leica Microsystems), operating software: Leica Application Suite.
3. pdV microscope (Applied Precision), operating software: softWoRx (Applied Precision).
4. 3D-SIM microscope (Applied Precision), operating software: softWoRx (Applied Precision).
5. Morgagni 268 transmission electron microscope (FEI), operating software: Morgagni Microscope Control and iTEM (Olympus Soft Imaging Solutions GmbH).
6. Phase contrast microscope (Zeiss).
7. Adobe Photoshop.
8. ImageJ/Fiji (public domain).

3 Methods

A flow chart of our setup of correlative microscopy is depicted in Fig. 1. Some basic notes on the alignment of images and on challenges arising when trying to compare the topography of individual structures in these images obtained with the different microscopes are given in Subheading 3.4.

3.1 4D (Space-Time) Imaging of Living Cells

1. Seed cells 12–24 h before starting the live cell observation into custom-made MatTek glass bottom dishes with 12 mm round gridded coverslips (Bellco) attached (*see* **Note 7**). For the live cell observation, cell density should be around 30–50 % (*see* **Note 9**). Incubate cells in normal medium containing 10% FCS, 1% penicillin/streptomycin and 250 μM Trolox (*see* **Note 1**). Here we used HeLa cells stably expressing H2B-mRFP to visualize chromatin in the nucleus. At least 5–10 min prior to mounting the cells on the microscope, exchange the normal medium to live cell medium (*see* **Note 2**).
2. Mount the sample on a live cell microscope with a well-preheated live cell chamber (37 °C) (*see* **Note 10**) and choose an appropriate area for observation. For easier relocalization in the subsequent steps, it is recommended to turn the grid into such a position that the numbers can be read in the normal way (not upside down, etc.). Take care to choose an area in the central 12 squares of the Bellco grid (numbers 33–34, 43–46, 55–58, 67–68), which is essential for the relocalization of a recorded cell in all successive steps. Write down the precise position (*see* **Note 11**).

3. Perform the live cell observation (*see* **Note 12**). We imaged the cells by acquiring stacks with 0.2 μm z-spacing using 20 % laser power and 250 ms exposure with SDLSM. For control cells, images were taken every 5 min for a total duration of 15 min (=4 stacks). HCC-treated cells were first imaged in normal medium (=stack 1), then HCC was induced stepwise by incubating in media with osmolaric concentrations of 340 mOsm, 425 mOsm, and 570 mOsm for 5 min each (=stacks 2–4). The medium was changed directly on the microscope (*see* **Note 13**), and for each step, images were taken after 4 min of incubation. With approximately 20 s of acquisition time for each step, this procedure left enough time for changing to the next higher osmolaric concentration within the 5 min time frame. After each change of medium, position and focus were checked and readjusted if necessary.

3.2 Imaging of Fixed Cells with CLSM and 3D-SIM

3.2.1 Cell Fixation

1. Fix the cells with 2 % PFA in 1× PBS for 8 min at room temperature (*see* **Note 14**). In contrast to the common fixation procedure, the HCC cells in our case were not washed with PBS prior to fixation as this would have caused the reversion of the HCC effect.
2. Replace the fixative with PBST in a stepwise manner: first add roughly the same amount of PBST to the fixative, then repeatedly remove a part of the mixture, and add new PBST (*see* **Note 15**).
3. Wash the cells twice with PBST.

3.2.2 Bleaching Experiments to Allow the Sequential Imaging with Fluorophores with Identical or Similar Emission Spectra

It may happen that in the subsequent experiment, target structures have to be detected by using the same fluorophore or a fluorophore with a similar emission spectrum as the one used for live cell imaging, which cannot be discriminated with the help of proper filter sets. Bleaching experiments can provide a solution to this problem. For example, in our experiment, we first recorded H2B-mRFP fluorescence, entirely bleached it thereafter, and used an Alexa594-coupled secondary antibody in the subsequent immunofluorescence experiment.

In such a case, remount and relocalize the cells on the live cell microscope (in our case SDLSM) and bleach the observed areas using FRAP settings (20 % of laser power for 488 and 561, 2 ROI intervals, 10 ms spot bleach time, 1 spot interval). Repeat this procedure several times until all fluorescence has vanished (we used three cycles) (*see* **Note 16**).

In case this protocol should only be continued on the next day, wash the samples twice with 1× PBS and keep them overnight at 4 °C.

3.2.3 Immunofluorescence of Specific Proteins

All steps are carried out at room temperature. An alternative procedure is described in **Note 4**. Care has to be taken that samples do not dry out during the washing steps (*see* **Note 17**).

1. Quench with 20 mM glycine in 1× PBS for 10 min to minimize background autofluorescence.
2. Wash twice with PBST.
3. Permeabilize the cells with 0.3 % Triton in 1× PBS for 10 min.
4. Incubate in blocking buffer for 1 h to 1 h 30 min.
5. Now two pathways are possible (*see* Fig. 1): For pathway II, which includes 3D-SIM, detach the coverslip from the glass bottom dish during blocking: incubate the dish in coverslip removal fluid for 45 min, and subsequently carefully apply pressure onto the coverslip from the inside of the dish using tweezers (*see* **Note 18**). Be very careful to touch only the periphery of the coverslip to avoid damage to the cells. Thoroughly remove the remaining glue from the coverslip using tweezers (*see* **Note 19**). If 3D-SIM is not included in the experiment (pathway I), it is better to move directly to confocal microscopy after immunofluorescence, without mounting the sample on a coverslip, as the quality of the TEM image will be better (compare Figs. 2C and 3D). In this case, skip this step and perform all further steps in the live cell dish (including confocal microscopy in 1× PBS instead of Vectashield).
6. Transfer the samples into a 12-well plate and perform all subsequent steps in that (unless otherwise indicated).
7. Prepare a dark humidified chamber (*see* **Note 20**) containing a strip of Parafilm well attached to a plain surface (e.g., plastic dish). Dilute the primary antibodies in blocking buffer (*see* **Note 21**) and incubate the samples for 1 h to 1 h 30 min on drops of antibody solution (typically, 30–60 μl per coverslip) on the Parafilm in the chamber. Blot dry the coverslip before placing it with the cells facing down on the drop to avoid further dilution of the antibodies.
8. Wash four times with blocking buffer.
9. Incubate for 45–60 min with the secondary antibodies equivalent to **step** 7.
10. Wash six times with blocking buffer.
11. Wash twice with PBST.
12. Postfix with 4 % PFA in 1× PBS for 10 min. Exchange the fixative to PBST in a stepwise manner as described in **step 2** of Subheading 3.2.1. Wash cells twice with PBST.
13. Stain for 7 min with DAPI equivalent to **step** 7.
14. Wash five times with PBS.
15. Mount the sample with Vectashield on microscope slides cleaned with ethanol and seal them with clear nail polish (*see* **Note 22**) (when following pathway II in Fig. 1) or proceed directly to confocal microscopy (for pathway I in Fig. 1).

3.2.4 3D Imaging with CLSM

Clean the coverslip with 80 % absolute ethanol, relocalize the cells on a Leica SP5 confocal microscope using transmission light (*see* **Note 23**), and scan them using standard settings (typically 700–1,000 Hz, line average 2, 50 × 50 × 210 nm voxel size). Be aware that possibly corrections for chromatic shift have to be done in post-processing. For samples that are intended for TEM straight after CLSM and are therefore not embedded in Vectashield but are scanned directly in 1× PBS, keep laser power at a minimum to avoid bleaching.

3.2.5 Imaging with 3D-SIM

Bright samples with low background intensities are crucial for 3D-SIM. This can be achieved by following the protocol given in Subheading 3.2.3. Alexa dyes turned out to be suited best for scanning with the harsh conditions required for 3D-SIM. In case GFP signals, e.g., from the live cell observation, should be imaged, we strongly recommend to enhance the signals using a GFP-Booster (ChromoTek).

1. Clean the coverslip thoroughly before scanning (*see* **Note 24**). As this can dissolve the nail polish, if necessary, reseal the samples before starting the relocalization to prevent the coverslips from moving. Clean also the objective of the microscope (*see* **Note 24**).
2. Choose the immersion oil with the best-matching refractive index. This depends on the sample, e.g., the thickness of the coverslip, the embedding media used, the thickness of the sample, and the temperature during acquisition. (*see* **Note 25**).
3. To allow relocalization of cells with 3D-SIM, an intermediate step at the pdV microscope is strongly recommended. The positioning of the slide on both microscopes is aligned towards each other. Mount the sample on the pdV microscope and relocalize the cells of interest using transmission light (compare **Note 23**). Save the positions of the cells.
4. In case you do not scan the samples at the 3D-SIM microscope within few hours, remove the oil and clean the coverslips again according to **step 1** before you scan them; otherwise the same oil can be used.
5. Transfer the slide to the 3D-SIM. Relocalize the cells using the stored positions from the pdV microscope and scan the samples (*see* **Note 26**).We typically use EMCCD 5 MHz mode for the red and green channel and Conv 5 MHz mode for DAPI. Care was taken that the settings for gain and laser transmission rates were set in such a way that exposure times range between 25 and 100 ms. Z-step size is always 125 nm.
6. Reconstruct super-resolution images using SoftWorx (*see* **Note 27**).

3.3 Procedures for TEM

3.3.1 Embedding of Cells in Resin

Incubation times indicated here are minimum times-longer incubations are no problem at all. As ethanol evaporates very quickly, great care has to be taken to replace the liquid immediately. It is better to leave a small amount of liquid on the sample than letting the cells dry out.

1. If the sample is mounted in Vectashield, put it in 1× PBS for about 20 min until the nail polish becomes grayish and can be peeled off from the coverslip. Take care not to apply shearing forces to avoid damage of the cells. Transfer the coverslip into a 12-well plate and incubate it in fresh 1× PBS for minimum 1 h. Change the solution every 10 min to wash out the Vectashield as good as possible.
2. Take the LR white resin out of the fridge at least 1 h prior to use.
3. Incubate twice for 5 min in cold (4 °C) 30 % ethanol on ice.
4. Incubate twice for 5 min in cold (4 °C) 50 % ethanol on ice.
5. Incubate twice for 10 min in cold (4 °C) 70 % ethanol on ice.
6. If it was not done before (this is the case when following pathway I of Fig. 1), detach the coverslip (as described in **step 5** of Subheading 3.2.3) from the glass bottom dish during the incubations with 90 and 100 % ethanol (**steps** 7 and **8**) (these steps can be prolonged to reach 45 min of incubation time in coverslip removal fluid). As the sample will not be mounted on a microscope slide, it is not necessary to remove the glue from the coverslip.
7. Incubate twice for 10 min in 90 % ethanol at room temperature.
8. Incubate three times for 10 min in 100 % ethanol at room temperature. Transfer the sample into an embryo dish for the last or one additional step, since the resin corrodes plastic. Label the dishes adequately if there is more than one sample (*see* **Note 28**).
9. Exchange the ethanol to LR white resin. To remove the remaining ethanol, move the coverslip around in the embryo dish using tweezers until no "clouds" of ethanol are observed anymore. Incubate at room temperature for a minimum of 30 min. Change the resin and move the coverslips around again (although typically no "clouds" should be observed anymore).
10. Seal the embryo dish with Parafilm and incubate overnight at 4 °C.
11. Let the sample and fresh LR white resin warm to room temperature for at least 1 h.
12. Prepare a box or plate with a plain surface covered with aluminum foil and label it well if there is more than one sample (preferably on the surface and on the sides since the resin can destroy the label).

13. Remove the lid of a capsule and make sure that it has a clear rim without protruding pieces of plastic on the inner side (*see* **Note 29**). Place the capsule into a capsule rack (stably but not too firm) and overfill it with resin (convex surface needed). Remove most of the resin from the embryo dish, take out the coverslip, and place it with the cells facing down on the capsule with the resin. If necessary, center it over the capsule. Take the capsule (with the coverslip on top) out of the rack using large tweezers and turn it upside down. Perform this turning without interruptions, not too slow and not too fast, to keep the coverslip in place. Place the capsule with the coverslip on the aluminum foil. To facilitate this, obtain a better angle by slightly turning the capsule in the tweezers before putting it down (without touching the coverslip). Carefully center the capsule on the coverslip. Process only one sample at a time.
14. Polymerize for 48 h at 60 °C (*see* **Note 30**).
15. Let the samples cool down for a minimum of 1 h (a few days are also possible).
16. Take the samples from the aluminum foil and remove the coverslips by dipping them as short as possible into liquid nitrogen (*see* **Note 31**). Process only one sample at a time.
17. The blocks are very stable and can be kept for extended periods (up to years) at room temperature. However, samples with fluorescent signals are better processed within 1–2 months.

3.3.2 Cutting Ultrathin Sections

1. Remove the capsule from the polymerized block (*see* **Note 32**). Be careful not to touch the surface of the block.
2. Relocalize the cells of interest on the ultramicrotome and cut a pyramid of maximum 1 × 1 mm in size using razor blades (*see* **Note 33**). For easier orientation in later steps, cut off one corner of the pyramid.
3. Prepare a dish with filter paper and label it (number grids from 1-X).
4. For cleaning the mesh grids, wash them briefly in 0.25 N HCl, put them on filter paper to remove most of the liquid, briefly wash them with acetone, and let them air dry on filter paper. Slot grids are very fragile, so it is not recommended to include them in these washing steps. Incubate mesh and slot grids for 20–40(−60) min in H_2O_{bidest} to reduce the charge of the grids. Only prepare grids to be used within the next 30–60 min (*see* **Note 8**). Take care that the side of the grids on which the sections will be in the end faces up on the filter paper and down on the water; for the mesh grids, this is the rough side; for the slot grids this is the side where also the membrane is located (darker side).

5. Mount the sample at the ultramicrotome and cut 100 nm sections (golden color) (*see* **Note 34**) using a diamond knife (*see* **Note 35**).
6. Take the slot grids out of the water using fine tweezers and let them air dry; do not put them down to prevent damage to the membrane. For mesh grids, it is sufficient to take them out immediately before use and drain the remaining water on filter paper.
7. Arrange the sections in groups of 2–4 (depending on section size and personal preference) using wooden sticks with a single eyelash attached and take them up with the grids by approaching the sections from the top. Make sure that the sections end up on the right side of the grid (*see* **step 4**). Try to keep the correct order of the sections (*see* **Note 36**).
8. Place the mesh grids on filter paper inside petri dishes (with the sections facing up) and let them air dry. As in **step 6,** the slot grids should be dried in the tweezers before putting them onto filter paper.

3.3.3 Confocal Microscopy of Ultrathin Sections

Fluorescent staining can be maintained throughout the embedding and ultrathin sectioning process (*see* Subheadings 3.3.1 and 3.3.2). This allows the comparison of a given physical section with a typical thickness of about 70–100 nm with the corresponding, yet much thicker (about 600–800 nm) light optical section obtained by 3D imaging of the whole cell nucleus with CLSM. Typically, using LR white resin, only DAPI fluorescence is maintained well enough in the sections to allow imaging (*see* **Note 37**). With our current setup, it is not possible to visualize DAPI fluorescence in the physical sections with 3D-SIM if TEM should be included in the analysis (*see* **Note 38**).

1. Clean microscope slides with ethanol.
2. Put one grid on each slide with the sections facing up and cover it with a big coverslip (e.g., 24 × 32 mm). To keep the coverslip in place, apply small drops of Fixogum on each corner and let it dry completely (*see* **Note 39**).
3. Relocalize the cells of interest (*see* **Note 40**) and record their exact positions on the grid in relation to the center of the grid.
4. Acquire images (only 1 plane, no z-stacks) using 400 Hz, the maximum gain setting and as little laser power as possible, typically ranging between 20 and 50 %. For overview images (e.g., 480 × 480 nm pixel size), use a line average of 8; for scans of individual cells/nuclei (50 × 50 nm or 30 × 30 nm pixel size), use a line average of 16. The “right” focal plane typically shows the brightest signal and the best signal-to-background ratio.
5. Unmount the sections by carefully removing first the Fixogum with tweezers and then the coverslip with a scalpel. Make sure that the immersion oil from the microscope does not come into contact with the grid. Put the sections back on filter paper.

3.3.4 Osmium Ammine Staining of DNA

There should be a minimum of one night between cutting and staining of the sections to make sure that they are completely dry. All steps are carried out in embryo dishes at room temperature. Make sure not to mix the grids during staining-this means that only 1–2 grids can be stained in one embryo dish (either one mesh/slot grid or one mesh and one slot grid).

1. Freshly prepare 5 N HCl (1.5 ml per embryo dish).
2. Freshly prepare osmium ammine staining solution (0.75 ml per embryo dish). Let the solution stand for 30–40 min.
3. Incubate the grids for 30–40 min in 5 N HCl (*see* **Note 41**). Make sure that the sections face down and that the grids float in the middle of the dish, not at the periphery (if necessary, remove surface tension by moving around the periphery of the dish with tweezers).
4. Drain the mesh grids on a tissue by carefully touching the tissue with the outer edge of the grid. Skip this step for slot grids.
5. Incubate the grids for 30 min in the osmium ammine staining solution equivalent to **step 3**.
6. Repeat **step 4**.
7. Wash for 5–10 min in H_2O_{bidest} equivalent to **step 3**.
8. Air dry on filter paper in petri dishes (*see* **step 8** of Subheading 3.3.2).

3.3.5 Imaging with TEM

There should be a minimum of one night between staining and observation of the sections to make sure that they are completely dry. Relocalize the cells (*see* **Note 42**) and image them with TEM. We typically acquire images with an average gray value of 50–60 % and take great care not to cut off signals on both sides of the spectrum (high and low values). Using different levels of magnification allows to obtain both overview images and high-resolution details. Be aware that the sample rotates with each step of magnification. This can be corrected in post-processing by putting the gray-scale images in pairs into a two-layered image in Photoshop, reducing the opacity of the upper one to 50–60 %, and aligning the images by eye using the rotate and resize and potentially also warp and distort functions (for details compare Subheading 3.4).

3.4 Image Alignment

After image acquisition and standard image processing, overlays of selected images from the different microscopes can be created to allow direct comparisons and complementary conclusions of the results obtained with each individual technique. We cannot provide a complete overview of options here as there exists a whole series of often highly sophisticated methods and software solutions. But we want to draw the attention to some basic problems and possibilities.

1. Find corresponding sections. According to our experience, it works best to start with the TEM image as a reference as here typically only a limited number of z-sections (often even only one) are available for each cell. Be aware that no perfect match can be found (*see* Subheading 1.2).
2. Align the images. As already mentioned, the images coming from the different microscopes are rotated and/or mirrored in comparison to each other. To allow proper comparison, this should be corrected. We do this manually by creating overlays of pairs of images with chromatin staining (*see* **Note 43**). Put the two images into a two-layered image in Photoshop, reduce the opacity of the upper one (typically colored in red) to 50–60 %, and align the two images by eye using the rotate and resize functions. Optionally increase the canvas size before doing this. If in addition to chromatin also other channels should be aligned, put them in separate layers of the Photoshop image, switch off layer visibility, and link them with the chromatin image. All transformations applied to the chromatin image will also be applied to these additional layers. For live cell (SDLSM), CLSM, and 3D-SIM images, the factor for resizing can be calculated from the pixel sizes. Embedding, however, causes some shrinkage-the degree varies from experiment to experiment or even from sample to sample-resulting in the cells on the ultrathin sections (i.e., both the CLSM image of the section and the TEM images) being smaller than calculated from the pixel sizes. Sometimes the alignment of size and especially rotation cannot be done in one step. In these cases, figure out the right parameters in as many steps as necessary, sum up all the transformations done, and repeat the alignment in one step. The reason behind this is that applying transformations practically always includes interpolation which results in blurry images when repeated several times. Due to the increased resolution in electron microscopy, the images are much bigger in comparison to the confocal or 3D-SIM images. In order not to lose the advantage of the higher resolution, we recommend to do the alignments as described before but to set the scale of the TEM image back to 100 % before applying the transformations. This maintains the option to generate high-quality blowups of the image. If needed, the image size can be reduced in a second step.

   However, due to the following reasons, the best possible alignment is typically not achieved by using the rotate and resize functions alone (exception: confocal–3D-SIM): (a) movements of the cell between the last acquisition of the live cell observation and fixation, (b) slight differences in the imaging plane due to an inevitably not absolutely precise alignment of the embedded block and the knife prior to sectioning, (c) shrinkage of the cell during embedding as mentioned above, and (d) deformation of the ultrathin section under the electron beam in the microscope (*see* also Subheading 1.2). Use the distort and warp functions to correct this as far as possible (*see* Fig. 2D).

Finally, set opacity back to 100 % for all channels, switch on visibility of additional channels again, and export all layers into single tiff images (*see* **Note 44**). If necessary, convert the pseudocolored images back into gray-scales.

3. Create overlays: for fluorescent images in the conventional way generating RGB or RGB plus gray images and for merges of fluorescent and TEM images according to the method described in **Note 43** (TEM image as a gray-scale and the fluorescent image inverted and with 50–60 % opacity in a second layer on top). As mentioned above, the alignments are done by hand and by eye. Therefore, drawing conclusions from these overlays is in our opinion only reasonable for relatively big structures like splicing speckles or nucleoli but very problematic for smaller structures as H3K4me3 or RNA polymerase II signals, for instance.

4 Notes

1. Trolox is a vitamin E analogue, which provides protection against radicals and thus reduces the effects of phototoxicity. For effective protection, Trolox should be added to the cells 12–24 h before imaging as it is supposed to be incorporated only during mitosis. We always add Trolox freshly at a final concentration of 250 μM and do not store Trolox-containing medium for more than 1 day. Prepare a stock solution of 250 mM in 100 % ethanol. For short-term live cell observations, it is not essential. However, using Trolox does not solve the problem of phototoxicity completely. Long-term observations with the acquisition of 3D stacks or with short intervals between exposures will still result in an increased fraction of apoptotic cells and/or impaired or even halted cell division rates (*see* also **Note 2**).
2. It is important to use medium without phenol red, since the latter is supposed to interfere with imaging and to increase the effect of phototoxicity by creating free radicals. Hepes-buffered medium in contrast to $NaHCO_3$-buffered medium helps to maintain a stable pH even outside a CO_2-enriched atmosphere [42]. It is recommended to aliquot the medium into smaller amounts (e.g., 50 ml) to avoid repeated warming up and cooling down. When starting a new aliquot, we add an additional 25 mM Hepes to the already contained 25 mM, as it seems to be not very stable. A clear indication for insufficient Hepes is given when cells stop dividing during the live cell observation or

are stuck in mitosis for hours. However, it was shown that Hepes generates cytotoxic products when exposed to light [43], providing a possible explanation for the phototoxic effects associated with intense imaging described in **Note 1**. Nevertheless, with this combination of Hepes and Trolox (*see* also **Note 1**), undisturbed cell cycles were observed up to 48 h (*see* ref. 44).

3. The fixative should either be prepared fresh (stir the solution on a heating plate for dissolving faster but take great care to remove the fixative from the heating plate as soon as all PFA is dissolved and avoid boiling of the solution) or aliquots stored at −20 °C can be thawed. Do not keep PFA at 4 °C and do not freeze/thaw the fixative more than once as this typically results in high autofluorescence background levels.
4. This blocking buffer is suitable for most applications. In special cases when samples show high background staining with this blocking buffer, an alternative buffer can be used, which consists of 150 mM NaCl, 15 mM Hepes/KOH pH 7.4 or 7.5, 2 mM $MgCl_2$, 0.1 mM EGTA pH 8.0, 0.2 % Triton, 0.5 % fish skin gelatin, 2 % BSA. When using this specialized buffer, it is recommended to apply the following changes to the protocol described in Subheading 3.2.3: do not wash with PBST but with PBS in **steps 2** and **11**, block for 1.5–2 h (**step 4**), and incubate with the primary antibodies for 1.5–2 h (**step 7**).
5. Store the DAPI stock solution at −20 °C. In our experience, repeated freeze/thaw cycles are not problematic.
6. For 2 ml osmium ammine solution, use 1.92 ml H_2O_{bidest}, 4 mg osmium ammine B, 80 μl 5 N HCl, 76 mg $Na_2S_2O_5$. It is recommended to dissolve the osmium ammine completely before adding HCl. $Na_2S_2O_5$ should be added last.
7. The MatTek glass bottom dishes are perfectly suitable for correlative microscopy as they combine several essential features: A hole is cut into a normal plastic petri dish, and a coverslip is attached from below, which allows high-resolution live cell observations with oil immersion objectives. The big advantage compared to other live cell systems is the fact that the coverslip can easily be detached from the dish using coverslip removal fluid (*see* **Note 18**), and thus the sample can be mounted on a microscopic slide or can be embedded for electron microscopy. Various predefined dishes are available including dishes with gridded coverslips (Bellco). If possible number 1.5 coverslips should be chosen as these are about 160–190 μm thick and match the refractive index of immersion oil and mounting medium best (ideal are 170 μm). Coverslips with other thicknesses can potentially be used (like in our case: the Bellco grids (*see* details below) are number 1.0 coverslips with a thickness of 130–160 μm)-in this case, for 3D-SIM, typically the refractive index of the immersion oil has to be adjusted accordingly.

We used custom-made dishes with 12 mm round gridded Bellco coverslips attached. This coverslip size is strongly recommended for moving to electron microscopy-in case square coverslips have to be used for a certain reason, they have to be cut to at least octagons using a diamond pencil to allow processing for TEM (compare **step 13** of Subheading 3.3.1). Be aware that cutting with a diamond pencil has the disadvantage of the coverslips getting more fragile and having a higher risk to break during the following treatments and of generating small pieces of glass which can spread all over the specimen. The Bellco grid fulfills two decisive criteria: (a) the square size of only 600 μm and the continuous numbering of the squares allow safe relocalization of all areas, and (b) the grid is carved into the glass by laser-etching and therefore leaves an imprint in the resin after polymerization which is important for relocalization with TEM.

As an alternative, self-gridded coverslips can be used, but several points are of great importance: (a) Do not apply too much pressure when drawing the grid onto the slide using a diamond pencil as this will result in breaking of the coverslips (in most cases not immediately when preparing them but later in the procedure)-only draw as you would do with a normal pencil. (b) Make the grid size as small as possible-e.g., 1 × 1 mm, which still allows freehand drawing of the grid without the need of a binocular. (c) The grid has to include a numbering to allow relocalization of the cells of interest. As with freehand drawing no exact pattern can be achieved and numbers tend to stick out of "their" square, it is recommended to number only every second square of the grid. (d) When choosing cells for observation, it is important to select an area which can easily be identified later-for example, an area that includes a specific part of a number.

No matter which type of grid is used, great care has to be taken to choose an area in the central part of the grid for the experiment as the procedure for TEM embedding allows only a small part of the coverslip being conserved in the block. For the 12 mm Bellco grid, this typically limits the usable area to the following numbers on the grid: 33–34, 43–46, 55–58, and 67–68.

8. Both grid types have advantages and disadvantages: mesh grids are very stable, but all parts of the sections located on the bars of the grid cannot be used for TEM imaging. In contrast, slot grids allow microscopy of the whole section but are very fragile and can easily break, making the whole grid useless. Based on this we typically combine both grid types (about 2:1 mesh:slot grids) for one experiment. For osmium ammine staining, gold grids are absolutely necessary as nickel or copper grids are not resistant to HCl treatment involved in the staining procedure.

9. Cell density for relocalization experiments should be around 30–50 %. Lower cell densities create problems in finding good observation areas for live cell imaging as in these cases the cells are usually too dispersed and therefore only few cells can be observed simultaneously. Higher cell densities make it (almost) impossible to see the grid and therefore make relocalization of the cells of interest in the successive steps very difficult.

10. Typically, the heating system of the microscope should be switched on about 4 h prior to use to allow equilibration of all components to the desired temperature, so that shifts in focus during acquisition are best possibly avoided. For stabilization, make sure that all doors and openings of the live cell chamber are closed.

    After mounting the sample (which of course involves opening the chamber), a certain drop in temperature is inevitable. This often leads to a change in focus around 30 min after reaching again the desired temperature. Therefore, it is recommended to adjust focus and/or z-volume settings only after this time.

    Despite these precautions, changes in focus during the acquisition cannot be avoided entirely, and reliable autofocus mechanisms are very helpful, in particular for long-term observations and cases where no stacks are acquired but only single planes. A contrast-based autofocus in transmission mode used to work fine for us and provides the advantage of limiting light exposure to the sample (transmission mode typically is operated by halogen lamps and not with strong laser light).

11. Be very thorough when registering the position of the cells of the live cell experiment. It is recommended to print a scheme of the grid and draw the area of interest into this-not only the number of the square but also the exact position. Otherwise an effective relocalization is hardly possible.

12. Start with imaging as many cells as possible in the live cell experiment and record all available cells with CLSM and 3D-SIM in the successive steps. Due to the many steps and especially relocalization steps involved in correlative microscopy, a number of cells will get lost during the process (especially at the TEM level).

13. For changing the medium directly on the microscope, make sure that the sample (i.e., the live cell chamber, here the MatTek dish) is firmly mounted on the microscope. If needed, fix it using sticky tape. Plastic pasteur pipettes turned out to be suited best for changing the medium. Try to touch the dish as little as possible with the pipettes to minimize the risk of moving the sample and therefore the risk of losing your selected position. If the schedule of the experiment allows it, it is strongly recommended to check position and focus before continuing. But despite careful handling, make sure to exchange most of the liquid and not only a minor part-particularly when working with dishes with small sunken observation areas, as in our case.

In these cases, also remove the liquid from the cavity as the exchange by diffusion seems to be minimal.

14. Fixation with 2 % PFA yields a satisfactory level of structural conservation of chromatin for 3D-SIM microscopy [45]. However, some circumstances might make it necessary to change the fixation to 4 % for 10 min instead (e.g., some antibodies work only or work much better with 4 % PFA).
15. The PFA fixative dries out very quickly and can create artifacts. The stepwise exchange of fixative against PBST helps to minimize this effect.
16. Using these FRAP settings, bleaching in one focal plane was enough to bleach the whole nucleus. Alternatively, bleaching can be performed by acquiring z-stacks with a z-spacing of, for example, 200 nm with SDLSM or CLSM by using long exposure times and/or continuous imaging modes in combination with high laser powers (50 % or above). In these cases, the samples showed bleaching in our hands, but even after 30 min, substantial signal intensities were still detectable. As it is not clear whether such a high and strong irradiation causes changes in the ultrastructure of the cells, these approaches were considered to be not applicable for our purposes.
17. Drying out of the cells has to be avoided by all means. For us plastic pasteur pipettes turned out to be most suitable to exchange buffers and solutions. It is recommended to have the next solution already in the pipette before removing the previous solution and to process one sample after the other (i.e., remove solution 1 from sample A, add solution 2 to sample A, remove solution 1 from sample B, add solution 2 to sample B). For timed incubation steps (e.g., 10 min), the timer was started when the solution of the first sample was changed, and then the subsequent samples were processed to keep the right timing.
18. The best system for detaching the coverslip is to put 1 ml of coverslip removal fluid into the lid of a 35 mm dish and place the dish with the sample inside. Take care that no air bubbles occur. If so, carefully lift the sample again on one side to remove the bubbles. During incubation in the coverslip removal fluid, the liquid inside the dish can be changed according to the needs of the protocol, but care should be taken not to dilute the coverslip removal fluid with any other liquid as this might reduce its effectivity. After 45 min, take the sample out of the lid and carefully wipe off the coverslip removal fluid to avoid cells getting into contact with it. Prepare a larger dish (e.g., a 50 mm petri dish) with the same solution the sample is currently incubated in (in the case of our protocol, this is blocking buffer). Hold the sample over this dish and apply pressure onto the coverslip from the inside-in this way, the cells will not dry out, as immediately after detachment they reside in solution again.

Do not apply much pressure onto the coverslip to avoid breaking it. It can happen that detachment is not possible after 45 min, for example, at low room temperature or if the coverslip removal fluid potentially got diluted by some other liquid. In this case, wipe the dish well from below and place it into fresh coverslip removal fluid. After some time, try again to remove the coverslip.

19. Removing the glue from the coverslip is essential for proper mounting of the sample on a microscope slide. Use tweezers to scratch off the glue and work from the inner rim towards the periphery of the coverslip. Work very thoroughly as any remnant of the glue can cause the invasion of nail polish under the coverslip when mounting the sample, which typically destroys it. Be very careful not to break the coverslip (therefore do not apply much pressure): Even after very thorough removal of the glue typically a small difference in height remains between the rim and the observation area. This does not play a role as long as the coverslip is intact but becomes a problem when a part of the rim is missing as this typically causes the invasion of nail polish when mounting.
20. For preparing the humidified chamber, ideally blocking buffer should be used, but also 1× PBS works fine. It is not recommended to use water as it has a clearly different osmolarity and possibly could influence the ultrastructure of the cell.
21. Typically it is enough to mix the antibodies by flicking prior to diluting them. In rare cases, a few minutes of centrifugation at high speed help to reduce precipitates. Also filtering of the diluted antibody through a 0.45 μm filter is possible.
22. Put a reasonably big drop (approximately 20–40 μl) of Vectashield in the center of the microscope slide-not in the center of the clear area but in the actual center of the slide-as the 3D-SIM microscope has a limited observation area. Take the coverslip out of the PBS and place it (with the cells up) on a soft tissue to remove all excess of PBS from the back. Drain the excess of liquid from the cell side with the edge of the soft tissue. Place the coverslip on the Vectashield (with the cells down). Work thoroughly but efficiently to avoid sample drying. Process only one sample at a time.

    Remove the excess of Vectashield either by preparing a stack of tissues and carefully pressing the sample against it from top or by putting soft tissue over the coverslip and carefully moving a finger over the edge of the coverslip. Repeat this several times by moving to fresh areas of the tissue stack/the soft tissue. In both cases, be very careful not to apply pressure onto the sample which would result in flattening of the cells and therefore in a change of morphology and possibly also ultrastructure. Finally, seal the sample well with clear nail polish.

23. Start with finding any number on the grid using transmission light. For easier relocalization, turn the dish in a position that allows reading the numbers in the normal way. Defocus to obtain better contrast of the grid. Once one number is identified "move" from this point to the area of interest-it is not necessary to be able to read all the numbers as long as track of the squares is kept (i.e., why it is recommended to have a printout of the grid at hand). Switch to the fluorescent mode and identify the cells of interest. For reliable relocalization, it is absolutely necessary to print the image of the area observed in the live cell experiment to make sure not to miss any cells and not to waste time and effort on cells that were not included before.

    Note that the image of the sample is now mirrored compared to the live cell image when not working with an inverted microscope.

24. Use either 80 % absolute ethanol or chloroform. Precision cotton swabs-pay attention that they are at least short-term resistant against chloroform-are a helpful tool for this. Avoid contact between the cotton swabs and the sealing during the cleaning procedure. Check slides thoroughly by eye. The same method is applied to clean the objective of the microscope-but reassure with the manufacturer and follow their recommendations.

25. Our standard oil has a refractive index of 1.514; possible alternatives range from 1.508 to 1.524 in steps of 0.002. A test sample can provide valuable indications on which oil is the best: prepare the test sample according to the current protocol and include fluorescently labeled beads. Regard the middle z-position of the beads as a center for the plane of symmetry. The correct oil will result in the brightest and the most symmetric signal with respect to small deviations in the z-position. Therefore, acquire a wide field image stack of ±2 μm around the center, create orthogonal projections in ImageJ, and check for symmetry. Rule of thumb: If you realize the Airy diffraction pattern only at the coverslip side, decrease the refractive index of the oil. Clear signs for suboptimal oil are if the reconstructions of the images are grainy and/or have a halo, e.g., around the nucleus.

    In our experiments with the number 1.0 gridded Bellco coverslips and the special morphology of the HCC-treated cells, the 1.514 oil seemed to be best.

26. Keep exposure times low to avoid bleaching and sample stress. Good reconstructions can be obtained already from images with maximum intensities of around 3,000 counts provided that the background is fairly dark, i.e., less than 200 counts, but most reliably good reconstructions are obtained with maximum intensities between 15,000 and 20,000. If the increase in illumination intensity implicates a strong increase of the background signal, this indicates sample-related issues, like unspecific

staining or autofluorescence. Change the sample preparation protocol in these cases.

Camera manufacturers always add an arbitrary (fixed) intensity offset to allow users statistical image evaluation. Use proper camera offset settings during the acquisition process in order to access the full dynamic range.

3D-SIM always requires the acquisition of an image stack. For optimal results, the physical thickness of the stack size should be greater than 1.5 μm (12 sections). The stack should extend more than 0.5 μm above and below the biological structure and ideally the stack "fades out," i.e., the mean intensity decreases from the images in the center to the images at the top or bottom of the stack. Otherwise, artifacts within the first and last optical section of the reconstruction (top and bottom image) are generated.

Make sure there is no movement of the sample or within the sample during acquisition. A global movement, e.g., produced by thermal drift, will constrict the overall image quality. Moving particles within a sample, e.g., moving filaments or particles moving through the field of view, can potentially create misleading artificial structures and local reconstruction artifacts. Movements can be determined and analyzed efficiently by the acquisition of small time-lapse series and the subsequent creation of a projection. This should already be done before performing 3D-SIM.

27. In most cases, there are only limited options in the reconstruction software, but typically the point-spread function (PSF; or optical transfer function, OTF) which will be used can be selected. It is very advantageous if different distinct PSFs for each channel can be used as the optical light path might slightly differ for different colors.

    To correct for the camera's background, an offset adjustment can be applied during reconstruction. Do not crop real background values created by photon and camera statistics as this will mess up the data and create image artifacts. If in doubt, keep the original background value and correct for brightness and contrast later.

    Due to the optical sectioning capabilities and the higher resolution, SI reconstructions will show pronounced chromatic aberrations. Single spots, e.g., in the green and the red channel, will appear at different x, y, and z locations. Unfortunately this effect is depth and sample dependent, i.e., you have to correct (align) your images later in post-processing.

28. Using colored sticker spots and labeling with pencil proved to be a save method for us.

29. Often the protruding pieces of plastic can be removed by running one's fingernail over it. Be careful not to overdo it and cause groovings.

30. In case the surfaces of the blocks are very small after polymerization, try to place a beaker with water in the 60 °C oven. This sometimes helps to reduce the decrease of the surface during polymerization.
31. Hold the capsule with big tweezers and already grab the coverslip with normal tweezers before dipping the sample into liquid nitrogen so that it is possible to immediately pull on the coverslip. Put the sample for approximately 2 s into liquid nitrogen-too long exposure to liquid nitrogen and too many repeated cycles of freezing and thawing cause cracks in the block. Cracks can potentially make the sample useless as the sections will fall apart along the cracks.
32. This works best if two razor blade cuts separated by a few mm are made on the long side of the capsule. Use the piece between the two cuts as a zipper and peel off the capsule.
33. Contrast on the block surface can be enhanced by casting a shadow on the block with a razor blade. Be careful not to touch and therefore cut the block. Note that the pattern is now mirrored in comparison to the live cell image. It is recommended to first mark an area a little larger than the surface of the final pyramid should be. This allows quick and rough removal of most of the surrounding material, leaves some space for small adjustments, and makes it easier to precisely prepare the pyramid. Register the cuts one by one in a drawing to ensure to keep oriented on the sample-typically most of the numbers of the grid will be cut off. Try to make the pyramid a perfect rectangle or square. It is especially important that the longer sides of the rectangle are indeed parallel-this longer side will be aligned to the knife later and therefore sectioning will take place perpendicular to it. For relocalization experiments, it is recommended to keep the size of the pyramid as small as possible to facilitate the identification of the cells of interest. Nevertheless, leave some space around the region of interest as cells on the very border of the section are difficult to image with TEM (due to deformations of the section in the electron beam if it does not cover one complete square of the mesh grid).
34. The thickness of the sections correlates very well with their color: silver = 70–80 nm, gold = 100 nm, purple/violet = 150 nm, blue > 150 nm. For correlative microscopy, 100 nm sections (golden color) are recommended as they typically yield sufficient DAPI signal for scanning the sections with CLSM but are also suitable for TEM. In case the microtome produces too thin/too thick sections despite being set at 100 nm, adjust section thickness on the microtome until golden sections are obtained.

    If the microtome has the option to calculate the number of sections and the total thickness of the cutoff segment, it is recommended to make use of it-this allows cutting of only as many sections as actually needed for the prepared number of grids and to adjust the number of sections to the thickness of the cells.

35. In principle also glass knives can be used. But they are of lower quality compared to the diamond knife and only rarely produce even sections without stripes. Therefore, the number of non-usable cells will increase dramatically if using glass knives.
36. Keeping the right order of sections-at least from grid to grid-has the advantage of providing a rough idea about the relative position of the section in the cell. For instance, if a very peripheral section of the sample is under view on the electron microscope, it is easy to know in which direction to move for obtaining more central sections.
37. If also the fluorescence of other fluorophores is to be maintained in the sections, we recommend other embedding media, for example, the Quetol 651 kit (Polysciences Europe). However, we have no experience with osmium ammine staining on Quetol-embedded sections. An increased section thickness improves signal quality for CLSM, but too thick sections cannot be imaged with TEM.
38. Imaging with TEM requires ultrathin sections on grids. To obtain images of the sections with 3D-SIM, the grids have to be mounted with Vectashield; otherwise, the image reconstruction process causes tremendous artifacts. Mounting, however, results in several problems: (a) the grid and especially the sections are relatively flexible which makes it almost impossible to mount them in a plane way, (b) the sections are very fragile and there is a high risk that they will break when unmounting the sections again for subsequent TEM imaging, and (c) it is not known whether Vectashield interferes with subsequent osmium ammine staining even if extensive washing steps are included after unmounting (compare discussions about osmium ammine staining quality in Subheading 1.2). CLSM, in contrast, does not require reconstruction of images and in addition provides worse resolution in *z*-direction, so that scanning of the ultrathin sections without mounting medium is possible.

    In case 3D-SIM should be performed on the sections without subsequent TEM imaging of the cell, we suggest the following: Perform sample preparation and ultrathin sectioning as described. Instead of taking the sections up with grids (Subheading 3.3.2, **step** 7), use a small metal loop (4 mm diameter) to transfer the sections onto well-cleaned coverslips prewarmed to 70–75 °C on a heating block: put down the loop on the coverslip (typically the water drop with the sections should get into contact with the glass), wait till the water is evaporated, and carefully remove the loop. After the sample dried completely, it can be mounted in Vectashield. The use of gridded coverslips (typically hand-gridded ones are sufficient) turned out very useful in facilitating later relocalization of the sections, as the position of the sections on the grid can be registered before mounting and the numbers can easily be identified at the microscope using transmission light.

39. Try to move the coverslip as little as possible to avoid damage to the sections. Use only small drops of Fixogum-otherwise the Fixogum can spread below the coverslip. If it touches the grid, it makes it useless for electron microscopy. For this reason, the use of a large coverslips is recommended. Nail polish can be used as an alternative to Fixogum, but as this is less viscous, greater care must be taken to avoid spreading under the coverslip and touching of the grid. In addition, using nail polish makes it more difficult to remove the coverslip again after scanning on the confocal microscope.
40. Find the sections using transmission light by starting from the center of the grid and moving to the periphery. Positions that are too peripheral on the grid (typically more than 10 squares away from the center on the 300 mesh grids) cannot be observed in the electron microscope. Try to orient yourself on the section based on the information from cutting the pyramid (*see* **step 2** of Subheading 3.3.2). Switch to fluorescent mode and locate the cells of interest. As with the previous relocalization steps, it is recommended to have a printout of the live cell observation at hand. Note that the sample is mirrored in comparison to the live cell experiments when working on an inverted microscope. Always be aware that now only a 2D section of the 3D volume of the cell is being viewed. This means that some cells might be cut in a midsection while others might be cut very peripherally, which makes identification of cells difficult. The best criterion for relocalizing cells is to focus more on comparing angles and distances of cells to each other rather than on the morphology of individual cells.
41. This HCl step serves to degrade RNA which otherwise will also be stained by osmium ammine. The incubation time in 5 N HCl varies from cell type to cell type. Typically 30 min should be enough, but the incubation can be extended to 40 min. A good measure for the efficiency of RNA degradation is the staining intensity in the nucleoli: they should have a color only slightly darker than the cytoplasm.
42. It is recommended to have a printout of the live cell observation at hand as well as the information of the positions registered when scanning the sections with CLSM (*see* **step 3** of Subheading 3.3.3). Note that the sample now is mirrored in comparison to the live cell experiments. Be aware that the angle of rotation is certainly different from what it was when scanning the sections with CLSM.

43. According to our experience, aligning two fluorescent images (e.g., SDLSM–CLSM, CLSM–3D-SIM) works best when focusing on the chromatin stainings only and when one or both of the two images are pseudocolored, resulting in one gray-scale and one red image or one green and one red image. For the alignment of one fluorescent image and the TEM image, the fluorescent image should be pseudocolored in turquoise (merge of green and blue) and inverted (resulting in all signals being red), while the TEM image remains in gray-scale.
44. The newer versions of Photoshop have an automatic function for this: file–scripts–export layers to files. Otherwise switch off visibility of all layers except for one, flatten the image, save it, and repeat these steps for all other layers.

Acknowledgements

This work was supported by grants to Thomas Cremer (DFG grant SFB684, CR-59/29-2).

We are indebted to Stanislav Fakan and Jacques Rouquette for introducing us into the techniques of osmium ammine staining for DNA and TEM and to Yolanda Markaki for helping with establishing immunofluorescence procedures for 3D-SIM. We thank our colleagues Dirk Eick for providing the RNA polymerase II antibodies, Otto Berninghausen for technical support with TEM, and Heinrich Leonhardt for continued support of our studies.

References

1. Caplan J, Niethammer M, Taylor RM 2nd, Czymmek KJ (2011) The power of correlative microscopy: multi-modal, multi-scale, multi-dimensional. Curr Opin Struct Biol 21(5):686–693
2. Giepmans BN (2008) Bridging fluorescence microscopy and electron microscopy. Histochem Cell Biol 130(2):211–217
3. Muller-Reichert T, Verkade P (2012) Introduction to correlative light and electron microscopy. Methods Cell Biol 111:xvii–xix
4. Svitkina TM, Borisy GG (1998) Correlative light and electron microscopy of the cytoskeleton of cultured cells. Methods Enzymol 298:570–592
5. Cremer C, Masters BR (2013) Resolution enhancement techniques in microscopy. Eur Phys J H 38(3):281–344
6. Toomre DK, Langhorst MF, Davidson MW (2012) Introduction to spinning disk confocal microscopy. http://zeiss-campus.magnet.fsu.edu/articles/spinningdisk/introduction.html. Accessed 29 Nov 2012
7. Cremer C (2012) Optics far beyond the diffraction limit. In: Träger F (ed) Springer handbook of laser and optics. Springer, New York, pp 1359–1397
8. Gustafsson MG (2000) Surpassing the lateral resolution limit by a factor of two using structured illumination microscopy. J Microsc 198(Pt 2):82–87
9. Heintzmann R, Cremer C (1998) Laterally modulated excitation microscopy: improvement of resolution by using a diffraction grating. Proc SPIE 3568:185–196
10. Gustafsson MG et al (2008) Three-dimensional resolution doubling in wide-field fluorescence microscopy by structured illumination. Biophys J 94(12):4957–4970
11. Schermelleh L, Heintzmann R, Leonhardt H (2010) A guide to super-resolution fluorescence microscopy. J Cell Biol 190(2): 165–175
12. Fiolka R, Shao L, Rego EH, Davidson MW, Gustafsson MG (2012) Time-lapse two-color 3D imaging of live cells with doubled resolution

using structured illumination. Proc Natl Acad Sci U S A 109(14):5311–5315

13. Shao L, Kner P, Rego EH, Gustafsson MG (2011) Super-resolution 3D microscopy of live whole cells using structured illumination. Nat Methods 8(12):1044–1046
14. Jones SA, Shim SH, He J, Zhuang X (2011) Fast, three-dimensional super-resolution imaging of live cells. Nat Methods 8(6):499–508
15. Vaughan JC, Jia S, Zhuang X (2012) Ultrabright photoactivatable fluorophores created by reductive caging. Nat Methods 9(12): 1181–1184
16. Denk W, Horstmann H (2004) Serial block-face scanning electron microscopy to reconstruct three-dimensional tissue nanostructure. PLoS Biol 2(11):e329
17. Zankel A, Kraus B, Poelt P, Schaffer M, Ingolic E (2009) Ultramicrotomy in the ESEM, a versatile method for materials and life sciences. J Microsc 233(1):140–148
18. Knott G, Marchman H, Wall D, Lich B (2008) Serial section scanning electron microscopy of adult brain tissue using focused ion beam milling. J Neurosci 28(12):2959–2964
19. Rouquette J et al (2009) Revealing the high-resolution three-dimensional network of chromatin and interchromatin space: a novel electron-microscopic approach to reconstructing nuclear architecture. Chromosome Res 17(6):801–810
20. Villinger C et al (2012) FIB/SEM tomography with TEM-like resolution for 3D imaging of high-pressure frozen cells. Histochem Cell Biol 138(4):549–556
21. Schroeder-Reiter E, Sanei M, Houben A, Wanner G (2012) Current SEM techniques for de- and re-construction of centromeres to determine 3D CENH3 distribution in barley mitotic chromosomes. J Microsc 246(1):96–106
22. Hayat M (2000) Principles and techniques of electron microscopy: biological applications. Cambridge University Press, Cambridge
23. Testillano PS et al (1991) A specific ultrastructural method to reveal DNA: the NAMA-Ur. J Histochem Cytochem 39(10):1427–1438
24. Vazquez-Nin GH, Biggiogera M, Echeverria OM (1995) Activation of osmium ammine by SO_2-generating chemicals for EM Feulgen-type staining of DNA. Eur J Histochem 39(2):101–106
25. Dubochet J, Sartori Blanc N (2001) The cell in absence of aggregation artifacts. Micron 32(1):91–99
26. Glaeser RM (2008) Cryo-electron microscopy of biological nanostructures. Phys Today 61: 48–54
27. Cremer T, Cremer C (2001) Chromosome territories, nuclear architecture and gene regulation in mammalian cells. Nat Rev Genet 2(4):292–301
28. Cremer T, Cremer M (2010) Chromosome territories. Cold Spring Harb Perspect Biol 2(3):a003889
29. Dixon JR et al (2012) Topological domains in mammalian genomes identified by analysis of chromatin interactions. Nature 485(7398): 376–380
30. Lieberman-Aiden E et al (2009) Comprehensive mapping of long-range interactions reveals folding principles of the human genome. Science 326(5950):289–293
31. Markaki Y et al (2012) The potential of 3D-FISH and super-resolution structured illumination microscopy for studies of 3D nuclear architecture: 3D structured illumination microscopy of defined chromosomal structures visualized by 3D (immuno)-FISH opens new perspectives for studies of nuclear architecture. Bioessays 34(5):412–426
32. Nora EP et al (2012) Spatial partitioning of the regulatory landscape of the X-inactivation centre. Nature 485(7398):381–385
33. Cremer T et al (2000) Chromosome territories, interchromatin domain compartment, and nuclear matrix: an integrated view of the functional nuclear architecture. Crit Rev Eukaryot Gene Expr 10(2):179–212
34. Mirny LA (2011) The fractal globule as a model of chromatin architecture in the cell. Chromosome Res 19(1):37–51
35. Rouquette J, Cremer C, Cremer T, Fakan S (2010) Functional nuclear architecture studied by microscopy: present and future. Int Rev Cell Mol Biol 282:1–90
36. Mor A et al (2010) Dynamics of single mRNP nucleocytoplasmic transport and export through the nuclear pore in living cells. Nat Cell Biol 12(6):543–552
37. Albiez H et al (2006) Chromatin domains and the interchromatin compartment form structurally defined and functionally interacting nuclear networks. Chromosome Res 14(7): 707–733
38. Albiez H (2007) Manipulation of global chromatin architecture in the human cell nucleus and critical assessment of current model views. Ludwig-Maximilians-University, Munich, Dissertation
39. Bornfleth H, Edelmann P, Zink D, Cremer T, Cremer C (1999) Quantitative motion analysis of subchromosomal foci in living cells using four-dimensional microscopy. Biophys J 77(5): 2871–2886
40. Strickfaden H, Zunhammer A, van Koningsbruggen S, Kohler D, Cremer T (2010) 4D chromatin dynamics in cycling cells: Theodor Boveri's hypotheses revisited. Nucleus 1(3):284–297
41. van Steensel B, Dekker J (2010) Genomics tools for unraveling chromosome architecture. Nat Biotechnol 28(10):1089–1095

42. Diaz G, Isola R, Falchi AM, Diana A (1999) CO_2-enriched atmosphere on the microscope stage. Biotechniques 27:292–294
43. Spierenburg GT, Oerlemans FT, van Laarhoven JP, de Bruyn CH (1984) Phototoxicity of N-2-hydroxyethylpiperazine-N′-2-ethanesulfonic acid-buffered culture media for human leukemic cell lines. Cancer Res 44(5):2253–2254
44. Hübner B, Strickfaden H, Müller S, Cremer M, Cremer T (2009) Chromosome shattering: a mitotic catastrophe due to chromosome condensation failure. Eur Biophys J 38(6):729–747
45. Markaki Y, Smeets D, Cremer M, Schermelleh L (2013) Fluorescence in situ hybridization applications for super-resolution 3D structured illumination microscopy. Methods Mol Biol 950:43–64

Chapter 22

Time-Lapse, Photoactivation, and Photobleaching Imaging of Nucleolar Assembly After Mitosis

Danièle Hernandez-Verdun, Emilie Louvet, and Eleonora Muro

Abstract

Nucleolus assembly starts in telophase with the benefit of building blocks passing through mitosis and lasts until cytokinesis generating the two independent interphasic cells. Several approaches make it possible to follow the dynamics of fluorescent molecules in live cells. Here, three complementary approaches are described to measure the dynamics of proteins during nucleolar assembly after mitosis: (1) rapid two-color 4-D imaging time-lapse microscopy that demonstrates the relative localization and movement of two proteins, (2) photoactivation that reveals the directionality of migration from the activated area, and (3) fluorescence recovery after photobleaching (FRAP) that measures the renewing of proteins in the bleached area. We demonstrate that the order of recruitment of the processing machineries into nucleoli results from differential sorting of intermediate structures assembled during telophase, the prenucleolar bodies.

Key words Live cell imaging, Dynamics, Photoactivation, FRAP, Nucleoli, PNB, Cell cycle

1 Introduction

In HeLa cells, transcription of the ribosomal genes (rDNAs) starts at telophase. During anaphase, the rRNA-processing proteins form a layer at the surface of all chromosomes in which 45S pre-rRNA foci are also present [1]. In telophase the rRNA-processing proteins and 45S pre-rRNA assemble into foci called prenucleolar bodies (PNBs). Two types of nucleolar processing proteins are defined based on the timing of their activity: (1) early-processing proteins localized in the dense fibrillar component (DFC) of the nucleolus and (2) late-processing proteins localized in the granular component (GC) of the nucleolus. The first set of proteins is involved in early-processing events of the pre-rRNAs and can bind to them co-transcriptionally. In this chapter, we have chosen Fibrillarin and Nop56 as markers of the early-processing machinery. For the second group, involved in later events of the pre-rRNA processing, we have chosen B23 and Nop52 that are involved in the formation of the large ribosome subunits. The biological

Yaron Shav-Tal (ed.), *Imaging Gene Expression: Methods and Protocols*, Methods in Molecular Biology, vol. 1042,
DOI 10.1007/978-1-62703-526-2_22,

question is how these two nucleolar processing machineries are recruited at the time of the activation of the rDNA transcription during the transition mitosis to interphase.

Live-cell imaging technology has made tremendous progress during the past decade using fluorescent molecules fused to proteins to observe and quantify dynamic processes in living cells. Three methods used to decipher the assembly of the nucleolus from mitosis to G1 phase are described. The principle of each method is the following:

1. 4-D time-lapse microscopy is based on rapid imaging of tagged fluorescent proteins in the volume of a living cell. The fluorescence signal gives information on the 3-D localization of the proteins over time and about the amount of the tagged proteins. When two different proteins are tagged, one in green and the other in red, the two signals provide information on the relative position and amount of these two proteins [2–4].
2. Photoactivation imaging is based on the possibility to render the photoactivatable GFP (PAGFP) fluorescent following photoactivation of the protein. The PAGFP is a mutated GFP protein that becomes fluorescent only after its activation at a specific wavelength (413 nm) and remains stable for days [5]. The trafficking of the pool of fluorescent proteins from the site of activation to the nuclear volume indicates the directionality of the movement [6].
3. Fluorescence recovery after photobleaching (FRAP) imaging is based on the bleach of the fluorescence of GFP-tagged proteins and the amount of fluorescence recovery in this bleached area [7]. The recovery of fluorescence measures the traffic of proteins tagged with non-bleached molecules [6] and gives information about the renewing of the protein pool in this area.

   These three imaging methods are used to analyze the order of recruitment, directionality of the trafficking, and dynamics of the nucleolar processing proteins at the time of nucleolar assembly.

2 Materials

2.1 Cells

1. Permanent human HeLa cell line (ATCC, CCL-2).
2. Medium for cell growth: minimum essential medium (MEM) containing Earle's salts and Glutamax without antibiotics and stored at 4 °C, and supplemented with 10 % fetal calf serum stored at −20 °C.
3. To detach the cells: trypsin–ethylenediaminetetraacetic acid (EDTA).

4. Medium for live imaging: Dulbecco's modified minimum essential medium (DMEM) "F12 Observation" (PAA Laboratories) without phenol red, without vitamin B12, without riboflavin, but with stable L-glutamine and stored at 4 °C (*see* **Note 1**).
5. Plastics for cell growth and observation: 75 and 25 cm^2 flasks with filter screw cap membrane with a pore size of 0.22 μm and 40 mm Petri dishes.
6. Imaging chamber (Ludin observation chamber; Life Imaging Services, Reinach, Switzerland) using glass coverslips of 32 mm diameter (Menzel-Glaser, Bioblock, France).

2.2 Overexpression of Tagged Proteins

1. Nucleolar protein sequences are inserted into the vector pEGFP-C2 or pDsRed2-C1 (both from BD Biosciences Clontech). In all cases, GFP and DsRed are fused to the NH_2 terminus of the proteins [3].
2. Stably transformed cell lines are established expressing GFP-Nop52, GFP-fibrillarin, GFP-B23, GFP-Bop1 [2], or DsRed-B23. Double-transfected cells are generated from stably transformed cells transiently transfected with DsRed-B23, DsRed-Nop52, mRFP-Nop56, or GFP-Nop52 [3].
3. Transfection with photoactivatable GFP (PAGFP) [5]-tagged nucleolar protein constructs (B23-PAGFP, Nop52-PAGFP or PAGFP-fibrillarin) is performed 4 h after seeding the cells on glass coverslips using liposome transfection reagent (*see* **Note 2**). The cells were doubly transfected with partner proteins tagged in red with DsRed-B23, DsRed-Nop52, or mRFP-Nop56 and observed 28–37 h after transfection [8, 9].

2.3 Time-Lapse Microscopy

1. Microscope: inverted wide-field microscope (Leica DM IRBE; Leica Microsystems, France). The stage plate is motorized to move in the *x* and *y* directions and *z* direction is controlled by a piezo-driven microscope objective nanofocusing/scanning device (PIFOC; Physik Instrumente, Karlsruhe, Germany) placed at the base of the objective.
2. Objective: ×100 PlanApo 1.4 numerical aperture, oil (Leica).
3. CCD camera: 5 MHz Micromax 872Y interline (Roper Scientific, Evry, France).
4. Microscope incubator with controlled temperature (Life Imaging Services).
5. Lamp: 175 W Xenon housed in a DG4 illuminator (Sutter Instruments, Novato, CA, USA) linked by an optical fiber to the microscope.
6. For imaging of GFP and DsRed, a dual narrow passband FITC/tetramethylrhodamine B isothiocyanate (TRITC) filter block is used.

7. In the DG4 illuminator, short-pass KP 500 and long-pass LP 515 filters are mounted in positions 1 and 2, respectively.
8. The acquisition software (MetaMorph) is set to trigger rapid wavelength changes to acquire two images (one for each channel) at each z-step (0.3 μm).

2.4 Photoactivation and Photobleaching Imaging

1. Confocal Microscope: Leica TCS SP2 AOBS Spectral confocal (Leica Microsystem).
2. Objective: ×63 PlanApo 1.32 numerical aperture, oil (Leica).
3. Lasers: Argon laser for green, Krypton laser for red, and a 50 mW laser diode GaN (Coherent) for PAGFP photoactivation.
4. Observation: the glass coverslips of 32-mm diameter with growing cells are mounted in the preheated imaging chamber and maintained at 37 °C.

2.5 Imaging Software for Time-Lapse Microscopy

1. For acquisition, MetaMorph (Molecular Devices Corporation, Sunnyvale, CA, USA).
2. For deconvolution, a custom-made software package is used [10].
3. For image analysis, ImageJ, a public domain Java image-processing program, is used that is downloadable from the Internet site: http://rsh.info.nih.gov/ij/.

3 Methods

Transcription of the rDNA genes starts during telophase on six chromosomes in HeLa cells [11]. The rRNA-processing proteins located around each chromosome should be targeted on to these six transcription sites. The processing proteins are progressively recruited on rRNA transcripts in an ordered manner. The dynamics of these recruitments were analyzed by time-lapse microscopy, photoactivation, and photobleach, from telophase to the end of early G1, the time frame that corresponds to the separation of the two daughter cells.

3.1 Time-Lapse Microscopy

3.1.1 Cells

1. The cells are seeded on a glass coverslip of 32 mm diameter, 22 h before observation, i.e., the period yielding the maximum mitotic cells. Typically 10^4 HeLa cells from exponentially growing batches are used.
2. The coverslip supporting the cells is mounted in the Ludin observation chamber (maintained at 37 °C).
3. The observation medium (*see* **item 4** in Subheading 2.1), DMEM F12 Observation, is added (2 mL at 37 °C).

4. The cells are maintained at 37 °C by placing the observation chamber and the microscope in the microscope incubator.
5. HeLa cells expressing both GFP-fibrillarin and DsRed-B23 or GFP-Nop52 and DsRed-B23 are selected (*see* **Note 3**).
6. The acquisition software is set to acquire two stacks of images (GFP and DsRed) of 0.3 μm at each *z*-step.

3.1.2 Observation

1. Search for round mitotic cells in the upper focal plane as compared to attached interphasic cells.
2. Start the acquisition from telophase to the formation of two daughter cells (Fig. 1a) (*see* **Note 4**).
3. Image a large part of the cell volume (surface of ≈10–20 μm^2) at each *z*-step of 0.3 μm because the size of the nucleus increases due to chromatin decondensation at this period and the proteins of interest are successively localized at the chromosome periphery, in PNBs, or in the six sites of rDNA transcription (Fig. 1b).
4. The frequency of the acquisition of the Z series is in general every 30 s, or every 10 s in particular cases.
5. Imaging at full overlapped speed of the CCD device assures that the two fluorescent tags are recorded sequentially at maximum speed, without movement of the filters.
6. To see the dynamics of the fluorescent proteins as a movie, the images are deconvolved [10] and assembled in stacks, one stack per time point. The stacks are projected along the *Z*-axis using the *Maximum Intensity* projection of ImageJ software. This step transforms each stack in one image per time point as described [4]. These images are then converted into a stack that can be saved as a QuickTime movie in ImageJ.
7. To measure the distribution of the fluorescent proteins, the fluorescence intensity is quantified using the sum of three consecutive non-deconvolved optical sections. This is necessary so that the PNB movement is taken into account (*see* **Note 5**). GFP and DsRed fluorescent signals are quantified in the same regions of interest (ROI) using ImageJ software.

3.1.3 Differential Timing for the Recruitment of the rRNA-Processing Proteins

During nucleolar assembly, time-lapse recording demonstrates that the recruitment of the processing proteins to the sites of rDNA transcription is rapid for early-processing proteins and takes longer for late-processing proteins (Fig. 1c). In addition, at the beginning of nucleolar assembly, both GFP-fibrillarin and DsRed-B23 are observed in the same PNBs (Fig. 1c). The concentration of both proteins measured in the same PNBs confirms these observations [3].

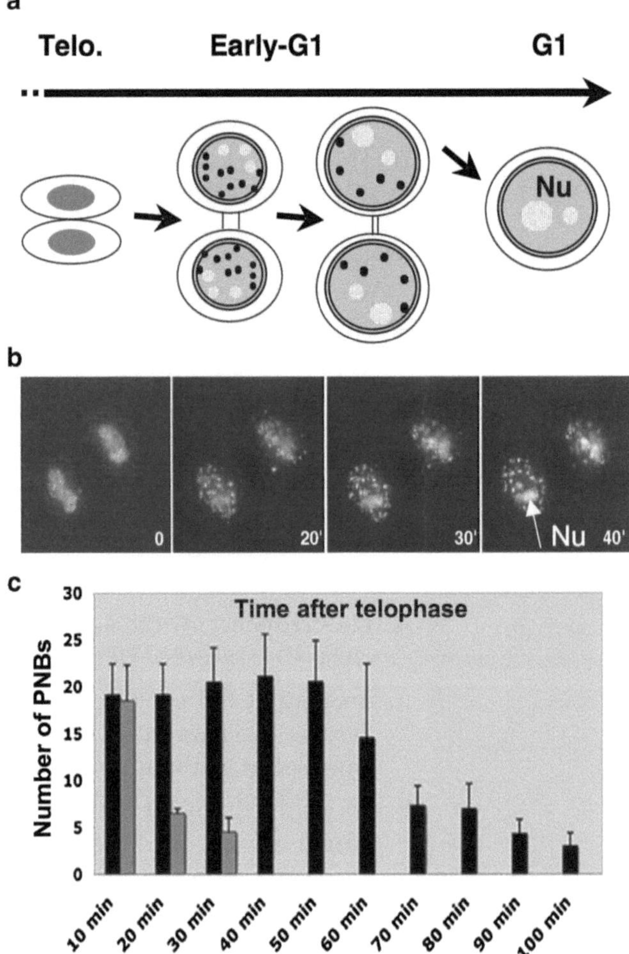

Fig. 1 From mitosis to interphase using time-lapse. (**a**) Scheme of the different steps organizing the nucleoli after mitosis. On the *left* of the cartoon, two daughter cells in telophase (Telo.) are presented; nucleus is *grey*. In early G1 the nuclear envelope (*double circles*) is assembled and between the two cells a cytoplasmic bridge is formed. Cells enter in G1 phase when the bridge is disrupted. During early G1 the nucleolus is assembled. In the nucleus, the three clear areas represent the sites of transcription-organizing nucleoli and the numerous *black spots* represent the PNBs (clusters of pre-rRNAs associated with rRNA-processing proteins). Note that the size of the nucleoli (Nu) increases during early G1. (**b**) Time-lapse imaging of one rRNA-processing protein during early G1. In telophase (0) the DsRed-B23-fluorescent (*white*) signal is visible in the two nuclei. As the cells progress to early G1 (20′–40′), the size of the nuclei increases. The DsRed-B23-fluorescent signal is visible in small foci, the PNBs, and is progressively recruited into the nucleolus (Nu) visible after 30′–40′. (**c**) Different timing of rRNA-processing proteins in PNBs. After telophase (10–30 min) both early (GFP-fibrillarin in *grey*) and late (Ds-RedB23 in *black*) rRNA-processing proteins localize in the same PNBs. Then only late-processing proteins can be detected in PNBs and the number of PNBs decreases, but they are still visible 100 min after telophase

3.2 Photoactivation Microscopy

The observation comprises four periods of recording: (1) before activation, one series of Z optical sections and single optical sections for reference of the initial state; (2) photoactivation; (3) recording of the fluorescence in single optical sections; and (4) distribution of the fluorescence in the volume by one Z series of images to analyze the distribution of the PAGFP-tagged proteins.

3.2.1 Observation

1. Prepare the HeLa cells expressing PAGFP-tagged nucleolar processing proteins (**item** 3 of Subheading 2.2) and observed as described above (Subheading 3.1.1, **steps 1–5**).
2. Before activation of the PAGFP, start imaging of 1–3 single optical sections and one Z series of images with 10 % of the power of a 20 mW argon laser at 488 nm (*see* **Note 6**).
3. Proceed to the activation of the PAGFP on a single section (0.6 μm thick) by a single laser pulse at 405 nm using 15–20 % of the power of a 50 mW laser diode GaN during 300 ms (Fig. 2a). To locate the activation site, a partner protein tagged in red is necessary in the case of small ROI (Fig. 2b) (*see* **Note** 7).
4. Choose a rapid recording of the images of the GFP (256 × 256 pixels, without averaging for rapid acquisition). The delay between activation and recording of the first activated GFP images is 121 ms. The photoactivation acquisition conditions may require an adjustment according to the protein abundance or dynamics (*see* **Note 8**).
5. After activation, record one image of the GFP signal at 488 nm with 10 % of the power of a 20 mW argon laser to minimize photobleaching.
6. The GFP signal is collected on single sections at 500 ms intervals during 2 or 5 min. The total duration of the recording may need to be modified according to the kinetics of the analyzed protein (Fig. 2c) (*see* **Note 9**).
7. Finally, Z series of GFP and either DsRed or mRFP signals are collected immediately or 5 min later to analyze the relative distribution of the activated pool of proteins with partner proteins in red.

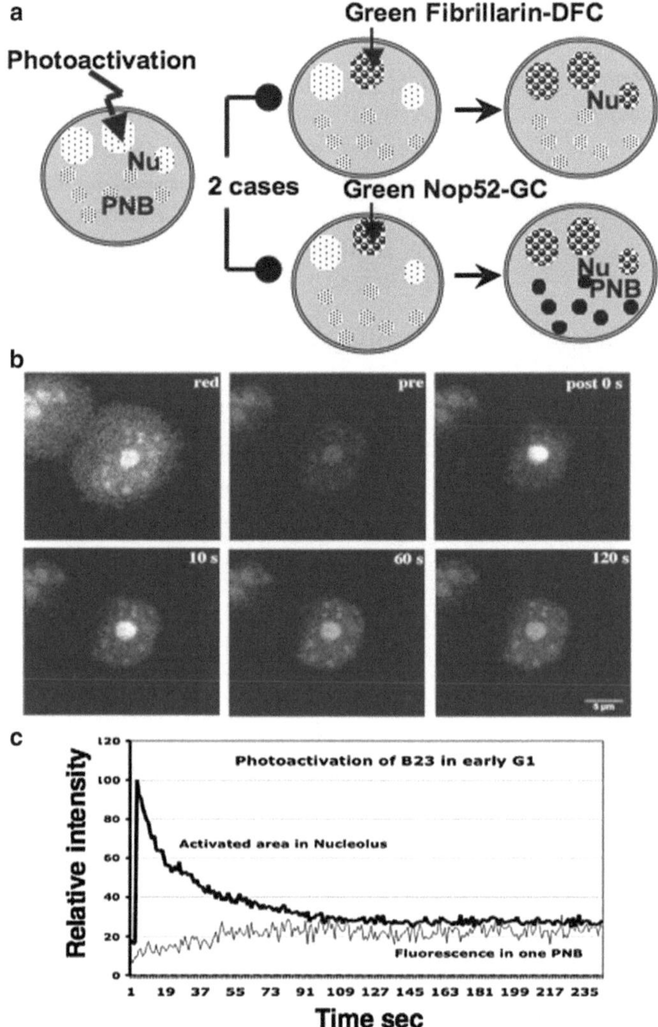

Fig. 2 Analysis of the traffic using photoactivation. (**a**) Comparison of early and late rRNA-processing proteins. The photoactivation in a nucleolus of PAGFP-fibrillarin (early processing) or PAGFP-Nop52 (late processing) renders the tagged molecules fluorescent at the site of activation (*black* nucleolus). From the site of activation, two types of migration of the fluorescent molecules (in *black*) are observed: (1) migration in each nucleolus but not into PNBs of PAGFP-fibrillarin and (2) migration into every nucleolus and all the PNBs of PAGFP-Nop52. (**b**) Images of photoactivation of PAGFP-B23 late rRNA-processing proteins. The fluorescent proteins are in *white*. To localize the site of activation, a protein partner of B23 tagged in red is overexpressed. The first image indicated with "*red*" shows the fluorescence of the red proteins. The nucleolus is the large structure in the center of nucleus and PNBs the small foci. At the same time, the second image indicated with "pre" shows the very low fluorescence of PAGFP-B23 before activation. The fluorescence is very high after photoactivation (post 0 s) and with time the distribution of the PAGFP-B23 in the PNBs is visible (10, 60, and 120 s). (**c**) Quantification of the PAGFP-B23 fluorescence. The relative intensity of the PAGFP-B23 in the activated area of the nucleolus (post 0 s) is set at 100 and is followed in this area for 2 min. The fluorescence decrease measures the redistribution of the activated molecules in the nuclear volume. In parallel the traffic of fluorescent molecules is visible in one PNB

3.3 FRAP Imaging

The imaging comprises four periods of recording: (1) before photobleaching, one series of Z optical sections to choose the best position of imaging and single optical sections for reference of the initial state; (2) photobleaching; (3) recording of the fluorescence in single optical sections; and (4) distribution of the fluorescence in the volume by one Z series of images. During nucleolar assembly, the traffic of proteins is measured in PNBs and nucleoli.

3.3.1 Observation

1. Prepare the HeLa cells expressing GFP-B23 or GFP-Nop52 nucleolar proteins (*see* **Note 10**) as described above Subheading 3.1.1, **steps 1–5**.
2. Search for early G1 daughter cells in the upper focal plane. Select the cell with the median focal plane parallel to the support (*see* **Note 4**).
3. Identify the fluorescent sites by their size; foci of 0.1–2 μm correspond to PNBs, and the larger fluorescent sites to nucleoli that are being formed (Fig. 3a, b). The PNBs are numerous compared to the six sites of active rRNA transcription where nucleolar assembly takes place.
4. The following recording setup is used: zoom 9.28; speed 800 Hz and image size 256 × 256 in standard conditions. For high-speed recording: zoom 16.52; speed 1,000 Hz and image size 128 × 128.
5. Three images are registered before the bleach as reference of the initial intensity level of the GFP-tagged proteins with the 488 nm line of the Argon laser at 10 % of power.
6. The bleach of GFP fluorescence in one PNB is carried out with the 488 nm line of the Argon laser by a single laser pulse at 100 % of power during 100 ms.
7. After bleaching, one optical section of GFP with the 488 nm line of the Argon laser at 10 % of power, every 500 ms for 1–2 min, is recorded (*see* **Note 9**).
8. Finally, one Z series of GFP signals is collected to visualize the 3-D distribution of the GFP-tagged proteins.
9. At least 10–15 cells should be bleached in the same conditions to analyze the recovery dynamics (Fig. 3c).

3.4 Image Analysis for Photoactivation and FRAP

The analysis of the images includes three consecutive steps: (1) correction of the bleach, (2) normalization of the intensity values, and (3) measurement of the dynamics with the curve of the fluorescence intensity against time.

3.4.1 Image Analysis

1. The bleach performed during the recording is measured in a stack of images of the neighboring cell that was not activated/bleached. The bleach coefficient is determined with ImageJ

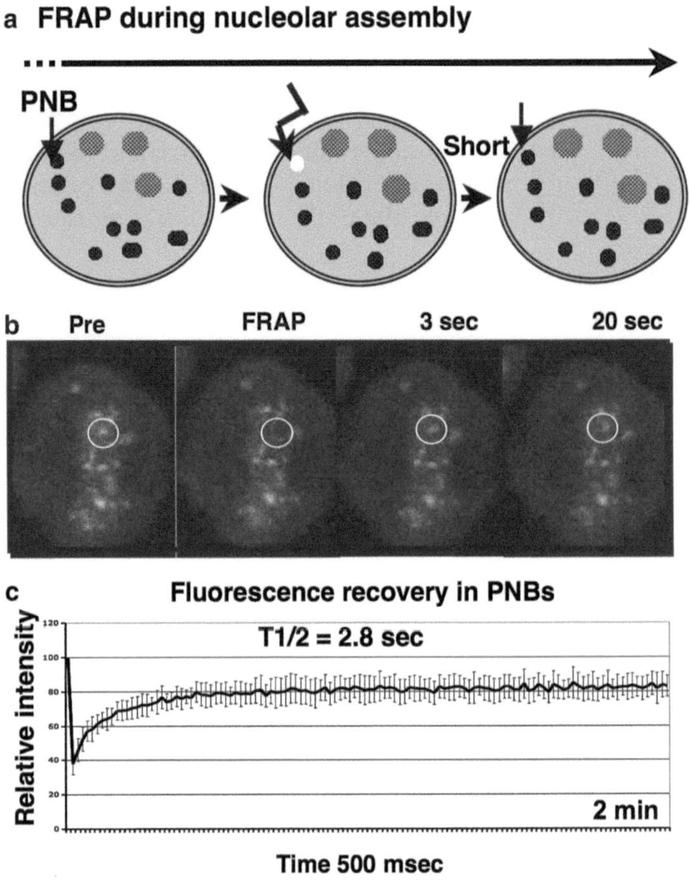

Fig. 3 Analysis of the dynamics using FRAP. (**a**) The dynamics of the nucleolar proteins in PNBs. The time of the fluorescence recovery after bleaching is short after photobleaching of GFP-B23 in one PNB (*arrow*) as visible in (**b**) and (**c**). (**b**) Images of the dynamics in PNBs. In this early G1 cell, the *circle* on the image marks the position of the bleach. Before the bleach (pre), the GFP-B23 fluorescence in one PNB is visible. The fluorescence is no longer visible after the FRAP (FRAP) and is visible in the same area 3 and 20 s after the FRAP. (**c**) Dynamics of the fluorescence recovery. The *curve* shows the fluorescence recovery of GFP-B23 in 12 PNBs during 2 min. The $T_{1/2}$ of 2.8 s required to reach the plateau indicates a rapid recovery

software (analyze/tools/curve fitting/exponential); the coefficient of bleach is applied to the images using the plug-in Stack-T-functions (bleach correction) and the corrected stack of images is registered (*see* **Note 11**).

2. The mean intensity values of activated/photobleached ROI are normalized respectively against the maximum/pre-bleach value at 100 and plotted against time. Using photoactivation, the dynamics are measured as a decrease of the fluorescence of the activated PAGFP proteins in the ROI (*see* **Note 12**). Using FRAP, the dynamics of the proteins are measured as a recovery of fluorescent molecules in the bleached area (ROI) (*see* **Note 12**).

3. The corrected and normalized measurements are fitted with the function $F(t) = a + \left[(b-a)^{*} \times t / (t+c)\right]$ proposed by Negi and Olson [12] to calculate the halftime ($t_{½}$) required to reach the plateau. In the function: t = time points; $F(t)$ = mean intensity in the ROI analyzed; a = intensity value immediately after bleaching/activation; b = intensity of the ROI when $t \rightarrow \infty$, it is the value of asymptotic trend of the function that it is ≈ the plateau; $c = t_{½}$.

3.4.2 Dynamics of Nucleolar Assembly

During nucleolar assembly, photoactivation allows measurement of the dynamics of a pool of proteins from nucleoli to PNBs in the nuclear volume. FRAP allows to quantify the recovery of fluorescent proteins into the bleached ROI either in nucleoli or PNBs. Using a combination of both approaches, we demonstrated that early-processing proteins are recruited to the incipient nucleoli and directly excluded from PNBs in late telophase, while late-processing proteins are still moving from nucleoli to PNBs. Moreover, this traffic analysis led us to demonstrate that the dynamics of these proteins is perturbed when inhibiting the CRM1-mediated nucleocytoplasmic transport; consequentially nucleolar assembly is blocked for a long time [9].

3.5 Trafficking of Processing Proteins During Nucleolar Assembly

Nucleoli assemble after mitosis, from telophase to the end of early G1. The signature of nucleolar assembly is based on the detectable resumption of rDNA transcription. However, active rDNA transcription does not possess the ability to organize a complete nucleolus. Nucleolar assembly also depends on the rRNA-processing complexes, ribosomal proteins (r-proteins), as well as the 45S rRNA produced during prophase and present in the perichromosomal compartment [13]. During telophase these nucleolar proteins and rRNAs assemble in PNBs [1, 14]. The time-lapse recordings demonstrate that during telophase PNBs contain both early- and late-processing proteins. The recruitment of early-processing proteins to sites of rDNA transcription lasts ≈20 min, while 90–120 min are necessary for the recruitment of late-processing proteins [3, 9]. When comparing the relative kinetics of two late-processing proteins (B23 and Nop52), similar dynamics are observed as well as synchronized departure from the same PNBs. These dynamics suggest formation of complexes between late rRNA-processing proteins in PNBs. This hypothesis is supported by FRET analysis that reveals B23-Nop52 complexes in nucleoli and in PNBs, while no complex is detectable around the chromosomes during anaphase [3].

PAGFP-fibrillarin, PAGFP-B23, or PAGFP-Nop52 can be photoactivated in one PNB or one incipient nucleolus, but the amount of proteins activated in one PNB is too low for analysis of the 3-D distribution of the proteins in each case. The PAGFP-fibrillarin activated in one incipient nucleolus migrates to the other incipient nucleoli and is not detectable in PNBs [9]. This illustrates the association of fibrillarin with the early-processing steps of pre-rRNAs [14] present in the six emerging nucleoli. In contrast, late rRNA-processing proteins (Nop52 or B23) migrate from one incipient nucleolus to the others and also to each PNB [9]. The recruitment of the late-processing proteins from PNBs to nucleoli implies bidirectional trafficking between these two structures and is achieved 2 h after telophase under normal conditions, when PNBs are no longer visible. This is the consequence of the increased number of binding sites in the incipient nucleoli, appearing when the first step of nucleolar assembly is accomplished. On the other hand, the persistence of PNBs and trafficking between PNBs and incipient nucleoli after this time is the signature of disturbed nucleolar assembly [9]. Thus, the recruitment of the processing complexes, first in the DFC (fibrillarin) and then in the GC (Nop52) (*see* **Note 13**) during nucleolar assembly after mitosis depends on the trafficking between PNBs and incipient nucleoli. Consequently, PNB formation is a means to control and regulate nucleolar assembly after mitosis and might explain the ubiquitous formation of PNBs in open mitosis.

4 Notes

1. In the case of a microscope incubator without CO_2, the pH 7.4 of the DMEM "F12 Observation" medium is maintained by adding 10 mM Hepes buffer.
2. It was not possible to generate a stable HeLa cell line for the expression of pPAGFP-B23, pPAGFP-Nop52, or fibrillarin-pPAGFP.
3. The time-lapse observations of individual proteins were also imaged. The couples were selected because it is possible to compare the trafficking of (a) an early-processing protein (fibrillarin) together with one late (B23) processing protein and (b) two late-processing (B23-Nop52) proteins during nucleolar assembly.
4. During anaphase the cells are round and easy to discriminate from the interphasic cells attached to the glass coverslip. When mitotic cells enter into interphase, they flatten and remain attached by the cytoplasmic bridge and the midbody during early G1. We observed that the two daughter cells cannot be imaged in the same focal plane during early G1 when they are still attached by the midbody.
5. In the analysis, different ROI(s) are included corresponding to diffuse or focal areas observed from telophase to early G1. Foci are defined as regions of local intensity greater than three times that of diffuse areas.
6. Verify before activation on a single section or Z series that imaging at 488 nm (argon laser power at 10 %) with a 4× frame average does not induce activation of PAGFP.

7. To localize the site of photoactivation for weak concentration of PAGFP proteins or in small bodies such as PNBs, the red signal from a protein partner expressing DsRed or mRFP is very useful. To localize the red fluorescence of DsRed-B23, DsRed-Nop52, or mRFP-Nop56, the Krypton laser wavelength at 558 nm is used. A Z series of images is recorded before the photoactivation to choose the plane containing PNBs and nucleolus within the nuclear volume.
8. If the trafficking is very rapid, shortening the photoactivation period (e.g., 100 ms) and increasing the power diode (e.g., 30 %) decrease the delay between photoactivation and recording. If the protein is not abundant, acquiring with the frame average mode (average of four scans for each image) combined with the high-speed recording mode (zoom 16.52; speed 1,000 Hz and image size 128 × 128) helps to increase the quality of the signal without slowing down the recording (when the spatial resolution allows it).
9. When possible, the reference would be the minimal time required to reach the equilibrium between the in and out traffic in the ROI.
10. The analysis of the protein dynamics in PNBs was performed using B23 and Nop52 because they are abundant proteins involved in late processing of the rRNAs of the large ribosomal subunits and were observed in PNBs during a long period using time-lapse imaging.
11. The Stack-T-functions plug-in can be downloaded from the following Website: http://www.macbiophotonics.ca/imagej/installing_imagej.htm. Download MBF ImageJ bundle plug-ins. To install the Stack-T-functions plug-in to an existing ImageJ installation, transfer its folder from the plug-ins folder of MBF to the plug-ins folder already present in the previously installed ImageJ application folder. If ImageJ is not yet installed, follow the instructions from this Website. Explanations for the Bleach Correction command are available at http://www.macbiophotonics.ca/imagej/t.htm#t_bleach.
12. Using photoactivation, the GFP-fluorescence intensity of the first image after the activation in the ROI is normalized to 100 %. Using FRAP, the intensity of fluorescence in the ROI before the bleach is normalized to 100 %.
13. In higher eukaryotes, the nucleolus is organized into three sub-compartments named fibrillar center (FC), dense fibrillar component (DFC), and granular component (GC). rDNA transcription occurs at the surface of the FC, the 47S rRNAs are present in the DFC where early-processing occurs, and the late rRNA processing takes place in the GC.

Acknowledgements

We thank the people of the Imaging Platform of the IJM and particularly A. Jobart-Malfait for her help. We are grateful to A-L Haenni for a critical reading of this paper. This work was supported in part by grants from the CNRS to UMR 7592; E. Louvet is funded by the Japan Society for the Promotion of Science.

References

1. Dousset T, Wang C, Verheggen C, Chen D, Hernandez-Verdun D, Huang S (2000) Initiation of nucleolar assembly is independent of RNA polymerase I transcription. Mol Biol Cell 11:2705–2717
2. Savino TM, Gébrane-Younès J, De Mey J, Sibarita J-B, Hernandez-Verdun D (2001) Nucleolar assembly of the rRNA processing machinery in living cells. J Cell Biol 153:1097–1110
3. Angelier N, Tramier M, Louvet E, Coppey-Moisan M, Savino TM, De Mey JR et al (2005) Tracking the interactions of rRNA processing proteins during nucleolar assembly in living cells. Mol Biol Cell 16:2862–2871
4. Louvet E, Tramier M, Angelier N, Hernandez-Verdun D (2008) Time-lapse microscopy and fluorescence resonance energy transfer to analyze the dynamics and interactions of nucleolar proteins in living cells. Methods Mol Biol 463:123–135
5. Patterson GH, Lippincott-Schwartz J (2002) A photoactivatable GFP for selective photolabeling of proteins and cells. Science 297:1873–1877
6. Lippincott-Schwartz J, Altan-Bonnet N, Patterson GH (2003) Photobleaching and photoactivation: following protein dynamics in living cells. Nat Cell Biol Suppl:S7–14
7. Phair RD, Misteli T (2000) High mobility of proteins in the mammalian cell nucleus. Nature 404:604–609
8. Muro E, Hoang TQ, Jobart-Malfait A, Hernandez-Verdun D (2008) In nucleoli, the steady state of nucleolar proteins is leptomycin B-sensitive. Biol Cell 100:303–313
9. Muro E, Gébrane-Younès J, Jobart-Malfait A, Louvet E, Roussel P, Hernandez-Verdun D (2010) The traffic of proteins between nucleolar organizer regions and prenucleolar bodies governs the assembly of the nucleolus at exit of mitosis. Nucleus 1:202–211
10. Sibarita JB (2005) Deconvolution microscopy. Adv Biochem Eng Biotechnol 95: 201–243
11. Roussel P, André C, Comai L, Hernandez-Verdun D (1996) The rDNA transcription machinery is assembled during mitosis in active NORs and absent in inactive NORs. J Cell Biol 133:235–246
12. Negi SS, Olson MO (2006) Effects of interphase and mitotic phosphorylation on the mobility and location of nucleolar protein B23. J Cell Sci 119:3676–3685
13. Hernandez-Verdun D (2011) Assembly and disassembly of the nucleolus during cell cycle. Nucleus 2:189–194
14. Fomproix N, Hernandez-Verdun D (1999) Effects of anti-PM-Scl 100 (Rrp6p exonuclease) antibodies on prenucleolar body dynamics at the end of mitosis. Exp Cell Res 251: 452–464

Chapter 23

Nucleation of Nuclear Bodies

Miroslav Dundr

Abstract

The nucleus is a complex organelle containing numerous highly dynamic, structurally stable domains and bodies, harboring functions that have only begun to be defined. However, the molecular mechanisms for their formation are still poorly understood. Recently it has been shown that a nuclear body can form de novo by self-organization. But little is known regarding what triggers the formation of a nuclear body and how subsequent assembly steps are orchestrated. Nuclear bodies are frequently associated with specific active gene loci that directly contribute to their formation. Both coding and noncoding RNAs can initiate the assembly of nuclear bodies with which they are physiologically associated. Thus, the formation of nuclear bodies occurs via recruitment and consequent accumulation of resident proteins in the nuclear bodies by nucleating RNA acting as a seeder. In this chapter I describe how to set up an experimental cell system to probe de novo biogenesis of a nuclear body by nucleating RNA and nuclear body components tethered on chromatin.

Key words Nuclear body, Cajal body, RNA transcription, Gene expression, Nuclear organization

1 Introduction

The dynamic spatial organization of the cell nucleus plays a primary role in genome function and maintenance [1, 2]. Within the nuclear environment, which is characterized by a lack of defining membranes, chromosomes occupy specific nonrandom territories. These chromosome territories harbor a variety of distinct nuclear bodies (NBs) involved in various aspects of genome activity, regulation and maintenance form in their closest association [3, 4]. NBs are highly dynamic structures, of which their components show rapid turnover with the surrounding nucleoplasm. Their structural integrity is mediated by transient low affinity protein–protein and protein–RNA interactions [5, 6]. Importantly, NBs do not assemble as preformed structural entities but rather emerge as a direct reflection of specific activities associated with gene expression and genome maintenance [7]. The molecular mechanisms of how NBs form and maintain their structural integrity are still

Yaron Shav-Tal (ed.), *Imaging Gene Expression: Methods and Protocols*, Methods in Molecular Biology, vol. 1042,
DOI 10.1007/978-1-62703-526-2_23, © Springer Science+Business Media, LLC 2013

poorly understood. More specifically, little is known regarding how an NB formation is initiated and how following building steps in an NB assembly is orchestrated.

Recently two opposite models of NB formation have been proposed [7]: (a) ordered assembly model in which an NB may form by a tightly controlled series of sequential building steps with very limited number of molecules which can act as an initiator of the assembly and, alternatively, (b) a stochastic self-organization assembly model in which an NB is built by the random interactions of individual components without a strict hierarchical order of assembly. In this model many components of an NB can initiate the formation. These two models were experimentally validated on the formation of the Cajal body (CB) [8], which is one of the major NB involved in biogenesis and recycling of many small nuclear RNAs [5, 9, 10]. By tethering individual CB components fused to the *Escherichia coli* Lac repressor (LacI) to a Lac operator (LacO)-repeat array gene locus in living cells, it was demonstrated that any CB component can initiate the formation of the entire NB. This finding provided conclusive evidence that an NB is formed by self-organization, but it does not address which initiation event triggers the physiological formation of a CB [8].

A critical step in NB formation is an initial nucleation event which serves to immobilize critical components to provide a seeding platform to trigger recruitment and retention of additional building blocks. Importantly, many NBs are formed at sites of active transcription at which their activities contribute to their formation. To investigate whether specific functionally related RNAs are sufficient to form major NBs, we developed an experimental cell system in which a specific RNA is tagged with a bacteriophage MS2 stem–loop sequence. When the MS2-tagged transcript is co-expressed with the LacI-NLS-GFP-MS2 coat protein, which selectively binds the MS2 loop, it is targeted to and accumulates on the 256 repeats of the LacO-binding sequence within the LacO array (Figs. 1 and 2a).

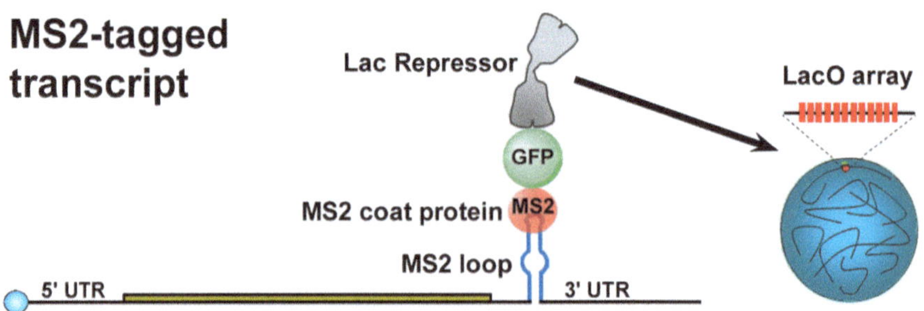

Fig. 1 Schematic representation of RNA-tethering system used to probe the contribution of specific RNAs in the formation of nuclear bodies. The hypothetical transcript is tagged with one bacteriophage MS2 stem loop in the 3′ UTR. The monomeric LacI-NLS-GFP-MS2 coat protein selectively binds the MS2 loop and targets the MS2-tagged transcripts to the LacO array where they accumulate

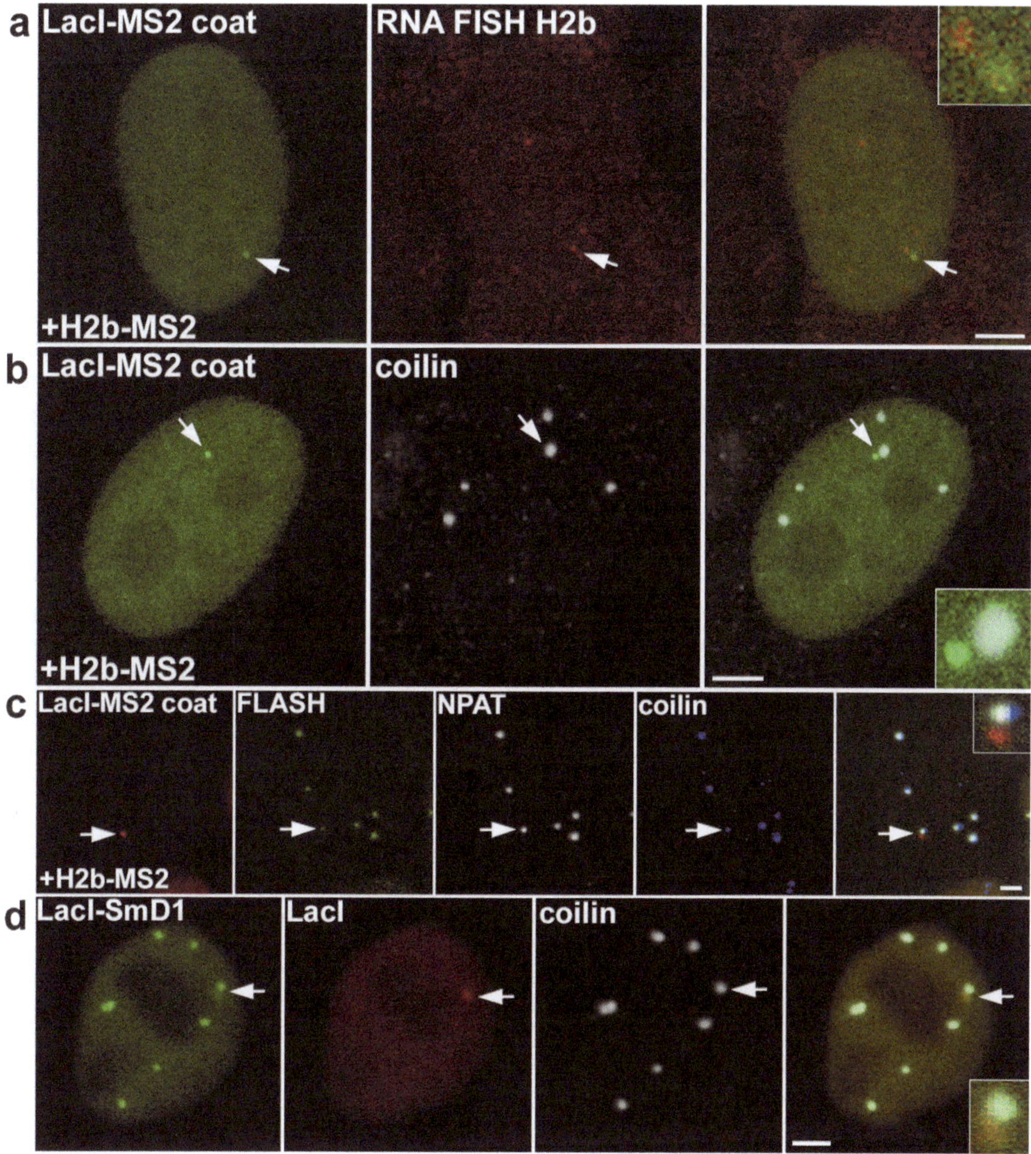

Fig. 2 Immobilization of specific RNAs to chromatin leads to the formation of nuclear bodies. (**a**) Histone H2b transcripts tagged with the MS2 loop were transiently co-expressed with the monomeric LacI-NLS-GFP-MS2 coat protein in HeLa cell containing a stably integrated LacO array. Tethering of H2b-MS2 transcripts on the LacO array was detected by RNA FISH using a specific DNA FISH probe against H2b pre-mRNA (*arrow* indicates higher magnification shown in *inset*). (**b**) Tethering of H2b-MS2 pre-mRNA to the LacO array in HeLa cells leads to de novo formation of a Cajal body (CB) detected by an antibody against a CB marker protein coilin (*arrow*; shown in *inset*). (**c**) Immobilization of H2b-MS2 pre-mRNA on the LacO array in HeLa cells leads to de novo formation of two nuclear bodies, a CB with a physically associated histone locus body (HLB). HLB is detected by GFP-FLASH (*green*) and with an antibody against NPAT (*white*), two HLB marker proteins, and histone pre-mRNA processing factors. CB is detected with anti-coilin antibody (*blue*) (*arrow*, *inset*). (**d**) Tethering of a single component of a CB, a core protein of the Sm ring of spliceosomal U snRNPs, GFP-LacI-SmD1, on the LacO array (detected by mCherry-LacI) in HeLa cell leads to de novo CB formation. The CB is detected with an anti-coilin antibody (*arrow*, *inset*). Scale bar = 2 μm

This system has been used to tether coding histone H2b or β-globin pre-mRNAs, or the noncoding NEAT1 or SatIII RNAs, to a LacO array in living cells, in order to generate a specific NB. This approach led to the de novo nucleation of a histone locus body with associated CBs (Fig. 2b, c), nuclear speckle, paraspeckle, and nuclear stress bodies, respectively, at these tethering sites [11]. Additional evidence was provided by the visualization of the de novo formation of paraspeckles by inducing transcription of noncoding RNA Men ε/β (NEAT1 in human) [12]. Overall, these data indicate that several types of coding and noncoding RNAs can initiate the formation of some NBs with which they are physiologically associated and function as structural elements and as a nucleator of NBs. Expanding the use of these approaches will undoubtedly uncover the total composition of individual NBs, which will lead to the understanding of an entire interactive network of NB components and physiological processes responsible for their formation and ultimately their functions.

In this chapter I explain the standard requirements and overall technical steps for developing an experimental cell system to probe the ability of specific RNAs and individual NB components to nucleate a NB. This chapter will describe (a) subcloning of coding and noncoding RNAs that specifically associate with NBs and tagging them with the MS2 sequence, (b) fusion of specific components of NBs with GFP-LacI, (c) tethering of MS2-tagged RNAs on a genomic LacO array and their detection by RNA FISH, (d) tethering of MS2-tagged RNAs and LacI-fused NB components to the array with the detection of de novo NB formation, and (e) live-cell imaging of de novo NB formation initiated by an RNA molecule.

2 Materials

2.1 Cloning of Coding and Noncoding RNAs Which Specifically Associate with NBs

1. Genomic DNA extraction kit.
2. pcDNA3.1 vector (Invitrogen) or other convenient PCR cloning vector for PCR subcloning.
3. Mammalian expression vector such as pcDNA3.3-TOPO (Invitrogen) driven by a modified enhanced CMV promoter.
4. Restriction enzymes.
5. T4 DNA ligase.
6. Competent strain(s) of *Escherichia coli* for DNA transformation.
7. GFP-LacI-NLS and mCherry-LacI-NLS sucloned in the vector derived from pEGFP-C1 (Clontech). These plasmids can be obtained from Addgene.
8. Site-directed mutagenesis kit.

2.2 Cell Culture

1. Dulbecco's Modified Eagle's Medium (DMEM) for cell culture.
2. Fetal bovine serum (FBS).
3. Trypsin for releasing attached cells from culture plates.
4. PBS (phosphate-buffered saline) for washing cells.
5. Adherent cell line of preference with a stably integrated LacO array with 256 repeats. It is highly recommended to use human-transformed cell lines which can be obtained commercially from various sources such as ATCC.
6. Lipofectamine 2000 (Invitrogen) and Opti-MEM Reduced Serum Medium (Invitrogen).
7. 0.1 M IPTG stock solution in sterile H_2O.

2.3 RNA FISH

1. PBS for washing cells.
2. 4 % paraformaldehyde in PBS: aliquot and freeze at −20 °C for later use.
3. 0.2 % Triton X-100 in PBS: 40 μl/20 ml PBS. Store at 4 °C for later use.
4. 20× Saline-sodium citrate buffer (SSC), pH 7.0.
5. Formamide and deionized formamide.
6. Yeast tRNA (Sigma): 10 mg/ml stock solution and freeze at −20 °C.
7. 50 % dextran sulfate in sterile H_2O: aliquot and store at −20 °C for later use.
8. Ice-cold absolute ethanol.
9. 3 M sodium acetate (pH 5.2).
10. Rubber cement.
11. Hybridization moisture chamber with paper towels soaked with 2× SSC in bottom.
12. Fluorescently labeled DNA FISH probe against the target gene sequence. The plasmid with the gene of interest should be labeled by nick translation with Cy3 (or other convenient fluorophores such as Alexa) to produce a FISH probe of size ~200 bp. Use 100 ng of probe per a 22×22 mm square coverslip.
13. Hybridization solution: 200 μl of 20× SSC, 200 μl of 50 % dextran sulfate, 200 μl of 0.1 M Tris–HCl (pH 7.2), and 700 μl of nuclease-free H_2O. It is possible to aliquot and freeze (−20 °C) the hybridization solution for later use.
14. Mounting medium with anti-fading agents (ProLong Gold or Vectashield) either with or without DAPI for DNA staining.

2.4 Microscopy

1. Fluorescent microscope of choice with either 63× or 100× Plan oil objectives. It is recommended to use a confocal microscope and collect vertical z-stacks to visualize de novo NB formation because the depth of field along *z*-axis is frequently important for NB detection on the chromatin locus. Since the nucleation of an NB is a very light-sensitive event, it is suggested to use a high-speed microscope equipped with a high-sensitivity CCD camera to reduce exposure time during live-cell imaging.
2. Microscope sample incubation chamber for live-cell imaging to maintain cells at 37 °C and 5 % CO_2.
3. Live-cell imaging chambers with thin glass bottom.
4. DMEM buffered with 15 mM HEPES without phenol red.
5. Image analysis software such as Metamorph (Molecular Devices), Image J (NIH), or Imaris (Bitplane).

3 Methods

3.1 Subcloning of Coding and Noncoding RNAs That Specifically Associate with NBs and Their Tagging with the MS2 Sequence

1. Isolate the gene of interest from genomic DNA derived from cells of interest using the genomic DNA extraction kit or use commercially available genomic libraries. Amplify the gene by PCR and subsequently subclone it into a convenient PCR vector (e.g., pcDNA3.1) using a PCR subcloning kit. The sequence needs to be verified by sequencing and compared with genome information available in genome databases.
2. Analyze for predicted RNA secondary structures in the transcript transcribed from the gene of interest by Mfold (http://mfold.rna.albany.edu/?q=mfold/RNA-Folding-Form). This computational approach will identify suitable predicted RNA stem–loop structure(s) for the introduction of one bacteriophage MS2 loop sequence into the gene of interest. However, it is always essential to analyze and critically evaluate positions of important functional domains present in the gene such as splice sites, cleavage sites, regulatory regions, and binding sites and to avoid using them for the MS2 loop insertion (*see* **Note 1**).
3. By site-directed mutagenesis, introduce one bacteriophage MS2 stem–loop sequence (CGTACACCATCAGGGTACG) as an extension of a predicted RNA stem–loop structure in the transcript as predicted by Mfold. The correct insertion of the MS2 loop must be verified by sequencing and the whole gene of interest should be re-sequenced.
4. To determine which sequence and structural elements of RNA of interest are necessary and sufficient for de novo nucleation of an NB, a series of processing and deletion mutants derived

from the original MS2 loop-containing wild-type transcript should be created at this stage by site-directed mutagenesis using the same technical approach as in **step 3**. Particularly, more stable (e.g., non-cleavable and unspliced) mutants of the transcript of interest should be generated and their efficiency to nucleate a de novo NB should be quantitatively compared to the wild-type-MS2 transcript.

5. Insert the wild-type MS2-tagged transcript and its processing and/or deletion mutants into a suitable mammalian expression vector such as pcDNA3.3-TOPO or others. Initially, a strong constitutive viral promoter such as the viral CMV should be used to achieve high expression levels. High levels of ectopic expression of RNA-MS2 provide functional evidence that tethering on chromatin leads to de novo NB formation. MS2-tagged transcripts can be expressed at a moderate level using an endogenous promoter to mimic endogenous levels of the native RNA of interest (*see* **Note 2**).

3.2 Fusion of NB Components with LacI

An alternative way to nucleate a de novo NB is to tether NB components fused with the LacI on the LacO array. This mimics a natural seeding event and thus initiates the assembly process of NB formation (Fig. 2d).

1. Clone or obtain a cDNA of an NB component of interest from various commercial sources (e.g., OriGene or Mammalian Gene Collection).
2. Obtain the GFP-LacI-NLS subcloned as an XhoI-EcoRI fragment into the pEGFP-C1 designed for fusion with proteins of interest from Addgene.
3. Using PCR add convenient available restriction sites present at the multicloning site of the pEGFP-C1 downstream of EcoRI, to the 5′ and 3′ ends of the NB component cDNA, for fusion with GFP-LacI-NLS. The prepared cDNA needs to have a stop codon to terminate the translation of the fusion product.
4. Insert the cDNA of the NB component into the GFP-LacI-NLS using the added restriction sites.
5. Verify the inserted clone by sequencing.

3.3 Cell Culture and Transfection

1. For preparation of cells before transfection, plate the cells containing a stably integrated LacO array (*see* **Note 3**) on 22 mm glass square coverslips, in medium supplemented with 10 % serum and 1 % glutamine without antibiotics, in a 6-well plate at 37 °C. Cells should reach 30–50 % confluency before transfection. Transfection efficiency is low when cell density is too high (*see* **Note 4**).

2. To transfect the cells on the following day, adjust the volume of medium in each well to 2 ml. Co-transfect the cells with plasmids encoding the MS2-tagged transcript and monomeric LacI-NLS-GFP-MS2 coat protein using Lipofectamine 2000. Briefly, dilute the plasmids (~1 μg in total) in 50 μl of Opti-MEM Reduced Serum Medium without serum and mix gently. The plasmid ratio of MS2-tagged transcript to LacI-NLS-GFP-MS2 coat protein should be 1:2 (*see* **Note 5**). In the second test tube, dilute 2 μl of Lipofectamine 2000 in 50 μl of Opti-MEM Reduced Serum Medium without serum, mix gently, and incubate for 5 min at RT. Then combine the diluted DNA with the diluted Lipofectamine 2000. Mix gently and incubate for 20 min at RT. Slowly add 100 μl of transfection complexes to each well containing cells and medium. Mix gently by rocking the plate back and forth. Incubate cells at 37 °C for 18–24 h.
3. Change the medium the next morning.
4. Fix cells for immunolocalization or visualize them live under the microscope 18–24 h after transfection.

3.4 Detection of Tethered MS2-Tagged RNA on the LacO Array by RNA FISH

To test whether MS2-tagged transcripts co-expressed with the LacI-NLS-GFP-MS2 coat protein are efficiently targeted and accumulated on the LacO array, perform RNA fluorescence in situ hybridization (FISH) using a specific DNA FISH probe against the MS2-tagged RNA (Fig. 2a) (*see* **Note 6**).

1. Wash cells briefly with PBS and fix with 4 % paraformaldehyde in PBS for 15 min at RT.
2. Wash cells three times with PBS at RT.
3. Permeabilize cells with 0.2 % Triton X-100 in PBS on ice for 5 min.
4. Wash cells twice with PBS and finally once with 2× SSC for 5 min at RT.
5. Prepare the hybridization solution.
6. Mix 100 ng of the DNA FISH probe against the target RNA (labeled with Cy3 by nick translation) and 40 μg yeast tRNA per coverslip and precipitate: 2 μl of probe, 4 μl of tRNA (10 mg/ml stock), 39 μl of sterile H_2O, 5 μl of 3 M sodium acetate (pH 5.2), 100 μl of ice-cold absolute ethanol. Put it in a –70 °C freezer for at least 20 min. Spin at maximum speed for 20 min at 4 °C. Remove the supernatant. The pellet may be clear so be very careful not to remove the pellet with the supernatant. Air dry the pellet and reconstitute in 6 μl of deionized formamide.
7. Denaturate the DNA FISH probe by incubating the tube for 10 min at 80 °C. Vortex several times during heating. Chill in water-ice slurry immediately for 5 min.

8. Add 18 μl of the hybridization solution from the **step 5** at RT, mix it, and spin it for 15 s at RT.
9. Use a petri dish to set up a moistened chamber. Place paper towels moistened with 2× SSC in the bottom of a plate.
10. Place 24 μl of the hybridization mixture from the **step 8** onto each coverslip, and seal coverslips to microscopic slides by rubber cement and hybridize in a moistened chamber overnight at 37 °C.
11. 30 min before washing, pre-warm the 2× SSC–50 % formamide to 37 °C.
12. Open the hybridization chamber, remove the rubber cement, and transfer the coverslips back to a 6-well plate with pre-warmed 2× SSC–50 % formamide.
13. Rinse coverslips twice for 30 min with 2× SSC–50 % formamide at 37 °C.
14. Rinse coveslips for 5 min with PBS at RT.
15. At this point the RNA FISH protocol is combined with indirect immunofluorescence to specifically visualize an NB or nuclear compartment functionally associated with immobilized MS2-tagged RNA. Incubate cells on coverslips with a primary antibody against a marker protein of the NB of interest, for 1 h at RT.
16. Wash cells three times with PBS at RT.
17. Incubate cells with an appropriate secondary antibody conjugated with Cy5 (or different convenient fluorophores such as Alexa) for 1 h at RT.
18. Wash cells three times with PBS at RT.
19. Mount coverslips in mounting medium with anti-fading agents either with or without DAPI. Leave slides overnight at 4 °C covered with aluminum foil to allow the medium to solidify.
20. Observe cells under a fluorescent microscope using either 63× or 100× Plan oil objectives with high numerical apertures. Check the localization of the LacO array as a single bright fluorescent spot in the nucleus with associated RNA FISH signal. Additional RNA FISH signals visible as bright foci, observed at the endogenous transcription sites of the gene of interest in the nucleoplasm, should be located and used as a positive control. The intensity of the RNA FISH signal at typical endogenous transcription sites should be approximately the same as the signal on the LacO array (Fig. 2a).

3.5 De Novo Nucleation of an NB by RNA or by Tethered NB Components

To test whether MS2-tagged transcripts or NB components immobilized on the LacO array are sufficient to de novo nucleate an NB, perform indirect immunofluorescence using the antibodies against marker protein(s) of NB of interest. To distinguish the position of the LacO array from endogenous NBs, where NB proteins fused with GFP-LacI should also be located, the cells need to be co-transfected with mCherry-LacI-NLS which marks the array specifically (Fig. 2d).

1. Wash transfected cells briefly with PBS and fix them with 2 % paraformaldehyde in PBS for 10 min at RT.
2. Wash cells three times with PBS at RT.
3. Permeabilize cells with 0.2 % Triton X-100 in PBS on ice for 5 min.
4. Wash cells three times with PBS at RT.
5. Incubate cells on coverslips with a primary antibody against a marker protein of NB for 1 h at RT.
6. Wash cells three times with PBS at RT.
7. Incubate cells with an appropriate secondary antibody conjugated with Cy5 for 1 h at RT.
8. Wash three times cells with PBS at RT.
9. Mount coverslips in mounting medium with anti-fading agents either with or without DAPI. Leave slides overnight at 4 °C covered with aluminum foil to allow the medium to solidify.
10. Observe cells under a fluorescent microscope using either 63× or 100× Plan oil objectives. Check the position of the LacO array as a single bright fluorescent spot in the nucleus where MS2-tagged RNAs are tethered. To distinguish the position of the LacO array from endogenous NBs labeled with GFP-LacI-NB proteins, check the localization of mCherry-LacI-NLS as a single bright red spot. Tethered GFP-LacI-NB fusion proteins are also clearly detectable on the array (Fig. 2d). The newly formed NB detected by specific antibodies against NB-specific components should be visible in strong association with the LacO array (Fig. 2b–d) but it is frequently located in different focal planes. Therefore vertical z-sections around the array should be acquired and all images should be combined and projected as a maximum intensity projection (*see* **Note 7**).

3.6 Kinetics of De Novo NB Formation by RNA

To establish the kinetics of de novo NB formation, treat cells with isopropyl-β-d-thiogalactopyranoside (IPTG) which selectively prevents binding of LacI-fusion proteins to the LacO array, which in turn leads to the disassembly of newly nucleated NB. When IPTG is washed out extensively from the cells, LacI-fusion binding is reestablished and de novo formation of a NB can be directly visualized and analyzed using time-lapse microscopy (*see* **Note 8**).

3.6.1 Time-Course of the Dynamics of De Novo NB Formation

1. Co-transfect cells containing a stably integrated LacO array growing on glass coverslips with LacI-NLS-GFP-MS2 coat protein and MS2-tagged RNA with Lipofectamine 2000 for 18–24 h.

2. Treat cells with IPTG (5 mM) for 16 h to prevent binding of LacI-NLS-GFP-MS2 coat protein with bound MS2-tagged RNAs to the LacO array.
3. Washout IPTG very extensively with PBS. This is time zero in the time-course experiment.
4. Perform a time-course series by fixing cells at: 15 min, 30 min, 1 h, 1 h 30 min, 2 h, 3 h and 4 h intervals. These time intervals would cover the assembly dynamics of a typical NB.
5. Perform indirect immunofluorescence localization using antibodies which detect marker proteins of an NB or nuclear subdomain of interest. Observe cells under a fluorescent microscope using either 63× or 100× Plan oil objectives. Check localization of the LacO array as a single bright green spot detected by the LacI-NLS-GFP-MS2 coat protein in the nucleus and analyze the assembly dynamics of an NB on the LacO array detected by a specific anti-NB antibody at each time point. LacI-NLS-GFP-MS2 coat protein with bound MS2-RNA accumulates on the LacO array in HeLa cells within ~30 min after IPTG withdrawal. A typical NB nucleated by MS2-tagged transcripts tethered on the LacO array likely forms within approximately 1–2 h [10].

3.6.2 Live-Cell 4D Imaging of De Novo Formation of a Nuclear Body Nucleated by MS2-Tagged RNAs on the LacO Array

1. Plate cells stably expressing the integrated LacO array in chambers for live-cell imaging.
2. Co-transfect cells with a GFP-tagged marker protein for an NB of interest with LacI-NLS-mCherry-MS2 coat protein and MS2-tagged RNA with Lipofectamine 2000 for 18–24 h. The plasmid ratio of MS2-tagged transcripts to LacI-NLS-mCherry-MS2 coat protein and to GFP-NB marker protein should be approximately 1:2:2.
3. Treat cells with IPTG (5 mM) for 16 h to prevent binding of LacI-NLS-mCherry-MS2 coat protein with tethered MS2-tagged RNA to the LacO array.
4. Wash out the IPTG extensively with PBS at RT.
5. Change medium to phenol red-free DMEM buffered with HEPES supplemented with 10 % serum.
6. Transfer the transfected cells in the live-cell imaging chamber to a high-speed fluorescent microscope equipped with a stage temperature-controlled incubation chamber to maintain cells at 37 °C and 5 % CO_2. Search for transfected cells with a strong but not overexpressed GFP signal detecting NBs and a diffuse nucleoplasmic red signal of LacI-NLS-mCherry-MS2 coat protein. The LacO array is not visible as a single bright spot in the nucleus immediately after IPTG withdrawal. LacI-NLS-

mCherry-MS2 coat protein with bound MS2-tagged transcripts accumulates on the LacO array in HeLa cells within ~30 min after IPTG withdrawal. Use a 63× Plan oil objective to simultaneously visualize several cells in the field. Adjust imaging conditions for high-speed image acquisition and low laser intensity to prevent photodamage of cells. During the search for optimal cells, minimize their exposure to reduce the bleaching of fluorescent signals.

7. Determine experimental conditions for time-lapse imaging including number of fields, number of vertical z-stacks, time intervals, and number of repeats. Recommended parameters are $z = 300–400$ nm and 5–10 min intervals for up to 5 h for a typical NB.
8. All collected images from the z-stack should be projected in three dimensions as combined maximum intensity projections for each time point using specific image analysis software.

4 Notes

1. *Position of MS2 loop insertion.* The right choice of site for MS2 loop insertion is critical for its interaction with a monomeric MS2 coat protein fused with the LacI and should not interfere with the proper processing and overall functionality of the MS2-tagged RNAs. It is recommended to insert one MS2 loop as an extension of a predicted helix region, or an RNA secondary structure located preferably outside of the coding region at the 3′ UTR, or portions of the coding region that are not involved in pre-mRNA processing or cleavage.
2. *Promoter.* The selection of a promoter will significantly affect the ectopic expression levels of tethered components and MS2-tagged RNAs. Therefore, it is essential at first to express both components at higher levels to assure that they would be sufficiently immobilized on the LacO array. Initially express these components under the strong constitutive viral CMV promoter with the ratio between plasmids encoding tethering components and MS2-tagged RNAs to 2:1. Once the tethering system proves that an NB of interest can be formed de novo, an endogenous promoter which typically produces moderate levels of expression can be used. This approach would mimic the endogenous levels of RNAs and thus more likely the biological conditions of NB formation.
3. *Stable cell lines expressing the LacO-repeat array.* In order to have a reproducible cell system for quantitative probing abilities of NB formation by different RNAs, a cell line with a stably integrated LacO array must be used. When use of special knockout cell lines or primary cells is essential, these cells can be transiently transfected with the plasmid encoding 256

repeats of the LacO. However, it is important to check the number of LacO repeats in the plasmid when transformed in bacteria because they tend to recombine and shorten the number of repeats. To minimize recombination this plasmid should be propagated in Stbl2 competent cells specifically designed for unstable inserts and cultured at 30 °C.

4. *Efficiency of transfection.* It is important to initially achieve a high level of transfection when using a transient transfection system. Therefore, it is always beneficial to use cells which are easy to transfect. It is recommended to use transformed cells such as HeLa or U2OS cells which are efficiently transfected with Lipofectamine 2000. As an alternative, it is possible to use electroporation which provides high transfection efficiency but requires a large number of cells for preparation. When primary cells, which are difficult to transfect, must be used it is recommended to use the Amaxa nucleofection system (Lonza) which has cell-type-specific reagents and good transfection efficiency for primary cells. The quality of plasmid DNA usually plays a critical role in high transfection efficiency. It is recommended to use a MidiPrep purification system from Zymo, Qiagen, Promega, etc. for high-purity plasmid DNA preparation.
5. *LacI-GFP-NLS-MS2 coat protein transfection.* To achieve successful targeting and accumulation of RNA-MS2 on the array, an appropriate level of LacI-GFP-NLS-MS2 coat protein needs to be expressed in the cells. It is suggested to use MidiPrep plasmid quality for high transfection efficiency and to adjust the plasmid ratio of MS2-RNA to LacI-GFP-NLS-MS2 coat protein to 1:2.
6. *Detection of MS2-RNA on the LacO array by RNA FISH.* Initially, it is important to verify whether MS2-tagged RNAs efficiently accumulate on the LacO array. LacI-NLS-GFP-MS2 coat protein marks the position of array as a single bright green spot in the nucleus, and the RNA FISH signal detecting MS2-RNA should be clearly visible in closed association with the array. Presence of endogenous and ectopic transcription sites visible as clear bright spots in the nucleoplasm detected by the FISH probe should be used as a positive control, and their signal intensity should be roughly equal to the FISH signal present on the array. Since the bacteriophage MS2 stem–loop sequence is relatively short (19 nucleotides), it is not a recommended target for RNA FISH.
7. *Detection of de novo formed NB of interest on the array.* For proper detection of de novo NB nucleation on the LacO array, endogenous NBs in the nucleus must be clearly detectable and used as a positive control. Suitable antibodies against the marker proteins of the NB of interest should be used. De novo NBs nucleated on the array are frequently positioned in different focal planes than the array. Therefore, it is important to

visualize cell nuclei vertically by z-stack sectioning and to project the 3D volume around the LacO array. It is necessary to evaluate whether ectopic expression of MS2-tagged RNAs and tethering components do not affect localization, number, size, shape, and topology of endogenous NBs and the behavior of their components. Furthermore, it is essential to check whether the de novo formed NBs are functional by several criteria such as size, shape, composition, and kinetic behavior of their components.

8. *Live-cell imaging of a NB formation by RNA.* The nucleation of an NB is a highly photosensitive event. Therefore, it is suggested to use a high-speed microscope such as a fast-spinning disc confocal microscope equipped with a high-sensitivity CCD camera to minimize cell exposure during live-cell visualization. The number of vertical z-sections depends on the size of NB of interest and cell shape. It is suggested to start with a number of z-sections which cover the entire cell nucleus from the top to the bottom and a step-width of 0.3–0.4 μm. The light intensity and exposure time should be extremely low to prevent photodamage of the cells and bleaching of the signals but still high enough for the signal detection over the background. An optimal signal-to-noise ratio should be adjusted by low laser intensity and fast image acquisition.

Acknowledgements

The laboratory of Miroslav Dundr is supported by NIH R01GM090156 grant from NIGMS.

References

1. Misteli T (2007) Beyond the sequence: cellular organization of genome function. Cell 128:787–800
2. Rajapakse I, Groudine M (2011) On emerging nuclear order. J Cell Biol 192:711–721
3. Mao YS, Zhang B, Spector DL (2011) Biogenesis and function of nuclear bodies. Trends Genet 27:295–306
4. Matera AG, Izaguire-Sierra M, Praveen K, Rajendra TK (2009) Nuclear bodies: random aggregates of sticky proteins or crucibles of macromolecular assembly? Dev Cell 17:639–647
5. Dundr M (2012) Nuclear bodies: multifunctional companions of the genome. Curr Opin Cell Biol 24:415–422
6. Caudron-Herger M, Rippe K (2012) Nuclear architecture by RNA. Curr Opin Genet Dev 22:179–187
7. Dundr M, Misteli T (2010) Biogenesis of nuclear bodies. Cold Spring Harb Perspect Biol 2:a000711
8. Kaiser TE, Intine RV, Dundr M (2008) De novo formation of a subnuclear body. Science 322:1713–1717
9. Nizami Z, Deryusheva S, Gall JG (2010) The Cajal body and histone locus body. Cold Spring Harb Perspect Biol 2:a000653
10. Machyna M, Heyn P, Neugebauer KM (2012) Cajal bodies: where form meets function. WIREs RNA. doi:10.1002/wrna.1139
11. Shevtsov SP, Dundr M (2011) Nucleation of nuclear bodies by RNA. Nat Cell Biol 13:167–173
12. Mao YS, Sunwoo H, Zhang B, Spector DL (2011) Direct visualization of the co-transcriptional assembly of a nuclear body by noncoding RNAs. Nat Cell Biol 13:95–101

Index

Yaron Shav-Tal (ed.), *Imaging Gene Expression: Methods and Protocols*, Methods in Molecular Biology, vol. 1042, DOI 10.1007/978-1-62703-526-2, © Springer Science+Business Media, LLC 2013

L

M

N

O

P

MIX
Papier aus verantwortungsvollen Quellen
Paper from responsible sources
FSC® C105338

If you have any concerns about our products, you can contact us on
ProductSafety@springernature.com

In case Publisher is established outside the EU, the EU authorized representative is:
Springer Nature Customer Service Center GmbH
Europaplatz 3, 69115 Heidelberg, Germany

Printed by Libri Plureos GmbH
in Hamburg, Germany